Serono Symposia USA
Norwell, Massachusetts

Springer

New York
Berlin
Heidelberg
Barcelona
Budapest
Hong Kong
London
Milan
Paris
Santa Clara
Singapore
Tokyo

PROCEEDINGS IN THE SERONO SYMPOSIA USA SERIES

Continued after Index

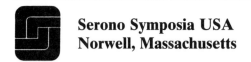

Serono Symposia USA
Norwell, Massachusetts

Barry B. Bercu   Richard F. Walker
Editors

Growth Hormone Secretagogues

With 138 Figures

Springer

Barry B. Bercu, M.D.
Richard F. Walker, Ph.D.
Pediatric Endocrinology
All Children's Hospital
St. Petersburg, FL 33701
USA

Proceedings of the International Symposium on Growth Hormone Secretagogues sponsored by Serono Symposia USA, Inc., held December 8 to 11, 1994, in St. Petersburg Beach, Florida.

For information on previous volumes, please contact Serono Symposia USA, Inc.

Library of Congress Cataloging-in-Publication Data
Growth hormone secretagogues/edited by Barry B. Bercu and Richard F.
Walker.
    p. cm.—(Proceedings of the Serono Symposia USA series;)
    "Proceedings of the International Symposium on Growth Hormone
Secretagogues sponsored by Serono Symposia USA, Inc., held December
8 to 11, 1994, in St. Petersburg Beach, Florida"—T.p. verso.
    Includes bibliographical references and index.
    ISBN 0-387-94707-8 (hardcover: alk. paper)
    1. Growth hormone releasing factor—Congresses. 2. Somatotropin—
Congresses. I. Bercu, Barry B. II. Walker, Richard F., 1939– .
III. International Symposium on Growth Hormone Secretagogues (1994:
Saint Petersburg Beach, Fla.) IV. Series: Proceedings of the Serono
Symposia.
QP572.G75G76   1996
612.6—dc20   96-632

Printed on acid-free paper.

Production coordinated by Chernow Editorial Services, Inc., and managed by Francine McNeill; manufacturing supervised by Jacqui Ashri.
Typeset by Best-set Typesetter Ltd., Hong Kong.
Printed and bound by Braun-Brumfield, Inc., Ann Arbor, MI.
Printed in the United States of America.

9 8 7 6 5 4 3 2 1

ISBN 0-387-94707-8 Springer-Verlag New York Berlin Heidelberg   SPIN 10524797

Preface

The traditional concept of a neuroendocrine mechanism for regulation of growth hormone (GH) secretion is based in large part on the work of Roger Guillemin. The work of Dr. Guillemin, who was awarded the 1977 Nobel Prize in Physiology and Medicine, supported the view that quantitative change in GH secretion was the net result of pituitary stimulation and inhibition by the hypothalamic neurohormones, GH releasing hormone (GHRH), and somatostatin (somatotropin release inhibiting factor; SRIF), respectively. During the 1970s, another endocrine research pioneer, Dr. Cyril Bowers, discovered that structural modification of enkephalin resulted in a family of peptides with GH releasing properties. These compounds, simply called GH releasing peptide (GHRP), were originally thought to mimic GHRH. However, upon subsequent investigation they were found to supplement the activity of the natural hormone through a different mechanism. Nearly two decades after their discovery, the differences between GHRP and GHRH have been described by many different laboratories throughout the world. The complementary GH secretagogues have different binding sites, second messengers, and effects on gene expression. Based on these differences, it has been suggested that expansion of the original two hormone mechanisms for GH regulation to include a third molecule may be appropriate, even though the naturally occurring analogue of GHRP has not yet been identified.

Despite our lack of knowledge concerning the natural product mimicked by GHRP, clinical development of the new family of GH secretagogues for diagnostic and therapeutic purposes has begun in earnest. The practical application of the GHRPs in medicine may be likened to the use of morphine for its therapeutic benefits, years before the endorphins were discovered. The rapid growth of research and development of GHRPs as drug products can be most likely attributed to the discovery that the prototype, GHRP-6 (growth hormone releasing hexapeptide; His-DTrp-Ala-Trp-DPhe-Lys-NH$_2$) was orally active, although its bioavailability was relatively low. Nonetheless, recognition of its oral activity led to the discovery of truncated peptides and nonpeptidyl mimics with significantly greater bioavailability. These compounds are now being actively developed for various and sundry purposes including such different maladies as growth retardation and osteoporosis. The broad range of potential applications for increased growth hormone secretion for treating disease in pediatric and geriatric populations makes GH secretagogues increasingly important pharmacotherapeutic entities.

Recognizing the significant advances in GH secretagogue research that have occurred over the past two decades, we felt that it was appropriate to organize a meeting specifically devoted to discussion of the GHRH and GHRP families of molecules. Because of their significant research contributions, Drs. Guillemin and Bowers began the meeting with historical perspectives on their independent discoveries of the different GH secretagogues. After an interesting and reflective session on the origins of GH secretagogue research, the meeting progressed to discussions of contemporary methods for chemical modification of GH secretagogues. The presentations focused on specific substitutions on the GHRH molecule that enhance its potency. In addition, theoretical models for designing xenobiotic GHRPs, and methods for practical application of these principles to create more active, truncated varieties of the prototypical molecules as well as to produce new, nonpeptidyl mimics of GHRP were presented. The structure, function, and regulation of the receptor for GHRH was also discussed to better understand how secretagogues elicit their specific activity at the level of the pituitary gland. Consistent with this topic, subsequent discussions dealt with cellular and molecular properties of the different GH secretagogues, focusing on topics ranging from second messenger utilization to gene expression. As a logical extension, the meeting progressed to evaluation of physiologic actions of GHRH and GHRP at the systems and organism level. These sessions included talks on basic and clinical aspects of GH secretagogue research dealing with such subjects as animal models for evaluating GH secretagogue potency and efficacy, pituitary mammosomatotroph transdifferentiation as influenced by GH secretagogues, behavioral effects of exposure to these compounds, and clinical utility of secretagogue administration in treating GH-deficient and obese children. The final session touched upon possible alternative uses for GH secretagogues, which included immunotherapy as well as treatment of osteoporosis and other degenerative and pathologic concomitants of aging.

This book, which presents specific important details of each author's oral presentation, provides the reader an opportunity to view the birth of GH secretagogue research from the eyes of the world-renowned scientists who pioneered the field to those of contemporary investigators using state-of-the-art techniques to bring theoretical concepts to clinical utility. Accordingly, we wish to thank the scientific committee, session chairs, speakers, and poster presenters for their outstanding contributions to the symposium and to the publication of this book. We are especially grateful to Leslie Nies and her superb staff at Serono Symposia USA, Inc., for their unflagging assistance with organization of the meeting, considerable financial assistance, and their patience throughout the process of organizing the symposium and publishing this volume.

BARRY B. BERCU
RICHARD F. WALKER

Contents

Contributors

JOSEPH P. ARENA, Department of Biochemistry and Physiology, Merck Research Laboratories, Rahway, New Jersey, USA.

EMANUELA ARVAT, Division of Endocrinology, Department of Clinical Pathophysiology, University of Turin, Torino, Italy.

MICHEL L. AUBERT, Department of Pediatrics, University of Geneva Medical School, Geneva, Switzerland.

MICHELE F. BELLANTONI, Department of Medicine, Johns Hopkins Bayview Medical Center, Johns Hopkins University School of Medicine, Baltimore, Maryland, USA.

BARRY B. BERCU, Departments of Pediatrics, Pharmacology, and Therapeutics, University of South Florida College of Medicine, Tampa, and All Children's Hospital, St. Petersburg, Florida, USA.

MICHAEL BERELOWITZ, Division of Endocrinology and Metabolism, State University of New York, Stony Brook, New York, USA.

MARC R. BLACKMAN, Department of Medicine, Johns Hopkins Bayview Medical Center, Johns Hopkins University School of Medicine, Baltimore, Maryland, USA.

J. EDWIN BLALOCK, Department of Physiology and Biophysics, University of Alabama, Birmingham, Alabama, USA.

PATRIZIA BORRELLI, Division of Endocrinology, Bambino Gesù Hospital, IRCCS, Rome, Italy.

CYRIL Y. BOWERS, Department of Medicine, Tulane University Medical Center, New Orleans, Louisiana, USA.

JOHN F. BRUNO, Division of Endocrinology and Metabolism, State University of New York, Stony Brook, New York, USA.

JAN BUSBY-WHITEHEAD, Department of Medicine, University of North Carolina, Chapel Hill, North Carolina, USA.

FRANCO CAMANNI, Division of Endocrinology, Department of Clinical Pathophysiology, University of Turin, Torino, Italy.

PAOLA CAMBIASO, Division of Endocrinology, Bambino Gesù Hospital, IRCCS, Rome, Italy.

MARCO CAPPA, Division of Endocrinology, Bambino Gesù Hospital, IRCCS, Rome, Italy.

PAOLO CARDUCCI, Institute of Anesthesiology and Intensive Care, Catholic University, Rome, Italy.

DANIELA CARTA, Service of Pediatric Endocrinology, Microcitemico Hospital, Cagliari, Italy.

FELIPE F. CASANUEVA, Department of Medicine, University Hospital, Faculty of Medicine, Santiago de Compostela, La Coruña, Spain.

MARIA ROSARIA CASINI, Service of Pediatric Endocrinology, Microcitemico Hospital, Cagliari, Italy.

GIAMPAOLO CEDA, Department of Geriatrics, University of Parma, and Stuard Hospital, Parma, Italy.

CHING H. CHANG, Department of Biochemistry and Physiology, Merck Research Laboratories, Rahway, New Jersey, USA.

HOWARD Y. CHEN, Department of Biochemistry and Physiology, Merck Research Laboratories, Rahway, New Jersey, USA.

KANG CHENG, Department of Biochemistry and Physiology, Merck Research Laboratories, Rahway, New Jersey, USA.

CHARLES J. COHEN, Department of Biochemistry and Physiology, Merck Research Laboratories, Rahway, New Jersey, USA.

LISA K. CONLEY, Department of Biological Sciences, University of Wisconsin, Milwaukee, Wisconsin, USA.

GIANLUIGI CONTE, Institute of Endocrinology, Catholic University, Rome, Italy.

EMILIANO CORPAS, Hospital General del Insalud, Guadalajara, Spain.

SALLIE B. COSGROVE, Department of Biochemistry and Physiology, Merck Research Laboratories, Rahway, New Jersey, USA.

VENITA DEALMEIDA, Department of Biochemistry, Molecular Biology and Cell Biology, Northwestern University, Evanston, Illinois, USA.

ROMANO DEGHENGHI, Europeptides, Argenteuil, France.

FRANCESCO DELLA CORTE, Institute of Anesthesiology and Intensive Care, Catholic University, Rome, Italy.

LAURA DE MARINIS, Institute of Endocrinology, Catholic University, Rome, Italy.

SUZANNE L. DICKSON, Anatomy and Human Biology Group, King's College, London, UK.

CARLOS DIEGUEZ, Department of Physiology, University of Santiago de Compostela, Santiago de Compostela, La Coruña, Spain.

JENNIFER DRISKO, Department of Biochemistry and Physiology, Merck Research Laboratories, Rahway, New Jersey, USA.

KEITH M. FAIRHALL, Division of Neurophysiology and Neuropharmacology, National Institute for Medical Research, The Ridgeway, Mill Hill, London, UK.

L. STEPHEN FRAWLEY, Department of Cell Biology and Anatomy, Medical University of South Carolina, Charleston, South Carolina, USA.

EASTER G. FRAZIER, Department of Biochemistry and Physiology, Merck Research Laboratories, Rahway, New Jersey, USA.

JENNY FRENKEL, Endocrinology and Diabetes Research Unit, Schneider Children's Medical Center of Israel, Petah Tikva, Israel.

ALAN R. FRIEDMAN, Animal Health Discovery Research, The Upjohn Company, Kalamazoo, Michigan, USA.

LAWRENCE A. FROHMAN, Department of Medicine, University of Illinois, Chicago, Illinois, USA.

MARIE C. GELATO, Department of Medicine, State University of New York, Stony Brook, New York, USA.

EZIO GHIGO, Division of Endocrinology, Department of Clinical Pathophysiology, University of Turin, Torino, Italy.

LAURA GIANOTTI, Division of Endocrinology, Department of Clinical Pathophysiology, University of Turin, Torino, Italy.

ANDREA GIUSTINA, Department of Clinical Medicine, University of Brescia, Brescia, Italy.

PAUL A. GODFREY, Department of Biochemistry, Molecular Biology and Cell Biology, Northwestern University, Evanston, Illinois, USA.

SILVIA GROTTOLI, Division of Endocrinology, Department of Clinical Pathophysiology, University of Turin, Torino, Italy.

ROGER GUILLEMIN, The Whittier Institute, La Jolla, California, USA.

S. MITCHELL HARMAN, Gerontology Research Center, National Institute on Aging, National Institutes of Health, Baltimore, Maryland, USA.

GERARD J. HICKEY, Department of Biochemistry and Physiology, Merck Research Laboratories, Rahway, New Jersey, USA.

BRUNO P. IMBIMBO, Mediolanum Pharmaceuticals, Milan, Italy.

LUCIEN ISRAËL, Department of Oncology, Avicenne Hospital, Bobigny, France.

TOM JACKS, Department of Biochemistry and Physiology, Merck Research Laboratories, Rahway, New Jersey, USA.

SIDNEY H. KENNEDY, Mood and Anxiety Division, Department of Psychiatry, University of Toronto, and Clarke Institute of Psychiatry, Toronto, Ontario, Canada.

TERESA M. KUBIAK, Animal Health Discovery Research, The Upjohn Company, Kalamazoo, Michigan, USA.

HAL LANDY, Growth and Metabolism, Serono Laboratories, Inc., Norwell, Massachusetts, USA.

Zvi Laron, Endocrinology and Diabetes Research Unit, Schneider Children's Medical Center of Israel, Petah Tikva, and Sackler Faculty of Medicine, Tel Aviv University, Tel Aviv, Israel.

Alfonso Leal-Cerro, Endocrinology Section, Department of Medicine, Virgin of the Rock Hospital, Seville, Spain.

Gareth Leng, Department of Physiology, The University Medical School, Edinburgh, UK.

Reid Leonard, Department of Biochemistry and Physiology, Merck Research Laboratories, Rahway, New Jersey, USA.

Sandro Loche, Service of Pediatric Endocrinology, Microcitemico Hospital, Cagliari, Italy.

Antonio Mancini, Institute of Endocrinology, Catholic University, Rome, Italy.

Kelly E. Mayo, Department of Biochemistry, Molecular Biology and Cell Biology, Northwestern University, Evanston, Illinois, USA.

Leslie A. McGuire, Department of Biochemistry and Physiology, Merck Research Laboratories, Rahway, New Jersey, USA.

Edoardo Menini, Institute of Biochemistry, Catholic University, Rome, Italy.

Vittorio Mignani, Institute of Anesthesiology and Intensive Care, Catholic University, Rome, Italy.

Teresa L. Miller, Department of Biochemistry, Molecular Biology and Cell Biology, Northwestern University, Evanston, Illinois, USA.

Frank A. Momany, Department of Medicine, Tulane University Medical Center, New Orleans, Louisiana, USA.

W. Michael Moseley, Animal Health Product Development, Growth and Performance, The Upjohn Company, Kalamazoo, Michigan, USA.

Anita Mynett, Division of Neurophysiology and Neuropharmacology, National Institute for Medical Research, The Ridgeway, Mill Hill, London, UK.

Michele Perrelli, Institute of Endocrinology, Catholic University, Rome, Italy.

PHILIPPE PLANCHON, Institute of Human Molecular and Cellular Oncology, Bobigny, France.

MANUEL POMBO, Department of Pediatrics, University Hospital, Faculty of Medicine, Santiago de Compostela, La Coruña, Spain.

SHENG-SHUNG PONG, Department of Biochemistry and Physiology, Merck Research Laboratories, Rahway, New Jersey, USA.

VERA POPOVIC, Institute of Endocrinology, Belgrade, Yugoslavia.

GRÉGOIRE P. PRÉVOST, Institute of Human Molecular and Cellular Oncology, Bobigny, France.

ERIC L. RICKES, Department of Biochemistry and Physiology, Merck Research Laboratories, Rahway, New Jersey, USA.

GUIDO RIZZI, Division of Medicine, Hospital of Saluzzo, Cuneo, Italy.

IAIN C.A.F. ROBINSON, Division of Neurophysiology and Neuropharmacology, National Institute for Medical Research, The Ridgeway, Mill Hill, London, UK.

MARC A. ROGERS, Department of Kinesiology, University of Maryland, College Park, Maryland, USA.

JESSE ROTH, Department of Medicine, Johns Hopkins Bayview Medical Center, Johns Hopkins University School of Medicine, Baltimore, Maryland, USA.

ALAN SCHWARTZ, Department of Medicine, Johns Hopkins Bayview Medical Center, Johns Hopkins University School of Medicine, Baltimore, Maryland, USA.

SAMIR SHAH, Growth and Metabolism, Ares Serono, Geneva, Switzerland.

MARY T. SHEEHAN, Department of Medicine, State University of New York, Stony Brook, New York, USA.

AVIVA SILBERGELD, Endocrinology and Diabetes Research Unit, Schneider Children's Medical Center of Israel, Petah Tikva, Israel.

PHILIP L. SMITH, Department of Medicine, Johns Hopkins Bayview Medical Center, Johns Hopkins University School of Medicine, Baltimore, Maryland, USA.

ROY G. SMITH, Department of Biochemistry and Physiology, Merck Research Laboratories, Rahway, New Jersey, USA.

RICHARD G.S. Spencer, Gerontology Research Center, National Institute on Aging, National Institutes of Health, Baltimore, Maryland, USA.

THOMAS E. STEVENS, JR., Department of Medicine, Johns Hopkins Bayview Medical Center, Johns Hopkins University School of Medicine, Baltimore, Maryland, USA.

KERRY J. STEWART, Department of Medicine, Johns Hopkins Bayview Medical Center, Johns Hopkins University School of Medicine, Baltimore, Maryland, USA.

GREGORY B. THOMAS, Division of Neurophysiology and Neuropharmacology, National Institute for Medical Research, The Ridgeway, Mill Hill, London, UK.

MICHAEL O. THORNER, Department of Medicine, University of Virginia, Charlottesville, Virginia, USA.

JORDAN D. TOBIN, Gerontology Research Center, National Institute on Aging, National Institutes of Health, Baltimore, Maryland, USA.

FRANCO J. VACCARINO, Department of Psychology, University of Toronto, and Clarke Institute of Psychiatry, Toronto, Ontario, Canada.

DOMENICO VALLE, Internal Medicine II, Catholic University, Rome, Italy.

NATHALIE VEBER, Institute of Human Molecular and Cellular Oncology, Bobigny, France.

JANET VITTONE, Department of Community and Internal Medicine, Mayo Clinic, Rochester, Minnesota, USA.

RICHARD F. WALKER, Department of Biochemistry and Molecular Biology, University of South Florida College of Medicine, Tampa, and All Children's Hospital, St. Petersburg, Florida, USA.

WILLIAM B. WEHRENBERG, Department of Health Sciences, University of Wisconsin, Milwaukee, Wisconsin, USA.

DOUGLAS A. WEIGENT, Department of Physiology and Biophysics, University of Alabama, Birmingham, Alabama, USA.

MATTHEW J. WYVRATT, Department of Medicinal Chemistry, Merck Research Laboratories, Rahway, New Jersey, USA.

JOSÉ L. ZUGAZA, Molecular and Cellular Endocrinology Laboratory, Faculty of Medicine, University of Santiago de Compostela, Santiago de Compostela, La Coruña, Spain.

Part I

Overview: Historical Perspective

1

Growth Hormone Releasing Factor: A Brief History of Its Time

ROGER GUILLEMIN

The first proposal for the existence of a growth hormone releasing substance of hypothalamic origin was probably made by Seymour Reichlin in 1959. Reichlin had shown that rats made obese by some (rather extensive) hypothalamic stereotaxic lesions had shorter long bones than controls and that these bones had narrower epiphyseal cartilages. It was to take 23 years eventually to validate the concept by the isolation, amino acid sequencing and reproduction by total synthesis of the hypothalamic growth hormone releasing factor (GHRF).

Why was that? Initially, there were problems of methodology: first, to design an adequate bioassay for such a substance; then to design methods of isolation of peptides present in picomolar amounts in kilogram quantities of fresh (brain) tissues; further, to design methods of sequencing the small amounts of peptides that would be involved.

There were also conceptual problems. At the time of Reichlin's seminal paper, the existence of hypothalamic hypophysiotropic releasing factors was still a hypothesis. So was that of their peptide nature; so was that of the size of such peptides, the favored idea being that they would be small like the 9-residues vasopressin and oxytocin.

Several laboratories started in earnest to look for a GHRF in the 1960s.

Using the width of the tibial cartilage of hypophysectomized rats as a bioassay for growth hormone released in the short-term incubation or tissue cultures of the (rat) pituitary, we could never demonstrate satisfactorily the existence of a GHRF in extracts of sheep hypothalamus in my laboratory at Baylor College of Medicine in Houston.

Similarly, we had no significant results when the same fluids were assayed by the early radioimmunoassays developed by Berson and Yalow and conducted in collaborative studies with Wilson Rodger and John Beck at the McGill University and with Sy Reichlin then at Farmington.

At the same time, nobody else was more successful.

I, for one, decided to forget looking for a GHRF and instead pursue the characterization of other hypothalamic releasing factors (anyone!) for which we had a reliable bioassay. And, in 1969, we isolated and established the primary structure of thyrotropin releasing factor (TRF), the first of the postulated hypothalamic hypophysiotropic releasing factors, as Schally's laboratory in New Orleans also concluded.

That fact, in itself, justified and validated efforts at isolating other releasing factors of hypothalamic origin. Indeed, in 1971, luteinizing hormone releasing factor (LHRF) was isolated and characterized by Schally's laboratory using pig brains, and in my laboratory at the Salk Institute in La Jolla, California, using sheep brains. The two molecules turned out to be identical and were small (10-residue) peptides.

That same year, the search for a GHRF was resumed. I asked a new postdoctoral fellow in my laboratory, Paul Brazeau, from the University of Sherbrooke in Quebec, Canada, to set up the radioimmunoassay (RIA) for rat growth hormone (rGH) in view of its sensitivity and specificity. I had decided to look for GHRF using the same tissue culture method (for rat pituitary cells) that Wylie Vale had designed earlier in the laboratory at Salk and which had been so successful in the isolation of LHRF. With Roger Burgus, Wylie Vale, and Paul Brazeau, we could never consistently show the presence of GHRF but instead and to our surprise, we isolated and characterized an inhibitor of the secretion of growth hormone. That was somatostatin.

Meanwhile, Schally's group in New Orleans had been reporting the isolation of growth hormone releasing hormone (GHRH) using as an assay the depletion of growth hormone (GH) content in rat pituitary halves in short-term incubations as well as release of GH from similarly incubated hemipituitaries, as I recall, the GH levels being measured by the same tibial cartilage assay with which we had been so unsuccessful before. I remember going with Roger Burgus and Wylie Vale to a meeting organized by the New York Academy of Science, at which Schally and Arimura were to present their results. I had slides that I intended to show to report our negative results using the very same methodology now used by Schally et al. The results presented at that meeting by Schally were so elegant, so persuasive (I think that a partial sequence was reported), that I decided not to say anything in public about our negative results, bowing, though somewhat incredulously, to what Schally called his "superior methodology."

In the review they published a few months later in *Science*, Schally et al. acknowledged the identity of their GHRH with a fragment of the α-chain of hemoglobin, thus recognizing it as an artifact, while, in the same review, casting doubt about the existence of a hypothalamic GH release-inhibiting factor.

To this day, I have remained puzzled at the sequence of events and the intellectual mechanisms that associated purification steps leading to the isolation of that fragment of hemoglobin and the bioassay results, in what appeared in Schally's laboratory as elegant chemistry and impeccable bioassay design and experiments. In my laboratory (and most others) the hemoglobin fragment GHRH did not stimulate release of GH in vitro or in vivo, when assayed by the now well-established radioimmunoassay for (rat) GH.

In June 1975, I learned from Henry Friesen on the occasion of a lecture I had given at the annual meeting of the Canadian Association for the Advancement of Science (CAAS) of John Hughes and Hans Kosterlitz in Scotland purifying a peptide of brain origin that they called enkephalin, which had biologic activities similar to those of opiates. I had never heard of the concept of these endogenous opioid peptides, but I knew that morphine had been reported to be a powerful stimulus for secretion of GH in man and animals. Could these opioid peptides be the endogenous releasers of GH? Using our large supplies of extracts of hypothalamic tissues, in a few months I had isolated three peptides with opiate-like activity, larger than the enkephalins α-, β-, and γ-endorphins, which turned out to be fragments of the N-terminal of what C.H. Li had called β-lipotropin. The endorphins indeed stimulated secretion of GH in vivo but not in vitro; the effect in vivo was inhibited by naloxone. But naloxone did not inhibit the spontaneous pulsatile release of GH in vivo. GHRF was not one of the endogenous opioid peptides.

At about the same time, Victor Mutt and collaborators reported that somatostatin existed in gut extracts in several molecular sizes, the 14 amino acid residue we had isolated earlier being the N-terminal to larger forms of 28 or 27 residues. All had biologic activity identical to or greater than that of the tetradecapeptide. We confirmed with Peter Bohlen and Nicholas Ling the presence of somatostatin-28 in the hypothalamus.

I thought that it was time to start again looking for GHRF in the hypothalamus. Paul Brazeau returned to the Salk Institute from McGill University where he then was. It was decided to use again a completely in vitro approach with the pituitary tissue culture method, and the RIA for rGH to assess the amounts of GH released in the incubated medium. We also started a fresh collection of fragments of rat hypothalamus. With Peter Bohlen in charge of the purification steps we rapidly purified GRF, knowing how to remove the several forms of somatostatin. With Richard Lubin we prepared monoclonal antibodies to GHRF using completely in vitro methodology. We could never, however, obtain the GRF component in pure enough form to show a single N-terminal residue for technical reasons that we never fully explained.

Over the preceding 3 or 4 years had appeared several papers from Larry Frohman's group reporting patients with acromegaly and no pituitary adenoma but with a peripheral tumor such as carcinoids, islet cell tumors, and others. Frohman had suggested that these tumors may be ectopic sources of GHRF or GHRF-like molecules. In 1980 I obtained a (very) small fragment of such a tumor through Jack Gerich and Bernard Scheithauer at the Mayo Clinic. We showed that it did not contain GH but that it contained, indeed, small amounts of a GH releasing substance. I asked Ron Evans at the Salk Institute whether he could clone the message for that molecule. Evans was not enthusiastic as the tissue had not been preserved adequately. And Frohman was not interested in collaborating in sending tissues from patients he was reporting about.

In 1981 I gave the plenary lecture at the annual meeting of the French Endocrine Society (Société Française d'Endocrinologie). The lecture was entitled, "Evidence for Central Nervous System Control of Somatic Growth." It took place in the same room, at the old Faculté de Médecine in Paris, where a hundred years earlier, Pierre Marie had described acromegaly. We still did not know what GHRF was. I closed the lecture by mentioning the reports by Frohman and a similar recent case of Michael Thorner. I told this mostly clinical audience of my interest in acquiring the tumor should they see such a patient in their practice or at the hospital.

A few months later, I received a letter from Geneviève Sassolas, a young assistant professor at the Medical School in Lyons, describing a patient with full-blown acromegaly, no pituitary adenoma, but with two rather large tumors in the pancreas. Arrangements were made to get the tumor to my laboratory at the Salk Institute. A few days later we had in the laboratory in La Jolla about 250 g of a rather heterogeneously looking tumor that Fusun Zeytin had gone to collect in the operating room in impeccable condition. In 48 hours Paul Brazeau had shown in his now well-standardized bioassay that one region of the tumor had very high GHRF activity, while another appeared to have somatostatin-like activity. Ten days later Peter Bohlen had completed the first step of high-performance liquid chromatography (HPLC) of the extract of 25 g of the part of the tumor containing GHRF activity. All the GHRF activity was in one peak, well separated from three zones containing ir-somatostatins. Two more HPLC and bioassays yielded three peaks, shown to be homogeneous, with GHRF activity. Amino acid compositions showed them to be respectively composed of 44, 40, and 37 amino acids. All three had the same N-terminal. Sequencing of the most prominent peak on the recently acquired ABI gas phase sequencer was started by Fred Esch while Nicholas Ling established the C-terminals. In just over four weeks from receiving the tumor in the lab, we had the amino acid sequence of the three forms of human pancreatic (hp)GHRF, the full sequence being 44-amide, the other two forms being, as the free acids, the 37 and 40 N-terminal sequences of the 44-amide. In the in vitro assay, 44-NH$_2$ was the most potent of the three. As the sequence came out from the ABI, Ling started the solid phase synthesis of the 44-amide hpGRF. Brazeau showed that the synthetic replicate was equipotent to the native hpGHRF-44 in the in vitro assay and Bill Wehrenberg showed the synthetic replicate to be fully active in vivo in the rat both in normal animals with or without Nembutal anesthesia and in animals pretreated with antisera to somatostatin. Brazeau started immunizing rabbits with the synthetic replicate of hpGHRF-44 to produce polyclonal antibodies. Within one month, we had prepared a small affinity chromatography column with those antibodies, and isolated by affinity chromatography the GHRF from the extraction of the hypothalamus-median eminence regions of three human brains. The sequencing by Esch of this human hypothalamic GHRF (hGHRF) showed it to be identical to hpGHRF-44 amide. The human hypothalamic extracts also contained GHRF-40.

Twenty-three years after the astute observations of Reichlin and his postulate of a humoral hypothalamic control of the secretion of growth hormone, we had finally established the primary structure of human GHRF and reproduced it in unlimited amounts by total synthesis.

In the 3 to 4 months that followed, the group in my laboratory at the Salk Institute performed all the fundamental biology of GHRF—physiology, mechanism of action at the cellular level, activity structure relationships, CNS localization, isolation and characterization from other species, radio-immunoassays, cloning of the cDNA for its precursor in collaboration with the group at the Roche Institute of Molecular Biology, mechanisms of antagonism by somatostatin, negative feedback by insulin-like growth factor (IGF)-1, -2, early clinical studies in normal subjects as well as hypothalamic dwarfs (in collaboration with Mel Grumbach and his group)—in a rather extraordinary series of reports. All of these conclusions were later confirmed by others, thus establishing the significance of that important and elusive molecule.

Somewhat before I obtained the GHRF-containing tumor from Lyons, I heard that Michael Thorner had also obtained from one of his patients an islet tumor that he suspected could be a source of GHRF. Eventually that tissue was split between my laboratory and that of Wylie Vale. Because the tumor from Lyons was so much larger and so rich in GHRF activity, we did not process the tissue from Michael Thorner until after we had completed the isolation and characterization of the GHRF from the tissue obtained from Geneviève Sassolas. The tumor from Thorner yielded essentially GHRF-40, i.e., the first 40-N-terminal residues of GHRF-44 amide, for reasons that were never explained. Wylie Vale, Joachim Spiess, Jean Rivier and Catherine Rivier found the same results independently.

I am writing this short historical review 12 years after the isolation of GHRF, on October 13, 1994, lying in bed at UCSD University Hospital, in the Clinical Research Center ward. I am here as one of the subjects of an extensive clinical investigation project conducted by Sam Yen to study the effects of GHRF in elderly men and women following the report of Daniel Rudman in 1989 of the beneficial effects of hGH in that same population. The early results appear quite interesting. I think that GHRF or one of its tailored analogues will be a major medication in clinical medicine. The next page of the history of GHRF is now to be written by Industry. I think that there will be a major market for that molecule. Because I am writing this short piece from memory, there are no references listed. Those should be easily available to whoever would want to go back to those historical papers.

2

Xenobiotic Growth Hormone Secretagogues: Growth Hormone Releasing Peptides

CYRIL Y. BOWERS

A number of different and still evolving concepts, strategies, approaches, and techniques underlie the story of the growth hormone releasing peptides (GHRPs). Even though the origin of the GHRPs was artificial, many investigators have validated these peptides as distinct chemical entities with novel actions on growth hormone (GH) release in animals and in humans. They appear to be valuable both diagnostically and therapeutically as well as basically because they may reflect the physiologic actions of a putative endogenous GHRP system as well as a natural GHRP-like ligand. A major challenge is to better understand the relationship between the actions of the GHRPs, growth hormone releasing hormone (GHRH), and somatotropin release inhibiting factor (SRIF) on GH release. Evidence strongly supports that GHRPs act on both the hypothalamus and pituitary to release GH. It is the hypothalamic component of the GHRP action that is still most incompletely understood as well as the relative importance of the hypothalamic and pituitary action in the in vivo release of GH. Most evidence indicates that the direct pituitary in vitro action of GHRP+GHRH on GH release is essentially additive and, thus, a direct pituitary action of these two combined peptides seemingly does not explain the not infrequent dramatic synergistic and marked effect on GH release in vivo in both animals and humans.

New dimensions in the release of GH are revealed by the in vivo actions of the GHRPs as well as by the combined actions of GHRP and GHRH. The complementary in vivo action of GHRP+GHRH on GH release that occurs in vivo at the pharmacologic level appears to result from a unique action of GHRP on the hypothalamus. In a primary way, the latter probably involves neither the increased release of GHRH nor the decreased release of SRIF but instead has been tentatively considered to involve release of a hypothetical factor designated the U-factor, or unknown factor. U-factor, released from the hypothalamus, acts together with GHRH and in part with GHRP on the pituitary to synergistically release GH. This is not envisioned

9

as a complicated explanation but it is convoluted and cumbersome and forecasts other new, still to be defined, hypothalamic factors that may be involved in the release of GH. Gradually evolving is how this primarily pharmacologic conceptual model might be related but yet different in specific ways to a physiologic conceptual model that involves the role and action of a putative GHRP-like ligand. In a physiologic model, the putative GHRP-like ligand would be envisioned as being secreted so that it acts first on the hypothalamus via a paracrine action before acting on the pituitary. Also, it is necessary to envision how U-factor+GHRH+GHRP act so uniquely to release GH synergistically. Our postulate is that this occurs because U-factor and GHRH attenuate the pituitary inhibitory action of SRIF on GH release.

To present an overview of this seemingly unusual and surprising story, basic and clinical findings of the GHRPs have been selected that arbitrarily and tentatively appear at this time to represent milestones. Rather than being discovered, the GHRPs were invented utilizing a closely integrated empirical-theoretical approach to design the peptides.

The empirical aspect of this approach emanated from the strategies, approaches, and techniques used over 15 years to chemically identify and biologically characterize the hypothalamic hypophysiotropic hormones as well as to design, synthesize, and biologically characterize small synthetic peptide agonist and antagonist analogues. From in vitro structure-activity relationships of small peptides we learned to expect the unexpected in regard to which amino acid sequence may release a particular single hormone or a certain pattern of pituitary hormones. Admittedly, at times large dosages of the small peptides may have been required to produce these effects; however, not infrequently this was offset by the peptide's specific action.

The theoretical aspect of the GHRP development approach included sophisticated, complicated, and continually evolving computational studies by Momany. By this theoretical approach, disparate findings that were revealed empirically were expanded into a working rationale and more meaningful conformational structures of the peptides. It was learned that considerably different amino acid sequences could assume similar conformations and that peptides with different sequences could release the same hormone. Importantly, the activity could be understood in terms of conformational analysis of the low-energy structures of the peptides. Also, sometimes different conformations appeared to release the same hormone, possibly indicating that certain peptides may bind at different sites on the receptor but nevertheless still activate the receptor presumably via a common mechanism.

TABLE 2.1. GHRP milestones—basic (1976–1985).

GHRP chemistry		GHRP-6 biologic effects	
1976	1. First GHRP	1980–82	1. In vitro and in vivo
1977–80	2. Intermediate GHRPs	1983	2. Desensitization
1979–80	3. Dlys³ GHRPs	1983–84	3. Synergism, in vivo
1980	4. GHRP-6 born	1983–84	4. Antiserum, GHRH, and SRIF in vivo
		1984	5. Perifusion
		1984	6. Repeated pulses, in vivo
		1985	7. Calcium dependent, in vitro
		1985	8. Different receptors

GHRP Milestones—Basic

1976–1985 (Table 2.1)

Key GHRPs

The GHRPs initially evolved in 1976 from assessment of the natural enkephalin pentapeptides and their analogues for GH releasing activity. The first key peptides, empirically obtained, were the C-terminal amidated DTrp² Met enkephalin in 1976 (1) and the DTrp³ peptides in 1977 (2, 3) listed in Table 2.2. These two peptides, which represent two different classes of pentapeptides, specifically release GH from the pituitary at high dosages in vitro but not in vivo and they do not have opiate activity (4). A number of closely related analogues were inactive for GH releasing activity. The data obtained from conformational energy calculations were used to search for structural features common to these two DTrp peptides and to various active and inactive peptides related to the key peptides listed in Table 2.2. (5). Conformational properties thought to be responsible for biologic activity were then incorporated into the design of the new peptides. Concomitantly, a linear empirical rather than a conformational theoretical approach was utilized in the design process. The linear approach consisted of substitution of amino acids with specific chemical properties at select

TABLE 2.2. Key GHRPs.

Active only in vitro		Inactive in vitro
1. TyrDTrpGlyPheMetNH₂	(DTrp²)	1. TyrGly²GlyPheMetNH₂
2. TyrAlaDTrpPheMetNH₂	(DTrp³)	2. Trp
3. TyrDTrpDTrpPheNH₂	(DTrp²,³)	3. Phe
4. TyrDTrpAlaTrpDPheNH₂	(Ala spacer)	4. Pro
		5. Sar
		6. DVal
		7. DAla
		8. DLeu

Dose: 100 μg/ml.

TABLE 2.3. Key GHRPs.

Active in vitro AND in vivo	
1. HisDTrpAlaTrpDPheNH$_2$	
2. HisDTrpAlaTrpDPheLysNH$_2$	(GHRP-6)
3. AlaHisDTrpAlaTrpDPheLysNH$_2$	
4. AlaHisDβNalAlaTrpDPheLysNH$_2$	(GHRP-1)
5. HisDβNalAlaTrpDPheLysNH$_2$	
6. DAlaDβNalAlaTrpDPheLysNH$_2$	(GHRP-2)

DβNal = D-2-naphthylalanine.

positions of the peptide. This was helpful in improving binding as well as intrinsic activity more than in providing a better understanding of the bioactive conformation of the peptides. Results of the empirical approach became the basis of the theoretical approach and vice versa.

The key peptides in Table 2.3 were sequentially developed and were increasingly more potent. Finally, after 4 years of developing only in vitro active peptides, new peptides including GHRP-6 with both in vitro and in vivo activity were discovered (6–8). Noteworthy is that during the first 4 years the basic model or blueprint for the design of future peptides kept changing. This reflected new empirical findings as well as new insight and understanding of the structure-activity relationship of these peptides by conformational studies. Advanced theoretical studies in the more recent understanding of the conformation of the GHRPs have been published by Momany (9). GHRP-6, GHRP-1, and GHRP-2 are examples of highly potent small peptides that specifically release GH in animals and, importantly, also in humans. Besides high potency and specificity of action, a concerted effort was made to keep these peptides small in size with physical chemical properties that impart favorable chemical and biologic stability, solubility, low toxicity, and rapid as well as cheap synthesis.

GHRP-6 was developed 2 years before the isolation of native GHRH (1-44)NH$_2$ in 1982 and had many of the characteristic actions of a hypothalamic hypophysiotropic hormone, presumably that of native GHRH, which led to the working hypothesis that GHRP-6 was acting via the putative native GHRH receptor. However, since GHRP-6 contained D amino acid residues, it was also immediately apparent that the amino acid sequence of native GHRH would not be directly related to GHRPs. Despite a number of similarities in the actions of GHRP-6 and GHRH—e.g., both peptide actions depend on Ca^{2+} in vitro (10), increase the GH response following pretreatment with SRIF antiserum, inhibit the GH response by pretreatment with GHRH antiserum (11), and are inhibited by SRIF—it became increasingly evident that GHRP-6 and GHRH act via different receptors to release GH (11, 12). Also, this evidence included full GH releasing activity of GHRH when the GH response was completely desensitized by repeated administration of GHRP-6 (12), selective inhibition of the GHRP-6 but not the GHRH GH response by the DLys^3 GHRP-6 antagonist (13), and the additive effects in vitro and synergistic effects in vivo by maximal dosages of GHRP-6+GHRH.

In the 1984 in vitro studies of Badger et al. (14) and in vivo studies of McCormick et al. (15) in which repeated pulses of GHRP-6 were given at 20- or 90-min intervals, GH was persistently released over 2- to 8-hour periods with or without small decreases in the GH responses. The studies of Badger et al. were performed using dispersed pituitary rat cells while the McCormick et al. studies were performed in adult male rats pretreated with a dopamine β-hydroxylase inhibitor and SRIF antiserum with iv pulse delivery of GHRP-6. These results emphasize that pulse delivery of GHRP-6 obviated the desensitization response both in vitro and in vivo. This is in contrast to our studies, in which the GHRP-6 in vivo response produced definite desensitization of the GH response after sc administration. In the perifusion studies of Badger et al., continuous administration of GHRP-6 desensitized the GH response. This again indicates that desensitization occurs at the pituitary level and that the desensitization probably occurs in vivo as a result of a direct pituitary rather than a hypothalamic action of GHRP-6, i.e., release of SRIF. Our crossover GHRP-6 induced desensitization studies in which GHRH is active at a time when GHRP-6 is inactive do not support the possibility that desensitization occurs because SRIF is released. If SRIF had been released, the GH responses of both GHRP-6 and GHRH would have been expected to be decreased. GHRP desensitization is an important point to understand and it may very likely be multifactorial in origin.

1986–1992 (Table 2.4)

From direct results of in vitro hypothalamic-pituitary incubate studies in 1987 and indirect results in which in vivo GH releasing activity did not appear to be explained by the in vitro effects of GHRP-6 and GHRH on GH release, GHRP-6 was postulated to act on both the hypothalamus and pituitary to release GH in vivo (16).

TABLE 2.4. GHRP milestones—basic (1986–1992).

GHRP chemistry		GHRP-6 biological effects
1988 1. GHRP-1	1986–87	1. H-P Incubates
1990 2. GHRP-2	1989	2. Continuous GHRP-6, GHRH pulses
	1989	3. Specific binding sites
	1989	4. Synergism, in vitro
	1990	5. Oral GHRP-6
	1991	6. Non–GHRH-mediated responses, monkeys
	1991	7. Protein kinase C, GHRP-6
	1991	8. Older female rats, chronic
	1992	9. GHRP-6, GHRH, SRIF
	1992	10. Depolarization, Ca^{2+} channels

In 1988 the second generation GHRP-1 heptapeptide, AlaHis-DβNalAlaTrpDPheLysNH$_2$, and in 1990, the third generation GHRP-2 hexapeptide, DAlaDβNalAlaTrpDPheLysNH$_2$, were developed. In rats, GHRP-2 is three times more potent than GHRP-1 and six times more potent than GHRP-6.

Results of other investigators performing various types of studies with GHRP-6 were starting to be published during this time period. Clark et al. (17) found that 6- to 8-hour infusion of GHRP-6 to rats enhanced the GHRH-induced GH responses and that GHRH antiserum pretreatment reduced the GH response of GHRP-6. They suggested that the potent effects of GHRP-6 in conscious rats primarily reflected its hypothalamic action, either to inhibit SRIF release or to stimulate GHRH release, with which it synergizes to release GH. These most provocative results still seem to need a special explanation because it seems very likely that the GH response during continuous GHRP-6 infusion should have been desensitized.

Codd et al. (18) at SmithKline Beecham described specific binding sites for GHRP-6 in membrane preparations of both the hypothalamus and pituitary, again supporting a dual action of GHRP-6 at these two anatomical sites.

A series of studies performed on GHRP-6 and other GHRPs by the Merck group (Cheng et al.) (19) revealed that GHRP-6 acts via different intracellular pathways than GHRH because GHRH, but not GHRP-6, raised adenosine $3',5'$-cyclic monophosphate (cAMP) levels in vitro. In addition, they reported that in vitro GHRP-6+GHRH released GH synergistically rather than additively, as we had previously reported, and that cAMP was synergistically increased by the two peptides in vitro. If the in vitro effect of the two peptides is synergistic, this would imply that in vivo synergism occurs because of a complementary combined action of GHRP-6 and GHRH on the pituitary via distinct receptors and distinct intracellular signal pathways. If the in vitro effect is additive rather than synergistic, the marked synergistic release of GH that occurs in vivo from the combined action of GHRP-6 and GHRH would be explainable only by an action of GHRP-6 on the hypothalamus.

A hypothalamic model to explain synergism seems more complicated mainly because evidence indicates that synergism probably is not mediated via an effect on the hypothalamus to release endogenous GHRH or to inhibit endogenous SRIF release. For example, GHRP produces synergism even with combined maximal dosages of GHRH and thus, in the presence of excess exogenous GHRH, any effect that GHRP may have on endogenous GHRH release would be insignificant. Also, when rats are pretreated with SRIF antiserum the GH response to GHRP-6 is further increased. If GHRP-6 release of GH had depended on inhibition of SRIF release or on inhibition of the pituitary action of SRIF, the SRIF antiserum would not have been expected to further increase the GH response to GHRP-6, but it did. Thus, we have concluded that if the synergistic action of GHRP+GHRH on GH release occurs via a hypothalamic action of GHRP, this hypothalamic action still remains unclear. Tentatively we have postulated that GHRP-6 acts on the hypothalamus to release U-factor and that U-factor, in combination with GHRH and GHRP, acts on the pituitary to induce synergism in part by attenuating the inhibitory action of SRIF on the pituitary to release GH (13).

In 1991 Malozowski et al. (20) found that propranolol, which inhibits SRIF release, augments the GH response to GHRP-6 in male rhesus monkeys. Also GHRP-6 was lower in potency than GHRH in the monkey but at maximal dosages GHRP-6 released greater amounts of GH than GHRH. They concluded that GHRP-6 did not reduce SRIF tone and that a mechanism other than release of endogenous GHRH must be involved in the release of GH by GHRP-6 since it can produce a greater maximal GH response than GHRH.

In 1991 Cheng et al. (21) obtained evidence for a role of protein kinase C in the induced release of GH by GHRP-6 from rat primary pituitary cells. The results of Pong et al. (22) in 1992 support that GHRP-6 elevates $[Ca^{2+}]_i$ by a depolarization mediated influx of Ca^{2+} via voltage-dependent Ca^{2+} channels.

1992–1994 (Table 2.5)

Hexarelin is a new hexapeptide GHRP developed by Deghenghi et al. (23) in 1992. It has the same amino acid sequence as GHRP-6 but is different from GHRP-6 by a single C-2 methyl substitution on the indole ring of the $DTrp^2$ residue. Although hexarelin was initially considered to be more potent and to have a longer duration of action than GHRP-6 our studies indicate the activities of these two peptides are essentially the same. Nevertheless, hexarelin does very effectively release GH in animals and in humans. Studies on the mechanism of action of hexarelin have been reported

TABLE 2.5. GHRP milestones—basic (1992–1994).

GHRP chemistry		GHRP biological effects		
1992	1. Hexarelin	1992	1.	Pit tumors, rat
1993–94	2. Nonpeptide GHRPs	1992	2.	GH mRNA
		1992	3.	GHRH+SRIF antiserum
		1992	4.	GH release and Pit cAMP, rat
		1993	5.	Protein kinase C, GHRP-1
		1993	6.	H arcuate neurons
		1994	7.	Portal blood GHRH, sheep
		1994	8.	Stalk-section, pig
		1994	9.	Transient, persistent $[Ca^{2+}]_i$ rise
		1994	10.	Receptor subtypes, species

by Conley et al. (24). Hexarelin was found to release GH more effectively than GHRH in conscious male rats during the time periods of higher somatostatin tone, suggesting that it may influence the release of SRIF and/ or the pituitary action more readily than GHRH.

Most noteworthy and indeed a seminal accomplishment has been the development of the first peptidomimetic GHRP (L-692,492) by Smith, Schoen, and Cheng et al. at Merck (25, 26). These fused benzolactam derivatives were found to release GH in vitro and in vivo. Several types of evidence strongly support that this new chemical class of organic compounds acts via the same receptor and intracellular pathway as GHRP. Additional chemical modifications of this basic nonpeptide GHRP have produced increasingly more potent compounds (L-692,585) (27).

Other studies of particular note during this time period were performed by Walker and Bercu. Chronic administration of GHRP-6+GHRH to Fisher rats for 60 days was found to prevent and/or decrease the high incidence of pituitary tumors that spontaneously occur in these rats (28). Understanding the mechanism involved in these surprising effects and the potential application of this approach is most exciting. Another important study by Walker and Bercu was the chronic administration of GHRP-6+GHRH to older female rats with a normal restoration of GH secretion (29).

Only a very limited number of studies involve the effect of GHRP at the transcriptional level. Chihara et al. (30) found that GHRH, but not GHRP-6, increased pituitary GH messenger RNA (mRNA) levels after administration of the peptide for 3 days. This again indicates that the GHRPs do not release GH by acting on the hypothalamus to release GHRH. More prolonged studies of this type are of obvious importance.

A basis for further understanding how GHRP, GHRH, and SRIF might act together or how their actions may be interrelated to regulate GH secretion are discussed in the two excellent critiques by Robinson (31) on the regulation of GH secretion by GHRH and SRIF and by Kraicer and Sims (32) on how SRIF acts to influence the ionic mechanisms involved in GH secretion.

Pertinent studies by Chihara et al. (30) were performed in rats concomitantly pretreated with antiserum to GHRH and to SRIF before GHRP-6, GHRH, and GHRP-6+GHRH were administered. A GHRH analogue (GHRH-A) was used in these studies that did not cross-react with the GHRH antiserum. GHRP-6 produced very little GH release again indicating that it depends on the presence of endogenous GHRH. Although a moderate release of GH was induced by the GHRH-A, a marked synergistic release of GH was induced by GHRP-6 and GHRH-A. A more direct interpretation of these results is that GHRP-6 and GHRH-A act directly on the pituitary to synergistically release GH. It is apparent that the synergism occurred in the absence of endogenous SRIF. Thus, if U-factor is involved in inducing synergism in combination with GHRH via attenuation of the pituitary action of SRIF, this action of the putative U-factor would not appear to totally explain how this occurs; however, U-factor may have other pituitary actions and GHRP may have other actions on the hypothalamus than to release U-factor.

The more recent carefully controlled study of Hickey et al. (33), in which synergism of the peptidomimetic GHRP(L-692,585,[L])+GHRH occurred in the pig after a pituitary-stalk section adds another new dimension. Interrelating the similarities and differences between studies of Chihara et al. (30) and Hickey et al. (33) seems instructive and once better understood could be helpful in revealing how the GHRP+GHRH synergism on GH release is induced. In both studies endogenous GHRH and SRIF are prevented from acting on the pituitary but in the pig studies the hypothalamus and pituitary are disconnected and thus the hypothalamic action of GHRP was prevented, while in the rat studies of Chihara et al. the antiserum did not prevent GHRP from acting on the hypothalamus. Regardless of these differences, the GHRP-6 or [L]+GHRH synergism induced in both studies suggests, especially in the pigs, that it occurs because both peptides are acting directly on the pituitary. Another possible interpretation of the results is that synergism occurs for different reasons in these two studies. In the pig study, it could be envisioned that the synergism might not have occurred if SRIF had been restored in physiologic amounts because the hypothalamic action of [L] would be prevented as a result of the stalk section and because the direct pituitary effect of [L] in attenuating the inhibitory action of SRIF on GH release would be projected to be minor. In the study of Chihara et al., the hypothalamus was not disconnected, and thus the hypothalamic action of GHRP-6 could occur and could induce the synergism. Of course, an alternate explanation for the GHRP+GHRH induced synergism could be a direct pituitary action of the two peptides. These are important issues for eventually understanding the synergism of GH release induced by GHRP+GHRH and also how GHRPs act to release GH.

Results of two studies performed in our laboratory seem important to better understand the in vivo action of GHRP (34). In the first study, the concomitant effects on GH release and on pituitary cAMP levels were determined after GHRP or GHRH administration to rats. After GHRP, the cAMP rise was small when the GH rise was large while after GHRH the rise of cAMP and GH were both large, especially cAMP. Despite the known dependency of GHRP on endogenous GHRH in order to release GH, the GHRP GH release probably did not occur because it released endogenous GHRH. If it had, the pituitary cAMP rise would have been much greater. Instead GHRP was postulated to have a permissive dependency on endogenous GHRH in order to release GH in vivo. In the second study, the concentration that inhibits 50% (IC_{50}) of SRIF was determined on the inhibition of the GH response to GHRP, GHRH, and GHRP+GHRH. These studies were performed in vitro and in vivo not only to determine whether any evidence could be obtained to indicate that GHRP, compared with GHRH, may more effectively attenuate the inhibition of SRIF on the GH response, but also to learn whether the two peptides together might have a special propensity to attenuate the action of SRIF, especially in vivo. From the results of these studies GHRP and GHRH alone were found to equally attenuate the action of SRIF on GH release in vitro and in vivo; however, the combined effects of GHRP+GHRH were additive in vitro but synergistic in vivo. Again, this unique in vivo effect has been envisioned to be the result of the action of GHRP on the hypothalamus to release U-factor.

Two studies by other investigators indicating that GHRP-6 acts on the hypothalamus are those of Dickson et al. (35), and Guillaume et al. (36). The studies of Dickson et al. revealed activation of the hypothalamic arcuate neurons by demonstrating an increase of c-*fos*–like immunoreactivity after systemic administration of GHRP-6 to rats and also an increased firing of the neurons in this area. The complementary studies of Guillaume et al. revealed that hexarelin acutely increased the irGHRH levels in the hypophyseal portal blood of sheep. Thus, the direct and indirect evidence that GHRPs act on the hypothalamus to release endogenous GHRH is substantial. Also apparent is that the relationships of the action on GH release of the GHRPs and GHRH can be multiple and that they depend on the experimental conditions. For example, there are well-established specific examples of when the GHRP-GHRH relationship is independent as well as dependent, additive as well as synergistic, and also permissive. To appreciate the full spectrum of the relationship of these two peptides on GH release a closely integrated in vitro and in vivo approach is essential.

Noteworthy are the studies of Herrington and Hille (37) on single rat somatotroph cells using the whole cell patch electrophysiologic and photometry technique. Eventually these techniques may aid in better understanding the SRIF and GHRP interactions at the pituitary level. GHRP-6 was found to elevate intracellular calcium by two mechanisms. Intracellular free calcium Ca^{2+} rose in two phases: a rapid transient phase, which results from release of Ca^{2+} from intracellular stores, followed by a persistent phase, which is consistent with Ca^{2+} entry through voltage-dependent Ca^{2+} channels. Since GHRP-6 induced a long-lasting depolarization by patch recordings, depolarization of the peripheral cell membrane was thought to be responsible for the persistent $[CA^{2+}]_i$ elevation. Herrington and Hille proposed the interesting possibility that the observations of Akman et al. (38) in which SRIF only partially inhibited the GH response to GHRP-1 in vitro may reflect the lack of SRIF to inhibit the intracellular release of Ca^{2+} transiently induced by GHRP-6. Since SRIF completely inhibits the GH response of GHRP-6 and, as reported in this study, SRIF does not inhibit the transient Ca^{2+} rise induced by GHRP-6, one could conclude that this transient rise of calcium is not related to the GHRP-6 GH release.

In 1993, Akman et al. (38) were unable to show a role for protein kinase C in the GHRP-1 induced release of GH from rat primary pituitary cells.

A possible milestone recently reported by Chen et al. (39) in Australia raises a number of potentially new fundamental issues about species and biologic differences possibly due to the type of GHRP. From in vitro dispersed pituitary cells of sheep and pituitary GH secreting tumor cells of patients with acromegaly, GHRP-2, but not GHRP-6, increased release of cAMP. Since neither GHRP-6 nor GHRP-2 released cAMP from dispersed pituitary cells of rats in vitro, the differences observed in sheep and rat pituitary cells may be due to species while the differences observed in the tumor cells or sheep cells may reflect GHRP receptor subtypes or possibly abnormalities of the tumor cells.

Chronic administration of GHRP-6 and GHRP-1 has been found to increase body weight gain of rats. More recently, GHRP-2 administered sc twice per day for 6 months has increased the body weight gain of adult male and female rats over the entire treatment period. No effect or only minor clinical, laboratory, or histopathologic adverse effects were produced by the prolonged treatment even with high dosages of GHRP-2 (unpublished).

A new unique approach has been developed for assessing the effect of GHRP-2 on body weight gain in immobilization stressed rats by Shimada et al. (40). In these studies GHRP-2 increased body weight gain, food utilization ratio, body length, and tibial length. Plasma insulin-like growth factor (IGF)-I levels also were increased as well as tibial bone mineral composition and density.

GHRP Milestones—Clinical

1989–1994 (Table 2.6)

The first report of the GH releasing activity of a GHRP in humans was in 1989 by Bowers et al. (41–43) and by Ilson et al. at SmithKline Beecham (44). GHRP-6 very effectively and specifically released GH in normal young men. GHRP-6+GHRH synergistically released GH. In similar types of studies over the next couple of years, GH was released and synergism

TABLE 2.6. GHRP milestones—clinical (1989–1994).

1989–94	1. GHRP-6, -1, -2
1989–94	2. Synergism, GHRP-6, -1, -2
1992	3. Continuous infusion, GHRP-6
1992–94	4. Acromegalic patients and somatotroph tumor cells
1993	5. Obese subjects, GHRP-6, -2
1993–94	6. Short-statured children, GHRP-6, -1, -2
1993–94	7. Hexarelin
1993–94	8. Nonpeptide GHRPs
1994	9. Sleep, GHRP-2
1994	10. Oral GHRP -2, chronic

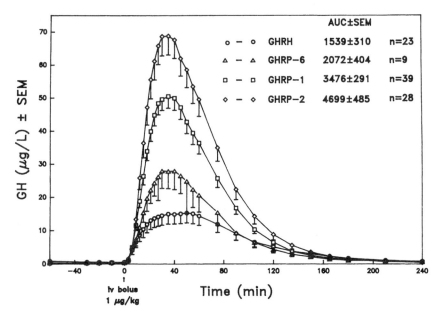

FIGURE 2.1. Comparative mean GH responses to 1.0 μg/kg GHRH(1-44) NH₂, GHRP-6, GHRP-1, and GHRP-2 in normal young men.

was induced by the second generation, GHRP-1 (45–47), and third-generation peptide, GHRP-2 (48–50). As recorded in Figure 2.1, each of these three GHRPs (-6, -1, and -2) after 1 μg/kg iv bolus administration to normal young men released more GH than GHRH(1-44)NH₂. GHRH was given at a maximal GH releasing dosage of 1 μg/kg. As found in rats, GHRP-6, GHRP-1, and GHRP-2 were progressively more active in humans by a factor of three. These GHRPs released GH by all routes of administration including iv bolus, sc, intranasal, and even orally (45–51). The large release of GH induced by 10-mg tablets orally of GHRP-2 to normal young men resulted in a mean GH rise of 54.4 ± 12.8 μg/L ($n = 9$). The actions of GHRP-6, -1, and -2 are highly specific. GH responses are large with no effect on thyroid stimulating hormone (TSH), luteinizing hormone (LH), and follicle stimulating hormone (FSH), and only a small rise of serum prolactin (PRL) and cortisol, which is still within the range of normal. In addition, large GH responses can be induced by intranasal GHRP-2 with little to no effect on serum cortisol or PRL. So far, only a few infrequently minor clinical or laboratory adverse effects have occurred in the rather large number of clinical studies performed with GHRP-6, -1, and -2 in all ages and both sexes.

Noteworthy are the results of GHRP-6 obtained during continuous iv infusion in dosages of 1 μg/kg/h over 24 to 36 hours. These studies were performed in 1992 by two different groups, Huhn et al. (52) and Jaffe et al.

(53). In each study, the amplitude but not the frequency of the normal spontaneous pulsatile secretion of GH was increased over the entire infusion period. At the end of the 36-hour infusion, iv bolus thyroid releasing hormone (TRH) did not release GH, while the GH response to iv bolus GHRH was increased but the GH response to iv bolus GHRP-6 was markedly decreased. Thus, it appears GHRP-6 may increase pulsatile GH secretion and the GH response even when the GH releasing action of the GHRP-6 is markedly decreased. The only equivocal rise of interpulse levels of GH during the infusion did not support that GHRP-6 had a special action in attenuating the inhibition of SRIF on GH release since SRIF secretion is supposedly responsible for the lower levels of GH during the interpulse periods. Extension of the infusion period is of obvious future importance.

At least six different clinical studies by several investigators [Bowers et al. (43), Laron et al. (54), Mericq et al. (55), Cassorla et al., Merriam et al., Argente et al. (56), Pihoker et al. (57), and Tiulpakov et al. (58)] have been performed demonstrating that GHRP-6, -1, and -2 release GH in normal and short-statured children. Various degrees of GH release were achieved in the short-statured children from no response in the completely GH-deficient children to relatively high responses. When GHRP+GHRH was administered, GH was released synergistically. Also noteworthy is that the GHRPs were effective after iv, sc, intranasal, and oral administration. Pihoker et al. (57) found that in the short-statured children who were below the third percentile for height had robust GH responses to 1 µg/kg iv bolus GHRP-2. The mean GH responses to 1 µg/kg iv bolus GHRP-2 and GHRH were 53.2 and 49.8 µg/L, respectively, and 142 µg/L when the peptides were administered together.

As reported by Chihara et al. (30) and Alster et al. (59), GHRP-6 released GH in vivo in patients with acromegaly and, as demonstrated by Hanew et al. (60), GHRP-6+GHRH synergistically released GH in acromegalic patients. Also GH was released in vitro from dispersed pituitary tumor cells of these patients [Chihara et al. (30), Hanew et al. (60), Renner et al. (61)]. Four of eleven of the acromegalic patients studied by Alster et al. had high GH responses to GHRP-6 and very low GH responses to GHRH. In one patient, this was reversed, i.e., high GHRH and low GHRP-6 response. Concomitant PRL levels were not increased by GHRP-6 or GHRH and there was no correlation between the GH and PRL rises induced by either GHRP-6 or GHRH.

In the in vitro studies of Renner et al. (61) GH was released by GHRP-6 from the dispersed tumor cells of all 12 of the human somatotroph adenomas patients studied while only six of them responded to GHRH. In addition, protein kinase C was involved in the intracellular signaling pathway of GHRP-6 but not GHRH. This again indicates GHRP and GHRH mediate effects via different receptors even in human tumor cells.

In the studies of Chihara et al. (30) one of the acromegalic patients responded both in vivo and in vitro, using dispersed tumor cells of the patient, to GHRP-6 but not GHRH. In this latter study, Chihara et al. postulated that GH-producing adenomas may represent a monoclonal proliferation of a single somatotroph cell type and suggested that subpopulations of this GHRP type of somatotroph cell may exist normally in the pituitary. In support of this conclusion, this group identified GHRP phenotypes distinct from GHRH phenotypes by the hemolytic plaque technique in subpopulations of somatotrophs from the normal rat pituitary.

The recent results of Chen et al. (39) appear to add another dimension to the GHRP story since GHRP-2, but not GHRP-6, released cAMP from dispersed human somatotroph adenoma cells in vitro. This is the first observation to indicate that different GHRPs may produce different effects. Important to determine is whether similar or different results will be observed with normal human pituitary cells in vitro.

The approach of Cordido et al. (62) may be helpful in the future in better understanding the pathophysiology of decreased GH secretion at the hypothalamic-pituitary level in obese subjects. It was shown by these investigators that GHRP-6 is the most potent GH stimulus so far known in obesity and that it probably acts through a non-GHRH, non–somatostatinergic-dependent mechanism. Particularly impressive was the large synergistic release of GH in these subjects by GHRP-6+GHRH. The reason why GHRP releases a relatively large amount of GH in obese subjects still needs elucidation. Clearly evident is that the pituitaries of these subjects certainly have the capacity to release large amounts of GH.

Hexarelin, a 2-methyl indole DTrp^{2}GHRP-6, has been shown to be active in the human as well as the rat. Dose-response relationships of hexarelin in humans have been reported by Imbimbo et al. (63). Ghigo et al. (64) recently demonstrated that hexarelin very effectively releases GH in normal young men and women after iv, sc, intranasal, and oral administration.

A major clinical milestone has been the studies of Gertz et al. (65) from the Merck group and Aloli et al. (66). They found that the peptidomimetic GHRP (L-692,429[L]) very effectively releases GH in humans after acute administration. The results were as predicted from their animal studies and the effects on GH release in humans paralleled those of the GHRPs. From results of these two important clinical studies, it was concluded that about 50% less GH is released by [L] in older than in younger subjects; however, this was two to four times more than that released by GHRH, indicating the potential value of [L] in restoring GH in elderly subjects. Effects on serum cortisol and PRL were similar in magnitude to physiologic pulsatile increases of these hormones. Detailed deconvolution analysis of the [L]-induced GH responses in the study of Aloi et al. (66) in the older subjects led to the provocative speculation and model that the observed shortened time interval between pulses occurs as a result of the action of [L] on the hypothalamus to increase the frequency of GHRH release, while the increased pulse amplitude occurs as a result of the direct pituitary action of [L] by attenuating the inhibitory pituitary action of SRIF on GH release.

Systemic administration of GHRH has been found to increase nocturnal sleep time in normal young men and more recently this also was found for systemically administered GHRP-2 (67). However, these two peptides appear to act differently to produce this effect, since the time of slow wave sleep was increased by GHRH and the time of rapid eye movement sleep was increased by GHRP-2.

Undoubtedly, new studies will include the effects of various types of GHRPs after chronic administration utilizing different dosages, routes, frequencies, and times of administration as well as different formulations of these agents given in different paradigms. Recently, Gonen et al. (68) found that GHRP-2 administered orally in various dosages twice per day for 14 days continued to release GH over the entire time period but the GH responses were partially attenuated by the higher dosages. Optimization of how to most effectively administer the GHRPs will require evaluation from a number of different viewpoints because of the substantial number of variables that potentially could influence the GH responses to the GHRPs during chronic administration.

Conclusion

New and novel appear to be hallmarks of the GHRPs and the peptidomimetic GHRPs.

Acknowledgment. This work was supported in part by National Institutes of Health grants DK-40202 and 5 M01RR05096 (GCRC). Special appreciation is expressed to Dr. Granda-Ayala, to the technicians and the fellows of the Endocrinology and Metabolism Section of the Department of Medicine, and to Robin Alexander for typing the manuscript.

References

1. Bowers CY, Chang JK, Fong TTW. A synthetic pentapeptide which specifically release GH, *in vitro*. 59th Annual Meeting of the Endocrine Society, Chicago IL, June 1977:232(abstr).
2. Bowers CY, Momany FA, Chang D. Growth hormone regulation at the pituitary level of hypothyroid and normal rats, *in vitro*. 60th Annual Meeting of the Endocrine Society, Miami FL, June 1978:379(abstr).
3. Bowers CY, Reynolds GA, Chang D, Hong A, Chang K, Momany F. A study on the regulation of GH release from the pituitary of rats, *in vitro*. Endocrinology 1981;108(3):1070–9.
4. Bowers CY, Momany F, Chang D, Hong A, Chang K. Structure-activity relationships of a synthetic pentapeptide that specifically releases GH *in vitro*. Endocrinology 1980;106(3):663–7.

5. Momany FA, Bowers CY, Reynolds GA, Chang D, Hong A, Newlander K. Design, synthesis and biological activity of peptides which release growth hormone, *in vitro*. Endocrinology 1980;108(1);31–9.
6. Bowers CY, Momany F, Reynolds GA. *In vitro* and *in vivo* activity of a small synthetic peptide with potent GH releasing activity. 64th Annual Meeting of the Endocrine Society, San Francisco CA, June 1982:205(abstr).
7. Momany F, Bowers CY, Reynolds GA, Hong A, Newlander K. Conformational energy studies and *in vitro* activity data on active GH releasing peptides. Endocrinology 1984;114:1531–6.
8. Bowers CY, Momany F, Reynolds GA, Hong A. On the *in vitro* and *in vivo* activity of a new synthetic hexapeptide that acts on the pituitary to specifically release growth hormone. Endocrinology 1984;114:1537–45.
9. Momany F. Computer-assisted modeling in xenobiotic growth hormone secretagogues. Growth Hormone Secretagogues, Serono Symposia, Dec 8–11, 1994.
10. Sartor O, Bowers CY, Chang D. Parallel studies of His-DTrp-Ala-Trp-DPhe-Lys-NH$_2$ and hpGKF-44NH$_2$ in rat primary pituitary cell monolayer culture. Endocrinology 1985;116:952–7.
11. Bowers CY, Sartor O, Reynolds GA, Chang D, Momany F. Evidence that GRF and GRP, His-DTrp-Ala-Trp-DPhe-Lys-NH$_2$, act on different pituitary receptors to release GH. 67th Annual Meeting of the Endocrine Society, Baltimore MD, June 1985:38(abstr).
12. Bowers CY, Momany F, Reynolds GA, Sartor O. Multiple receptors mediate GH release. 7th International Congress of Endocrinology, Quebec Canada, July 1984:464(abstr).
13. Bowers CY, Sartor AO, Reynolds GA, Badger TM. On the actions of the growth hormone releasing hexapeptide GHRP. Endocrinology 1991;128:2027–35.
14. Badger RM, Millard WJ, McCormick GF, Bowers CY, Martin JB. The effects of growth hormone (GH) releasing peptides on GH secretion in perifused pituitary cells of adult male rats. Endocrinology 1984;115:1432–8.
15. McCormick GF, Millard WJ, Badger TM, Bowers CY, Martin JB. Dose-response characteristics of various peptides with growth hormone-releasing activity in the unanesthetized male rats. Endocrinology 1985;117:97–105.
16. Reynolds GA, Bowers CY. In vitro studies with GH releasing peptides. 69th Annual Meeting of the Endocrine Society, Indianapolis IN, June1987:49(abstr).
17. Clark RG, Carlsson LMS, Trohnar J, Robinson ICAF. The effects of a growth hormone-releasing peptide and growth hormone-releasing factor in conscious and anesthetized rats. J Neuroendocrinol 1989;1:249–55.
18. Codd EE, Shu AYL, Walker RF. Binding of a growth hormone releasing hexapeptide to specific hypothalamic and pituitary sites. Neuropharmacology 1989;28:1139–44.
19. Cheng K, Chan WWS, Barreto A, Convey EM, Smith RG. The synergistic effects of His-D-Trp-Ala-Trp-D-Phe-Lys-NH$_2$ on growth hormone (GH)-releasing factor-stimulated GH release and intracellular adenosine 3',5'-monophosphate accumulation in rat primary pituitary cell culture. Endocrinology 1989;124:2791–8.
20. Malozowski S, Hao EH, Ren SG, Marin G, Liu L, Southers JL, et al. Growth hormone (GH) responses to the hexapeptide GH-releasing peptide and GH-

releasing hormone (GHRH) in the cynomolgus macque: evidence for non-GHRH-mediated responses. J Clin Endocrinol Metab 1991;73:314–7.

21. Cheng K, Chan WWS, Butler B, Barreto A, Smith RG. Evidence for a role of protein kinase-C in His-D-Trp-Ala-Trp-D-Phe-Lys-NH$_2$ induced growth hormone release from rat primary pituitary cells. Endocrinology 1991;129:3337–42.

22. Pong S, Chaung L, Smith RG, Ertel E, Smith M, Cohen CJ. Role of calcium channels in growth hormone secretion induced by GHRP-6(His-D-Trp-Ala-Trp-D-Phe-Lys-NH$_2$) and other secretagogues in rat somatotrophs. 74th Annual Meeting of the Endocrine Society, San Antonio TX, June 1992:255(abstr).

23. Deghenghi R, Cananzi M, Battisti C, Locatelli V, Müller EE. Hexarelin (DP23905)—a superactive growth hormone releasing peptide. J Endocrinol Invest 1992;15(4):45.

24. Conley LK, Stagg LC, Giustina A, Imbimbo BP, Deghenghi R, Wehrenberg WB. The mechanism of action of hexarelin and GHRP-6: analysis of the involvement of somatostatin. 75th Annual Meeting of the Endocrine Society, Las Vegas NV, June 1993:1451(abstr).

25. Smith RG, Cheng K, Schoen WR, Pong SS, Hickey G, Jacks T, et al. A novel nonpeptidyl growth hormone secretagogue. Science 1993;260:1640–3.

26. Schoen WR, Pisano JM, Prendergast K, Wyvratt MJ, Fisher MH, Cheng K, et al. A novel 3-substituted benzazepinone growth hormone secretagogue (L-692,429). J Med Chem 1994;37:897–906.

27. Jacks T, Hickey G, Judith F, Taylor J, Chen H, Krupa D, et al. Effects of acute and repeated intravenous administration of L-692,585, a novel nonpeptidyl growth hormone secretagogue, on plasma growth hormone, ACTH, cortisol, prolactin, thyroxine (T4), insulin and IGF-I levels in beagles. 76th Annual Meeting of the Endocrine Society, Anaheim CA, June 1994:365(abstr).

28. Walker RF, Yang SW, Masuda R, Hu CS, Bercu BB. Effects of growth hormone releasing peptides on stimulated growth hormone secretion in old rats. Serono Symposia, 1992, Dec 3–6. In: Bercu B, Walker R, eds. Growth hormone II: basic and clinical aspects. New York: Springer-Verlag 1994:167–92.

29. Walker RF, Yang S-W, Bercu BB. Robust growth hormone (GH) secretion in aged female rats co-administered GH-releasing hexapeptide (GHRP-6) and GH-releasing hormone (GHRH). Life Sci 1991;49:1499–504.

30. Chihara K, Kaji H, Hayashi S, Yagi H, Takeshima Y, Mitani M, et al. Growth hormone releasing hexapeptide: basic research and clinical application. Serono Symposia, 1992, Dec 3–6. In: Bercu B, Walker R, eds. Growth hormone II: basic and clinical aspects. New York: Springer-Verlag 1994:223–30.

31. Robinson ICAF. Regulation of growth hormone output: the GRF signal. Serono Symposia, 1992, Dec 3–6. In: Bercu B, Walker R, eds. Growth horomone II: basic and clinical aspects. New York: Springer-Verlag 1994:47–65.

32. Kraicer J, Sims SM. Ionic mechanisms governing the control of growth hormone secretion by somatostatin. Serono Symposia, 1992, Dec 3–6. In: Bercu B, Walker R, eds. Growth hormone II: basic and clinical aspects. New York: Springer-Verlag 1994:17–32.

33. Hickey GJ, Baumhover J, Faidley T, Chang C, Anderson LL, Nicolich S, et al. Effect of hypothalamo-pituitary stalk transection in the pig on GH secretory activity of L-692,585. 76th Annual Meeting of the Endocrine Society, Anaheim CA, June 1994:366(abstr).

34. Bowers CY, Veeraragavan K, Sethumadhavan K. Atypical growth hormone releasing peptides. Serono Symposia, 1992, Dec 3–6. In: Bercu B, Walker R, eds.

Growth hormone II: basic and clinical aspects. New York: Springer-Verlag 1994:203–22.

35. Dickson SL, Leng G, Robinson ICAF. Systemic administration of growth hormone-releasing peptide activates hypothalamic arcuate neurons. Neuroscience 1993;53(2):303–6.

36. Guillaume V, Magnan E, Cataldi M, Dutour A, Sauze N, Renard M, et al. Growth hormone (GH)-releasing hormone secretion is stimulated by a new GH-releasing hexapeptide in sheep. Endocrinology 1994;135:1073–6.

37. Herrington J, Hille B. Growth hormone-releasing hexapeptide elevates intracellular calcium in rat somatotropes by two mechanisms. Endocrinology 1994; 135:1100–8.

38. Akman MS, Girard M, O'Brien LF, Ho AK, Chik CL. Mechanisms of action of a second generation growth hormone-releasing peptide (Ala-His-D-βNal-Ala-Trp-D-Phe-Lys-NH$_2$) in rat anterior pituitary cells. Endocrinology 1993; 132:1286–91.

39. Chen C, Wu D, Bowers CY, Clarke IJ. The mechanism of action of synthetic GH-releasing-peptides: a comparison between sheep, rat and human pituitary cells. Growth Hormone Secretagogues, Serono Symposia USA, Norwell, MA, 1994, Dec 8–11:31(abstr).

40. Shimada O, Kiyofuji T, Hanada K, Chihara K, Bowers CY. KP-102 administration overcomes growth retardation of rats subjected to immobilization stress. Growth Hormone Secretagogues, Serono Symposia USA, Norwell, MA, 1994, Dec 8–11:41(abstr).

41. Bowers CY, Thorner MO, Reynolds GA, Durham D, Barrera CM, Pezzoli SS. Efficacy of the hexapeptide GHRP and GHRP+GRF synergy in release of GH in normal men. 71st Annual Meeting of the Endocrine Society, Seattle WA, June 1989:216(abstr).

42. Bowers CY, Reynolds GA, Durham D, Barrera CM, Pezzoli SS, Thorner MO. Growth hormone releasing peptide stimulates GH release in normal men and acts synergistically with GH-releasing hormone. J Clin Endocrinol Metab 1990;70:975–82.

43. Bowers CY, Alster DK, Frentz JM. The growth hormone releasing activity of a synthetic hexapeptide in normal men and short statured children after oral administration. J Clin Endocrinol Metab 1992;74:292–8.

44. Ilson BE, Jorkasky DK, Curnow RT, Stote RM. Effect of a new synthetic hexapeptide to selectively stimulate growth hormone release in healthy human subjects. J Clin Endocrinol Metab 1989;69:212–4.

45. Bowers CY, Newell DC, Alster DK. Small GH releasing peptide. Proceedings of the Ninth International Congress of Endocrinology, 1992 Aug 30–Sept 5; Nice, France, Mornex, Caffiol, Leclere, 1992:256–62.

46. Bowers CY. GH releasing peptides—structure and kinetics. Workshop on Growth No. 5, Novo Nordisk A/S, Copenhagen, Denmark, 1992, Sept 24–26; J Pediatr Endocrinol 1992;6(suppl 1):21–31.

47. Bowers CY, Newell DC, Granda-Ayala R, Garcia M, Barrera C. Comparative studies on GH release in younger and older men and women. 74th Annual Meeting of the Endocrine Society, San Antonio, TX, June 1992:172(abstr).

48. Bowers CY. Novel GH releasing peptides. In: Melmed S, ed. Molecular and clinical advances in pituitary disorders. Los Angeles: Endocrine Research and Education, 1993:153–7.

49. Bowers CY, Reynolds GA, Granda-Ayala R, Garcia M. Dimensions of GH releasing peptides (GHRPs). 75th Annual Meeting of the Endocrine Society, Las Vegas, NV, June 1993:413(abstr).
50. Bowers CY, Reynolds GA, Servera S, Reyes Y, Baquet T, Granda-Ayala R. GHRP-2 and GHRH actions and interactions on GH release in humans. 76th Annual Meeting of the Endocrine Society, Anaheim, CA, June 1994: 336(abstr).
51. Hayashi S, Okimura Y, Yagi H, Uchiyama T, Takeshima Y, Shakutsui S, et al. Intranasal administration of His-D-Trp-Ala-D-Phe-LysNH$_2$ (growth hormone releasing peptide) increased plasma growth hormone and insulin-like growth factor-I levels in normal men. Endocrinol Jpn 1991;38(suppl 1):15–21.
52. Huhn WC, Hartman ML, Pezzoli SS, Thorner MO. 24-h growth hormone (GH)-releasing peptide (GHRP) infusion enhances pulsatile GH secretion and specifically attenuates the response to a subsequent GHRP bolus. J Clin Endocrinol Metab 1993;76:1201–8.
53. Jaffe CA, Ho PJ, Demott-Friberg R, Bowers CY, Barkan AL. Effects of a prolonged growth hormone (GH)-releasing peptide infusion on pulsatile GH secretion in normal men. J Clin Endocrinol Metab 1993;77:1641–7.
54. Laron Z, Bowers CY, Hirsch D, Almonte AS, Pelz M, Keret R, et al. Growth hormone-releasing activity of growth hormone-releasing peptide-1 (a synthetic heptapeptide) in children and adolescents. Acta Endocrinol 1993; 129:424–6.
55. Mericq V, Cassorla F, Garcia H, Avial A, Bowers CY, Merriam G. Growth hormone responses to growth hormone releasing peptide (GRP) and to growth hormone releasing hormone in growth hormone deficient children. J Clin Endocrinol Metab 1995;80:1681–4.
56. Argente J, Pozo J, Barrios V, Munoz MT, Gonzalez R, Bowers CY. Growth hormone-releasing peptide 2 (GHRP-2) selectively stimulates growth hormone (GH) secretion orally administered. 33rd Annual Meeting of the European Society for Paediatric Endocrinology, 1994 June, Maastricht (Netherlands):82.
57. Pihoker C, Bowers CY, Reynolds GA, Badger TM. Growth hormone (GH) responses to intranasal GH releasing peptide-2 (GHRP-2) and combined GHRP-2/GH releasing hormone (GHRH). 76th Annual Meeting of the Endocrine Society, Anaheim, CA, June 1994:337(abstr).
58. Tiulpakov AN, Bulatov AA, Peterkova VA, Elizarova GP, Volevodz NN, Bowers CY. Growth hormone (GH) releasing effects of synthetic peptide GHRP-2 and GH releasing hormone (GHRH 1-29NH$_2$) in children with GH insufficiency and idiopathic short stature. Metabolism, 1995;44:1199–1204.
59. Alster DK, Bowers CY, Jaffe CA, Ho PJ, Barkan AL. The GH response to GHRP (His-DTrp-Ala-Trp-DPhe-Lys-NH$_2$), GHRH and TRH in acromegaly. J Clin Endocrinol Metab 1993;77:842–5.
60. Hanew K, Utsumi A, Sugaware A, Shimizu Y, Abe K. Enhanced GH responses to combined administration of GHRP and GHRH in patients with acromegaly. J Clin Endocrinol Metab 1994;78:509–12.
61. Renner U, Brockmeier S, Strasburger CJ, Lange M, Schopohl J, Muller OA, et al. Growth hormone (GH)-releasing peptide stimulation of GH release from human somatotroph adenoma cells: interaction with GH-releasing hormone, thyrotropin-releasing hormone, and octreotide. J Clin Endocrinol Metab 1994;78:1090–6.

62. Cordido F, Penalva A, Dieguez C, Casanueva FF. Massive growth hormone (GH) discharge in obese subjects after the combined administration of GH-releasing hormone and GHRP-6: evidence for a marked somatotroph secretory capability in obesity. J Clin Endocrinol Metab 1993;76:819–23.
63. Imbimbo BP, Mant T, Edwards M, et al. Growth hormone releasing activity of hexarelin in humans: a dose-response study. Growth hormone II: basic and clinical aspects. Serono Symposia USA, Norwell, MA, 1992 Dec 3–6:37(abstr).
64. Ghigo E, Arvat E, Gianotti L, Imbimbo BP, Lenaerts V, Deghenghi R, et al. Growth hormone-releasing activity of hexarelin, a new synthetic hexapeptide, after intravenous, subcutaneous, intranasal, and oral administration in man. J Clin Endocrinol Metab 1994;78:693–8.
65. Gertz BJ, Barrett JS, Eisenhandler R. Growth hormone response in man to L-692,429, a novel nonpeptide mimic of growth hormone releasing peptide (GHRP-6). J Clin Endocrinol Metab 1993;77:1393–7.
66. Aloi JA, Gertz BJ, Hartman ML, Huhn WC, Pezzoli SS, Wittreich JM, et al. Neuroendocrine responses to a novel growth hormone secretagogue, L-692,429, in healthy older subjects. J Clin Endocrinol Metab 1994;79:943–9.
67. Moreno-Reyes R, Kerkhofs M, L'Hermite-Baleriaux M, Bowers CY, Thorner MO, Van Cauter E, et al. The sleep-promoting effects of GHRH in normal men are not mimicked by GHRP. 76th Annual Meeting of the Endocrine Society, Anaheim, CA, June 1994:204(abstr).
68. Gonen B, Meng X, Orczyk G, Bowers CY. Growth hormone releasing peptide: clinical studies. International Symposium on Growth Hormone Secretagogues, 1994, Dec 8–11, St. Petersburg, FL.

Part II

Chemistry of Growth Hormone Secretagogues

3

Synthetic Analogues of Growth Hormone Releasing Factor (GHRF) with Improved Pharmaceutical Properties

Teresa M. Kubiak, Alan R. Friedman, and W. Michael Moseley

The protein hormone, somatotropin or growth hormone (GH), is essential for normal somatic growth. The biologic activity of GH is species specific. Currently, recombinant human GH is used for increasing growth in children with short stature, while recombinant bovine growth hormone is registered for enhancing lactation in cows. The secretion of GH from the anterior pituitary is controlled by two hypothalamic peptides, somatostatin (inhibitory) and growth hormone releasing factor (GHRF, stimulatory). GHRF offers a potential alternative to GH in applications where increased circulating concentrations of GH are desirable.

Although GHRF peptides are largely of hypothalamic origin, the sequences of three human GHRF peptides, hGHRF(1-37)NH$_2$, hGHRF(1-40)OH, and hGHRF(1-44)NH$_2$, were determined from material isolated from human pancreatic tumors (1, 2). Subsequently, the latter two were shown to be present in human hypothalami (3). Presently, the sequences of GHRFs from various species, pig (4), bovine (5), goat and sheep (6), rat (7), and mouse (8, 9) have been determined. Most GHRFs have a high degree of identity with human GHRF [hGHRF(1-44)NH$_2$]. Rat GHRF [rGHRF(1-43)OH] (7), and mouse GHRF [mGHRF(1-42)OH] (8, 9) are exceptions since they share only 62% to 67% homology with hGHRF(1-44)NH$_2$ and exhibit 67% sequence homology between themselves.

In contrast to GH, GHRF peptides are active across species and, unlike GH treatments, administration of exogenous GHRF does not abolish the pulsatile and physiologic nature of GH release in vivo (10–14). Additionally, the high potency and smaller size of GHRF or its analogues (44 amino acids or smaller vs. 190 residues for GH) may provide drugs that are easier to formulate for sustained delivery.

The C-terminus of GHRF (beyond residue 30) is highly diverse across species while the N-terminal (1-29) fragment represents the highest degree of sequence conservation (Fig. 3.1) (15–18). Moreover, GHRF(1-29)NH_2 is the shortest GHRF fragment displaying GH-releasing activity (18) and receptor binding affinity (19) similar to GHRF(1-44)NH_2 indicating that residues 30-44 are not required for biologic activity. Rat and mouse GHRFs have an amino terminal His rather than Tyr, which is present in other nonrodent mammalian GHRFs. With the exception of mGHRF(1-42)OH, in GHRFs from all other species identified to date, residue 2 is occupied by Ala (8, 9, 17). The presence of Val^2 in mGHRF(1-42)OH is unique (8, 9). As shown by us (20), this substitution renders mGHRF resistant to dipeptidylpeptidase-IV (DPP-IV), the main enzyme responsible for rapid hydrolysis of the Ala^2-containing GHRFs from other species (21–25). As first shown for human GHRF(1-44)NH_2 in elegant pioneering studies by Frohman et al. (21, 22), DPP-IV–mediated hydrolysis of the Ala^2-Asp^3 peptide bond constitutes the main metabolic pathway of GHRF degradation in blood plasma. Subsequently, the same DPP-IV-dependent metabolism of GHRF has been found in bovine (23, 26–28), porcine (23, 25), and rat plasma (24) in vitro. These DPP-IV–catalyzed cleavages yield truncated des(Tyr^1-Ala^2)-GHRF fragments virtually devoid of inherent GH-releasing activity (21, 23).

Our main goal has been to find proprietary and potent GHRF analogue(s) with superior characteristics to the native hormone for applications in production animals to enhance growth and lactation. However, since the highly conserved (1-29) GHRF fragment was taken as a template for further modifications, superior analogues identified in our studies could have a potential for human use as well.

Since 1982, when the sequence of human GHRF (hGHRF) was determined (1, 2), hundreds of synthetic analogues have been made for

```
            DPP-IV
            1   ⇓  5        10        15        20        25        30        35        40
hGRF        Y A D A I F T N S Y R K V L G Q L S A R K L L Q D I M S R Q Q G E S N Q E R G A R A R L-NH2

pGRF        Y A D A I F T N S Y R K V L G Q L S A R K L L Q D I M S R Q Q G E R N Q E Q G A R V R L-NH2

bGRF, cGRF  Y A D A I F T N S Y R K V L G Q L S A R K L L Q D I M N R Q Q G E R N Q E Q G A K V R L-NH2

oGRF        Y A D A I F T N S Y R K I L G Q L S A R K L L Q D I M N R Q Q G E R N Q E Q G A K V R L-NH2

rGRF        H A D A I F T S S Y R R I L G Q L Y A R K L L H E I M N R Q Q G E R N Q E Q R S R F N-NH2

mGRF        H V D A I F T T N Y R K L L S Q L Y A R K V I Q D I M N K Q - G E R I Q E Q R A R - - L S-OH
```

FIGURE 3.1. Comparisons of the amino acid sequences of GHRFs from different species. Residues that differ from those in the hGHRF sequence are underlined. The arrow indicates a DPP-IV clevage site at the 2-3 peptide bond in GHRFs with N-terminal Tyr-Ala or His-Ala sequences. hGHRF, human GHRF; pGHRF, porcine GHRF; bGHRF, bovine GHRF; cGHRF, caprine GHRF; oGHRF, ovine GHRF; rGHRF, rat GHRF; mGHRF, mouse GHRF.

structure-activity relationship (SAR) studies yielding important insights into the nature of residues crucial for receptor binding, biologic activity, and metabolic and chemical stability.

The performance of a drug in vivo is determined by its inherent potency and pharmacokinetic and pharmacodynamic properties (metabolic stability, uptake, and elimination rates). From a commercial point of view, it is important to find analogues with improved metabolic and chemical stability, enhanced potency, and ease of production combined in a single molecule. Herein, we present our perspective and a summary of findings both from our laboratory and from other laboratories that helped identify modifications in GHRF that yielded highly potent GHRF analogues with improved pharmaceutical properties.

Enzymatic Degradation of GHRF and Modifications Leading to Analogues with Improved Metabolic Stability

As mentioned above, in blood plasma, DPP-IV–mediated cleavage of the Ala^2-Asp^3 bond of GHRFs with the N-terminal Tyr^1-Ala^2 or His^1-Ala^2 sequences is the main route for GRF inactivation. In vitro, the half-life of unmodified GHRFs ranges from ca. 8 minutes in porcine plasma (23) to 17 minutes in human plasma (21) and 20 to 50 minutes in bovine plasma (23, 26–28). In vivo, in humans, hGHRF(1-44)NH$_2$ disappears from the circulation with a half-life of 6.7 minutes resulting mostly from the DPP-IV–mediated conversion of the intact hormone into its almost biologically inactive hGHRF(3-44)NH$_2$ fragment (21). In addition to DPP-IV–mediated cleavage, human (21, 22), rat (24), and mouse (20) GHRFs have been reported to undergo hydrolysis and inactivation by plasma trypsin-like endopeptidases. However, this metabolic pathway is minor when compared with the rate and extent of the DPP-IV–mediated metabolism. Trypsin-like cleavages have not been identified when bGHRF was incubated in bovine or porcine plasma in vitro (23).

DPP-IV is reported to cleave N-terminal dipeptides from proteins or peptides having either an X-Ala, X-Pro, or X-Hyp as the N-terminal sequences with a preference for X being an aromatic amino acid. It has also been shown that DPP-IV absolutely requires a free and protonated amino group at the N-terminus for effective enzyme binding and action. DPP-IV does not cleave (a) substrates having a D–amino acid residue either as the N-terminal (P$_2$) or penultimate (P$_1$) position or (b) peptides with either proline or hydroxyproline at P$_1'$, which is position 3 from the N-terminus (for review of DPP-IV substrate specificity see res. 29 and 30).

In retrospect, after the identification of DPP-IV as the main enzyme metabolizing GHRF (21), and based on the DPP-IV substrate specificity, it was not surprising that the desNH$_2$Tyr^1, D-Tyr^1, D-Ala^2, N-MeTyr^1, and N-

AcTyr[1] modifications in GHRF prevented cleavage by DPP-IV (22). However, it was somewhat unexpected that cyclization between the β-carboxyl group of Asp[3] (or substituted Asp[8]) and ε-amino group of Lys[12], with or without Ala[15] modification (31), resulted in only limited degradation by DPP-IV (22), although these cyclic peptides had the same N-terminal Tyr-Ala sequence as the native hormone. Potentially, the intramolecular bridging of the molecule could have restrained the Asp[3] (or Asp[8]) and Lys[12] residues and led to a possible global conformational change that rendered the GHRF N-terminus no longer easily accessible for DPP-IV recognition and cleavage. These cyclic analogues were also found to be resistant to trypsin-like degradation (22).

One of our early ideas was to take advantage of the DPP-IV present in plasma to achieve a temporary protection and a gradual release of the core GRF peptide from N-terminally extended GHRF analogues (pro-GHRFs), designed as substrates for DPP-IV (32, 33). Bovine [Leu[27]]bGHRF(1-29)NH$_2$ was chosen as a model GHRF to be N-terminally extended since a similar 29-amino acid human GHRF analogue was reported to have almost full growth hormone releasing activity (18). Leu[27] was introduced to replace easily oxidizable Met[27] in the model peptide with the intent of making GHRF more chemically stable and compatible with recombinant DNA synthesis and cyanogen bromide (CNBr) processing for future large-scale peptide preparation by biosynthetic methods (34). We have shown that, in bovine plasma in vitro, GHRF analogs with N-terminal extensions specific for DPP-IV recognition, e.g., [Ile^{-1}Pro^{0}Leu[27]]bGHRF(1-29)NH$_2$ or [Ile^{-2}Pro^{-1}Ile^{0}Leu[27]]bGHRF(1-29)NH$_2$ were processed by DPP-IV present in plasma. However, only [Ile^{-1}Pro^{0}Leu[27]]bGHRF(1-29)NH$_2$ generated the core [Leu[27]]bGHRF(1-29)NH$_2$ while [Ile^{-2}Pro^{-1}Ile^{0}Leu[27]]bGHRF(1-29)NH$_2$ was converted to [Ile^{0}Leu[27]]bGHRF(1-29)NH$_2$, a peptide with a wrong "reading frame" for DPP-IV recognition that was not further processed to [Leu[27]]bGHRF(1-29)NH$_2$ (32). In bovine plasma in vitro, the half-life of [Leu[27]]bGHRF(1-29)NH$_2$ generated from either Ile-Pro, Val-Pro, or Tyr-Ala extended pro-GHRFs was almost doubled as compared with the half-life of the core peptide incubated directly in bovine plasma (32, 33). The same pro-GHRFs were as active as the core peptide in releasing serum GH in vivo in steers despite their low in vitro GH releasing potencies of only 0.5% to 13% of that of [Leu[27]]bGHRF(1-29)NH$_2$ (33). It has therefore been concluded that the core peptide must have been generated from the DPP-IV–cleavable pro-GHRFs to effectively release GH in vivo. This notion was additionally supported by the fact that a non–DPP-IV–cleavable analogue with desNH$_2$Tyr-Ala in the extension had low GH-releasing potency in vitro (3%) and did not induce GH-release over the control levels in vivo in steers (33). Taken together, our results confirmed the prodrug concept of N-terminally extended GHRF analogues as targets for the DPP-IV–mediated release of the core GHRF and extension of its half-life in bovine plasma in vitro. However, in vivo, probably due to high organ and tissue levels of DPP-IV, these pro-GHRFs could have been rapidly processed to [Leu^{27}]bGHRF(1-29)NH$_2$ as evidenced by no significantly delayed, prolonged, or enhanced pattern of GH-release observed in steers (no advantage over the core peptide) (33).

Another approach we took was to focus on naturally occurring amino acid position 2 modifications of GHRF. We hypothesized that substituting Ala^2 with any amino acid other than Pro or Hyp could render GHRF an inactive substrate for DPP-IV (26–28). Our goal was not only to make GHRF resistant to DPP-IV but also to maintain high potency for GH release in a format that retains all natural amino acids. This last requirement was for future large-scale peptide preparation by biosynthetic methods. Using the [Leu^{27}]bGHRF(1-29)NH$_2$ template, initially Gly, Ser, or Thr substitutions at position 2 were explored because of their perceived structural similarity to the native Ala (26). Subsequently, 11 additional DNA coded amino acids were added to the series as substitutes at position 2. These substitutions were made in a template that included a Gly^{15} → Ala^{15} modification (27, 28) that had been reported to enhance GH-releasing potency (35, 36).

While the 3-29 fragment was quickly generated from both, [Leu^{27}]bGHRF(1-29)NH$_2$ and [Ala^{15}Leu^{27}]bGHRF(1-29)NH$_2$ during incubation of the peptides in bovine plasma in vitro, the Ser^2- and Thr^2-modified analogues were cleaved slowly at the 2-3 peptide bond. All the other position 2 substitutions yielded peptides resistant to plasma DPP-IV as judged by the absence of the corresponding 3-29 fragments in their plasma incubation mixtures (23, 26–28). Furthermore, it has been concluded that the DPP-IV stabilizing position 2 modifications did not lead to the formation or exposure of any new sites vulnerable to hydrolysis by other plasma enzymes (28). The DPP-IV–catalyzed cleavages of the Ser^2- and Thr^2-analogues observed in bovine plasma were somewhat unexpected. They were later confirmed in a separate study with a purified DPP-IV preparation derived from porcine kidney (37). This finding led to a proposal to also include N-terminal X-Ser and X-Thr in the extended family of sequences cleavable by DPP-IV (37). Independently, hydrolysis of a Ser^2-modified human GHRF analogue by human placental DPP-IV was reported by Bongers et al. (38).

The single position 2–modified analogues with Gly, Ser, and Thr were three to eight times more stable than [Leu^{27}]bGHRF(1-29)NH$_2$, while their corresponding position 2/Ala^{15}-GHRF analogues were six to nine times more stable than [Leu^{27}]bGHRF(1-29)NH$_2$, depending on the bovine plasma pool (Table 3.1). The Ala^{15} substitution alone in [Leu^{27}]bGHRF(1-29)NH$_2$ afforded an analogue about twice as stable in bovine plasma as the parent compound, with a concurrent slower formation of the 3-29 fragment. The partial protection of the 2-3 peptide bond in [Ala^{15}Leu^{27}]bGHRF(1-29)NH$_2$ was surprising because this analogue was not modified at the N-terminus (27, 28). A similar observation was made by Su et al. (25) for [Ala^{15}]hGHRF(1-29)NH$_2$ incubated with porcine plasma in vitro. The effect

TABLE 3.1. Stability of bGHRF analogs in bovine plasma in vitro

| Peptide | Plasma pool A | | Plasma pool B | |
	% Intact peptide after 1 hour[a]	$t_{1/2}$ (min)	k^a (min^{-1})	$t_{1/2}^b$ (min)
[Leu^{27}]bGHRF(1–29)NH$_2$	39.8^g	50	0.0171^f	41 (+)
[Ala^{15}Leu^{27}]bGHRF(1–29)NH$_2$	56.2^h	96	0.00797^d	87 (+)
[Gly^{2}Leu^{27}]bGHRF(1–29)NH$_2$	73.3^c	nd	0.00224^h	309 (−)
[Gly^{2}Ala^{15}Leu^{27}]bGHRF(1–29)NH$_2$	89.3^{e,f}	nd	0.00184^g	377 (−)
[Ser^{2}Leu^{27}]bGHRF(1–29)NH$_2$	76.1^{c,d}	nd	0.00266^c	260 (+)
[Ser^{2}Ala^{15}Leu^{27}]bGHRF(1–29)NH$_2$	84.0^f	nd	0.00193^g	360 (+)
[Thr^{2}Leu^{27}]bGHRF(1–29)NH$_2$	75.6^{c,d}	147	0.00207^{g,h}	334 (±)
[Thr^{2}Ala^{15}Leu^{27}]bGHRF(1–29)NH$_2$	91.1^e	297	0.00185^g	375 (±)
[desNH$_2$Tyr^{1}D-Ala^{2}Ala^{15}]hGHRF(1–29)NH$_2$	80.0^{d,f}	217	nd	nd (−)
bGHRF(1–44)NH$_2$	nd	nd	0.0110^e	63 (+)

[a] Data are presented as means from three experiments. [b] (3–29)-fragment was either present (+) or absent (−) in the plasma incubation mixtures. For the two Thr^2-analogues, the (3–29)-GHRFs were not seen for the first 1 hour of incubation but could be detected at later time points (±). The (3–29)-GHRFs were generated from the Ser^2-analogues faster than from the Thr^2-peptides, but their formation rates were much slower than those from the GHRF peptides with the native Ala^2. nd, not determined. [c–h] Means within a column lacking a common superscript letter are different at $p < .05$. Table reproduced with permission from Kubiak et al. (28). © American Chemical Society.

of the Gly^{15} to Ala^{15} modification on increased plasma stability of GHRF analogues is not well understood. However, it may be indirectly related to the increase in helicity and hydrophobicity imparted by this modification.

The in vitro plasma half-lives of the Ile^{2}Ala^{15}-, Val^{2}Ala^{15}-, and Thr^{2}Ala^{15}-modified analogues (best GH-secretagogues identified later in the study) were respectively 16, 11, and 7 times longer than that of [Leu^{27}]bGHRF(1–29)NH$_2$, and 9, 6, and 4 times longer than that of the Ala^{15}-analogue (Table 3.2). The greater stability of the Val^{2}Ala^{15} and Ile^{2}Ala^{15} GHRFs seems to be due to the fact that these analogues, in contrast to their Thr^{2}Ala^{15} counterpart, were not cleaved by bovine plasma DPP-IV (27, 28). The increased stability of the Ile^2 analogue over the Val^2 analogue may additionally be related to its increased hydrophobicity. This characteristic may also serve to protect the analogue from other plasma proteases, perhaps by hydrophobic interaction with serum components. Both helicity and hydrophobicity were also correlated with increased in vitro plasma stability in a series of GHRF analogues additionally modified at position 19 (39).

Shortly after we had identified Ile^2, Val^2, and Thr^2 as superior substitutions in bGHRF, the primary structure of mouse GHRF with its N-terminal His-Val sequence was elucidated and revealed (8, 9). It was then gratifying to discover, as expected, that Val^2- and Ile^2-modified analogues from the human GHRF series, with either His or Tyr at position 1, synthesized by the Hoffmann-La Roche group, also turned out to be resistant to plasma and

TABLE 3.2. Stability of GHRF analogues in bovine plasma in vitro

Compound	% Intact peptide after 1 hour (plasma pool C)		Plasma pool (D)	
			$k \times 10^4$ (min^{-1})	$t_{1/2}$ (min)
[Leu^{27}]bGHRF(1–29)NH$_2$	46.8 ± 4.1^a	(+)	179 ± 5.6	39
[Ala^{15}Leu^{27}]bGHRF(1–29)NH$_2$	$54.1 \pm 1.5^{a,b}$	(+)	100 ± 4.4	69
[Thr^{2}Ala^{15}Leu^{27}]bGHRF(1–29)NH$_2$	$91.1 \pm 3.1^\#$	(−)	27 ± 3.6	257
[Val^{2}Ala^{15}Leu^{27}]bGHRF(1–29)NH$_2$	$83.4 \pm 5.4^{d,e,f}$	(−)	16 ± 1.2	433
[Ile^{2}Ala^{15}Leu^{27}]bGHRF(1–29)NH$_2$	$80.3 \pm 17.5^{c,d,e,f}$	(−)	11 ± 0.6	630
[Leu^{2}Ala^{15}Leu^{27}]bGHRF(1–29)NH$_2$	$86.3 \pm 4.0^{e,f}$	(−)	nd	nd
[Glu^{2}Ala^{15}Leu^{27}]bGHRF(1–29)NH$_2$	$83.9 \pm 1.5^{d,e,f}$	(−)	nd	nd
[Phe^{2}Ala^{15}Leu^{27}]bGHRF(1–29)NH$_2$	88.1 ± 3.4^f	(−)	nd	nd
[Gln^{2}Ala^{15}Leu^{27}]bGHRF(1–29)NH$_2$	$71.0 \pm 4.2^{c,d}$	(−)	nd	nd
[Asn^{2}Ala^{15}Leu^{27}]bGHRF(1–29)NH$_2$	$71.7 \pm 8.0^{c,d}$	(−)	nd	nd
[Lys^{2}Ala^{15}Leu^{27}]bGHRF(1–29)NH$_2$	$82.7 \pm 5.4^{d,e,f}$	(−)	nd	nd
[Asp^{2}Ala^{15},Leu^{27}]bGHRF(1–29)NH$_2$	89.4 ± 3.2^f	(−)	nd	nd
[Arg^{2}Ala^{15}Leu^{27}]bGHRF(1–29)NH$_2$	88.0 ± 10.5^f	(−)	nd	nd
[Tyr^{2}Ala^{15}Leu^{27}]bGHRF(1–29)NH$_2$	107.0 ± 21.8^g	(−)	nd	nd

$^{a-g}$ Within a given assay, means lacking a common superscript letter were different at $p < .05$. nd, not determined. ^{h} C, D denote two different plasma pools; each experiment was done in triplicate. $^\#$ One hour stability index taken from Table 3.1, plasma pool A. (+)/(−) Indicates the (3–29)-fragment either generated (+) or not generated (−) in plasma. Table reproduced with permission from Kubiak et al. (28). © American Chemical Society.

purified DPP-IV while displaying high GH-releasing activity both in vitro and in vivo (40, 41).

Secondary Structures of GHRF and Inherent GH-Releasing Activity

In entirely aqueous solutions hGHRF(1-29)NH$_2$ (42) does not have well-defined structures and is present mostly in a random disordered conformation with a low α-helical content as determined by circular dichroism (CD) studies (42). However, ordered, mostly helical structures can be induced by low concentrations (30–35%) of trifluoroethanol (TFE) (42) or 75% to 80% methanol in aqueous media (36, 43). For hGHRF(1-29)NH$_2$ in 30% TFE, two distinct α-helical regions have been identified: the first extending from residues 6 to 13 and the second from 16 to 29. The region from residues 13 to 16 has been referred to as a half-turn (42). As shown by nuclear magnetic resonance (NMR) in 75% aqueous methanol, the Ala^{15} substitution in hGHRF(1-29)NH$_2$ resulted in an extended helix (residue 4 to 29) without the turn observed in the Gly^{15} analogue (43). The finding that the amino acid substitution at position 15 can either stabilize (Ala) or disrupt (Sar) the global helical structure of GHRF and the fact that these changes in helical conformation correlated well with inherent (in vitro)

GRF bioactivity (35, 36) led to the hypothesis that the presence of organic solvents might mimic the hydrophobic environment at the peptide-receptor interface (43). Also, NMR and CD studies on mono and bicyclic [Ala^{15}]hGHRF analogues, constrained via lactam bridges between Asp at position 3 (or 8) and Lys^{12} and/or Lys^{12} and Asp^{25}, proved the existence of stable α-helical structures present in these molecules in aqueous media even in the absence of organic solvents (43). However, this stabilization of helical conformation was accompanied by somewhat decreased GH releasing activity, indicating that the introduced constraint could have reduced peptide flexibility and either prevented the proper orientation of all the side chains that interact with the receptor or hindered the formation of the bend or kink in the helix that might be required for optimal binding (43).

The results of our CD and/or NMR studies indicate that the high propensity for α-helix formation and similar secondary structures may be necessary but not sufficient for eliciting similar GH releasing activity (39, 44, 45). Specific examples include [Leu^{27}]bGHRF(1-29)NH$_2$, its D-Ala^2-substituted analogue and [Leu^{27}]bGHRF(3-29)NH$_2$ fragment, which in 35% TFE had very similar, highly helical secondary structures in the 8 to 29 region, with only subtle differences at the N-termini as shown by NMR (44). In bovine anterior pituitary cell cultures, the D-Ala^2 analogue was as potent as the unsubstituted peptide while the des(Tyr-Ala) 3 to 29 fragment was virtually inactive in inducing GH release (44). This lack of correlation between secondary structures in solution and in vitro bioactivity suggests that not only highly helical structures but also residues 1 and 2 in GHRF are important for effective GHRF-receptor interaction. Also, NMR and CD studies of [Leu^{27}]bGHRF(1-29)NH$_2$ analogues with replacements of Lys^{12,21} → Arg^{12,21} or Arg^{11,20} → Lys^{11,20} identified similar conformations induced by 35% TFE (45). However, the double Arg-modified analogue was as potent as the unsubstituted peptide, while the double Lys-substituted peptide displayed only 0.03% potency of the model peptide in the in vitro bovine anterior pituitary cell assay (45). It might be that either different conformations are induced at the receptor level, or that arginines at position 11 and 20 are important contact residues crucial for the GHRF-receptor interaction, which cannot be replaced by lysines without loss of the hormone bioactivity.

In contrast to the modifications at position 15, for position 19-modified analogues from the [Thr^{2}Ala^{15}X^{19}Leu^{27}]bGHRF(1-29)NH$_2$ series (X^{19} = Leu, Val, or Ile), α-helicity in aqueous buffer was inversely correlated to the analogue inherent bioactivity (39). The Ala^{15}Val^{19}-analogue, which was the least helical of the three in the entirely aqueous medium, showed the highest relative GH-releasing activity in rat anterior pituitary cell cultures (39). Therefore, it can be that it is not only the global helicity that decides about analogue interaction with the receptor. Other important factors include spatial arrangements of individual amino acid side chains, hydrophobicity, local secondary structures in specific regions of the molecule, and an overall peptide flexibility that might be vital for a perfect fit to the receptor.

Despite some discrepancies between secondary structures observed in solution and bioactivity, in general, the highly beneficial effects of the α-helix enhancing Ala[15] modification in the central part of the GHRF molecule on both in vitro and in vivo GH releasing activity have been well documented. This was also confirmed in our studies, especially when the Ala[15] substitution was combined with DPP-IV stabilizing replacements at position 2 (26–28).

Interplay Between Metabolic Stability, Inherent (In Vitro) and Apparent In Vivo GH Releasing Activity

While inherent (in vitro) potency of an analogue (drug) largely reflects how well a given molecule interacts with its receptor, the in vivo performance is a function of inherent potency and clearance rates (metabolism and excretion). When in vivo GHRF-induced GH release was computer simulated, using a theoretical one compartment model of Wagner (46), it became apparent that the most efficient way to potentiate GHRF action and the intensity of its biologic response in vivo, without increasing the drug dose, would be via an analogue with a prolonged half-life, other factors remaining the same (26).

Historically, GHRF analogues have been reported as superpotent, based mostly on the results of GH releasing activity determined in rat anterior pituitary cell cultures. However, when we used a bovine pituitary cell assay (47) to test a set of over 80 various GHRF analogues, including a few so called superpotent compounds, none of these analogues performed any better than the native bGHRF(1-44)NH$_2$ (unpublished data). Perhaps, the superior behavior of some GHRFs represented a species-dependent unique feature of the rat system that did not apply to the bovine cell cultures. In retrospect, it has been suggested that the reported inherent potency of some GHRF analogues, e.g., with an Ala^2 → D-Ala^2 substitution (48), had been greatly overestimated and the major increase in in vitro and in vivo potency in the rat was subsequently shown to be due to enhanced resistance to DPP-IV rather than increased receptor affinity (49).

Early evaluations of the in vitro plasma stability and in vitro and in vivo GH releasing activities of [Thr^{2}Ala^{15}Leu^{27}]bGHRF(1-29)NH$_2$ let us realize the importance of metabolic stability on the in vivo bioactivity. This analogue, estimated to be only about 25% to 30% as potent as the native hormone in vitro in bovine pituitary cells, in steers turned out to be twice as active (single iv injection) and fourfold more potent (dose-response study) than bGHRF(1-44)NH$_2$. We view the improved activity in vivo to be due mostly to its improved proteolytic stability (26, 28). Although other DPP-

IV protected analogues were weakly active in vivo (even those with low inherent potency) in order to have *improved* activity in vivo, inherent activity at the receptor must be above a certain threshold value. In that light, despite effective protection against DPP-IV, the analogues either inactive at 1 to 5 nM (Glu²Ala¹⁵, Arg²Ala¹⁵, or Lys²Ala¹⁵) or only weakly active (Asp²Ala¹⁵ or Tyr²Ala¹⁵) in the rat pituitary cell cultures in vitro (Fig. 3.2) (27, 28) were only slightly active in steers (Fig. 3.3) (28). In these cases, stabilization against DPP-IV, which possibly led to a longer half-life of the intact peptides in the circulation, did not offset deleterious effects of their low inherent GH releasing activity. On the other hand, compounds (e.g., Gly², Gly²Ala¹⁵, Ser², Ser²Ala¹⁵, Thr², Leu²Ala¹⁵, or Thr²Ala¹⁵) with moderate in vitro bioactivities, combined with their enhanced plasma stability, produced serum GH responses in steers either comparable to, or higher than, the treatment with the native bGHRF(1-44)NH₂ (Fig. 3.3) (26–28). Here, the improvement in metabolic stability was able to override lower inherent potency and resulted in full or even enhanced performance of the GHRF analogues in vivo. A similar observation applies to the Ile²Ala¹⁵ and Val²Ala¹⁵ analogues, which were comparable to the unsubstituted parent peptide in their in vitro bioactivity (Fig. 3.2) but turned out to be five times

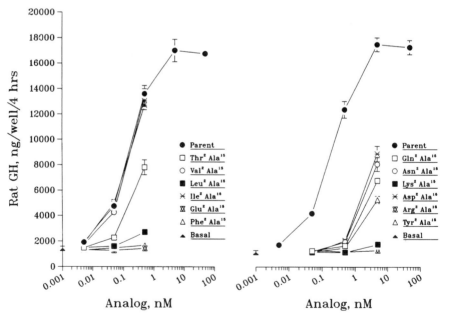

FIGURE 3.2. Effects of bGHRF analogues from the [X²Ala¹⁵Leu²⁷]bGHRF(1-29)NH₂ series on GH release in rat anterior pituitary cell cultures in vitro. Reproduced with permission from Kubiak et al. (27). © 1992 ESCOM Science Publishers B.V.

as potent as $bGHRF(1-44)NH_2$ in vivo and were identified as the most potent analogues from the X^2/Ala^{15} series (28).

$[DesNH_2Tyr^1D-Ala^2Ala^{15}]hGHRF(1-29)NH_2$ has been reported by others to be 10 to 15 times more potent than the $hGHRF(1-44)NH_2$ in releasing GH (50) and to be more potent than $hGHRF(1-29)NH_2$ in stimulating GH release and milk production of dairy cows (51). The lower estimated GH-releasing potency of this analogue in our in vivo model in steers (52) [only four times higher than that of $bGHRF(1-44)NH_2$ (28)] may be due to a different experimental design, different animals (dairy cows versus steers) as well as different standards used in these in vivo experiments. The lower stability in bovine plasma of $[Thr^2Ala^{15}Leu^{27}]bGHRF(1-29)NH_2$ and $[desNH_2Tyr^1D-Ala^2Ala^{15}]hGHRF(1-29)NH_2$, determined in our study, may partially explain the slightly lower GH responses to these two analogues in vivo in steers as compared with the more metabolically stable $[Val^2Ala^{15}Leu^{27}]bGHRF(1-29)NH_2$ and $[Ile^2Ala^{15}Leu^{27}]bGHRF(1-29)NH_2$. These results once more emphasize the significant contributions of both improved metabolic stability and inherent potency in increasing the GH releasing activity observed in vivo.

It is worthwhile to comment on the dose-dependent GH releasing activity in vivo of the Ser^2Ala^{15} analog. This peptide was not different from $bGHRF(1-44)NH_2$ when tested in steers at a single iv dose of 0.01 nmol/kg (26, 28) but as shown in Figure 3.3, a dose 10 times higher resulted in a 67% greater serum GH response in vivo than was induced by the native hormone (28). The DPP-IV related degradation of $[Ser^2Ala^{15}Leu^{27}]bGHRF(1-29)NH_2$, which was observed in bovine plasma in vitro, could have been accelerated under the in vivo conditions, where overall organ and tissue DPP-IV activity may be much higher than the plasma DPP-IV levels. The DPP-IV activity of the bovine plasma might represent only a small fraction of the overall organ and tissue DPP-IV activity, as has been reported for the rat system (53). It is possible that a higher dose of $[Ser^2Ala^{15}Leu^{27}]bGHRF(1-29)NH_2$ used in vivo led to a greater fraction of the intact peptide remaining in serum above a threshold level needed to increase serum GH. Therefore, it may be concluded that even a partial protection of GHRF against DPP-IV cleavage, like in the Ser^2Ala^{15} analogue, is sufficient to improve peptide performance in vivo (28).

An interesting C-terminally extended GHRF analogue with relatively low intrinsic potency but superior performance in vivo was reported by Smith et al. (54). This construct, designed for effective expression in *Escherichia coli*, consisted of the 44 amino acids of mature hGHRF followed by a 33 amino acid extension, identical to the C-peptide in hGHRF precursor protein. The protein was expressed with an extra Met-Ala extension at the N-terminus from which the N-terminal Met was then removed by the *E.coli* enzymatic system leaving the final 78 amino acid product N-terminally extended with a single Ala^0 residue (54). Despite relatively low in vitro GH releasing potency (about 10-fold lower than that of native

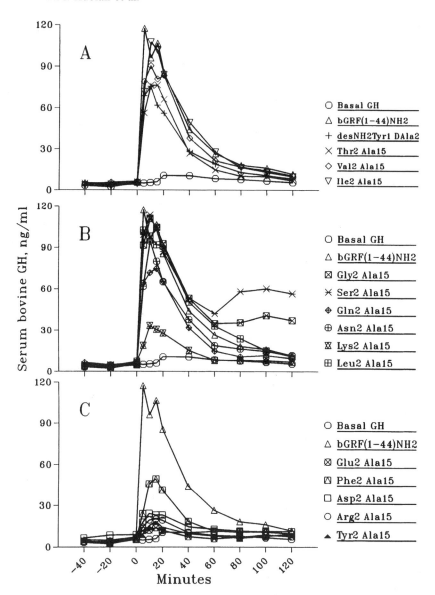

FIGURE 3.3. Serum bGH levels after a single iv injection of position 2/Ala¹⁵-substituted analogues of [Leu²⁷]bGHRF(1-29)NH₂ in the meal-fed steers (52). Doses: Panel A: 0.02 nmol/kg, except for bGHRF(1-44)NH₂, which was at 0.1 nmol/kg; Panels B and C: 0.1 nmol/kg. All analogues were tested in the same 17 × 17 Latin square design experiment but the data are split for clarity of presentation. dNH₂Tyr¹D-Ala² denotes [desNH₂Tyr¹D-Ala²Ala¹⁵]hGHRF(1-29)NH₂.

hGHRF), subcutaneous injections of this [Ala^0]pro-GHRF in wether lambs resulted in plasma GH responses equal to or greater than the responses typically seen in previous infusion studies with similar dosages of human GHRF (54). It is interesting that this truncated hybrid GHRF propeptide, unlike other prohormones, showed remarkable bioactivity in vivo. According to the authors' hypothesis (no metabolic stability data provided), the apparent increase in pro-GHRF potency in vivo might have been due to the increase in its half-life over that of mature GHRF in animal serum (54). It seems very likely that the presence of Ala^0 could have stabilized this proGHRF against DPP-IV proteolysis. In our studies, the half-life of a bGHRF analogue N-terminally extended with a single Ile residue (Ile^0-bGHRF) was greatly prolonged in bovine plasma in vitro due to its demonstrated resistance to plasma DPP-IV (32). This may be the case with [Ala^0]pro-GHRF, which should also be out of phase for DPP-IV recognition and, additionally, due to its increased size, might be cleared from the circulation at a slower rate than the native hormone.

A fully synthetic, pegylated GHRF analogue, recently described by Felix et al. (55), might be the most effective GHRF analogue reported to date. [desNH$_2$Tyr^{1}D-Ala^{2}Ala^{15}]hGHRF(1-29)-Gly-Gly-Cys(NH$_2$)-S-Nle-PEG$_{5000}$ was found to be 33.6 and 54.5 times more potent than hGHRF(1-44)NH$_2$ in mice and pigs, respectively (55). The C-terminal GHRF modification with polyethylene glycol MW 5,000 (PEG$_{5000}$) did not alter the in vitro potency relative to the parent GHRF nor had any effect on peptide conformation determined in 75% aqueous methanol as shown by CD (55). The superior performance of this analog in vivo was proposed to be linked to its longer half-life in the peripheral circulation due to decreased metabolic clearance (55).

Chemical Degradation of GHRF and Analogues with Improved Chemical Stability

Chemical rearrangement and deamidation of Asn- and, to a lesser extent, Gln-containing biologically active peptides and proteins is often accompanied by a substantial loss of their bioactivity. At neutral or basic pH, the generally accepted mechanism of deamidation proceeds via a reactive succinimide intermediate formed by intramolecular nucleophilic attack of the backbone amide nitrogen of the $i + 1$ residue onto the Asn side chain carbonyl with a subsequent elimination of NH$_3$. Hydrolysis of the cyclic intermediate yields Asp and β-Asp (also referred to as iso-Asp) products, with the latter being the favored product at physiologic pH. Also, partial racemization through the succinimide intermediate and peptide cleavage between i and $i + 1$ residue have been reported (56).

Deamidation of Asn^8 has been shown to be a major chemical degradation pathway of bovine (47) and human (57) GHRF. Deamidation is known to be sequence and conformation dependent and this is probably why Asn^{28} in bovine GHRF is degraded much slower than Asn^8 (47) and no Asn^{35} deamidation in hGHRF was observed (57).

Incubation of [Leu^{27}]hGHRF(1-32)NH$_2$ at pH 7.4 led to extensive peptide degradation (47). The major degradation products were identified as [β-Asp^{8}Leu^{27}]hGHRF(1-32)NH$_2$ and [Asp^8 Leu^{27}]hGHRF(1-32)NH$_2$. When tested for GH release in bovine anterior pituitary cell cultures, the β-Asp^8 and Asp^8-substituted peptides were, respectively, 400 to 500 and 25 times less potent than the parent Asn^8 peptide (47). Although the rate of chemical degradation of [Leu^{27}]bGHRF(1-32)NH$_2$ in phosphate buffer at pH 7.4 and 37°C ($t_{1/2}$ = 150 hours) was about 200 to 500 times slower than the peptide metabolism in bovine plasma in vitro, the fact that the Asn^8 deamidation/isomerization is accompanied by such a drastic loss in bioactivity implied a need for Asn replacements to prevent such rearrangements in sustained release formulations.

Asn^8 → Ser^8, Asn^{28} → Ser^{28}, and Asn^{8}Asn^{28} → Ser^{8}Ser^{28} substitutions in [Leu^{27}]bGHRF(1-32)NH$_2$ resulted in much more chemically stable analogues with the respective half-lives of 746 hours, 202 hours, and 1550 hours in phosphate buffer at pH 7.4 (47). These Ser for Asn replacements also resulted in full retention of the in vitro GH releasing activity of [Leu^{27}]bGHRF(1-32)NH$_2$ (47). Increased solution stability and enhanced GH releasing activity were also reported for Asn^8 → Gln^8, Asn^8 → Thr^8, and Asn^8 → Ser^8 modifications in the [His^{1}Val^{2}X^{8}Ala^{15} Leu^{27}]hGHRF(1-32)OH series (17, 41). A recently described Asn^8 → Ala^8 replacement in hGHRF, beneficial from the bioactivity point of view (58–60), should also lead to analogues with improved chemical stability due to the elimination of deamidation sites.

Isomerization of Asp^3 to β-Asp^3 and cleavage at Asp^3-Ala^4 under acidic conditions were also revealed as another potential chemical degradation pathway of GHRF that resulted in a considerable decrease in biologic activity of (57).

Comparison of Various Superpotent GHRF Analogues in a Single Bioassay In Vivo

Many pharmaceutical companies and individual laboratories have been making analogues of GHRF for the past 12 years. The literature discloses the in vivo GH releasing activity or potency of several GHRF analogues with responses ranging from 5- to 50-fold greater than either bGHRF(1-44)NH$_2$ or hGHRF(1-29)NH$_2$. With the wide range of responses to putative superpotent GHRF analogues, we were interested in a head-to-head comparison of one of our better analogues, [Ile^{2}Ser^{8}Ala^{15}Leu^{27}Ser^{28}Hse^{30}]-bGHRF(1-30)NHEt, against other reported superior analogues in the same in vivo bioassay system. The meal-fed steer model of Moseley et al. (52) was used for the study with analogues administered as a single iv dose of

0.2 nmol/kg. GH releasing activities were expressed as a 0 to 3-hour area under the serum GH curve and compared against bGHRF(1-44)NH$_2$ (Fig. 3.4). The in vivo bioactivity of our analogue (222%) was not statistically different from that of [desNH$_2$Tyr^{1}D-Ala^{2}Ala^{15}Leu^{27}]-hGHRF(1-29)NH$_2$ (192%), a superior GHRF analogue first reported by Felix et al. (36). The relative activities of [D-Ala^{2}Ala^{8,9,15,27}D-Arg^{29}]hGHRF(1-29)NH$_2$ (a gift from Dr. David Coy) and [N-MeTyr^{1}Ala^{8,9,15,22}Nle^{27}Ala^{28}]hGHRF(1-29)NH$_2$ (60; a gift from Dr. Jean Rivier) were, respectively, 151% and 139% of that of bGHRF(1-44)NH$_2$ (100%) and were statistically different from the activity of our Ile^2 analogue ($p < .05$). All four analogues were significantly more active than bGHRF(1-44)NH$_2$ ($p < .05$). Although in this case dose-response curves were not generated, our previous experience with GH releasing activity measured at a single iv dose versus potency determined from a full dose-response curve (28) allows us to estimate that the analogues from this study were probably three- to fivefold more potent than the native bGHRF(1-44)NH$_2$ under the conditions used for testing. We cannot comment on the relative GH releasing activity of the pegylated GHRF from Hoffmann-LaRoche (55) or the Eli Lilly [Ala^0]pro-GHRF (54) in our steer assay because we were not aware of the analogue's existence at the time of the in vivo experiment described here.

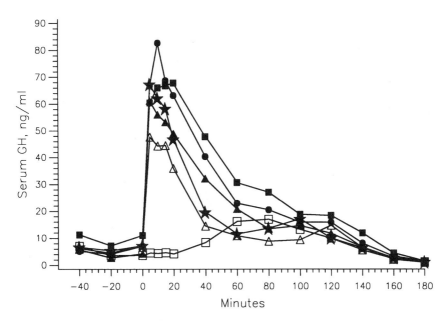

FIGURE 3.4. Serum bGH levels after a single iv dose (0.02 nmol/kg) of various GHRF analogues in the meal-fed steer model (52). △ bGHRF(1-44)NH$_2$; ● [Ile^{2}Ser^{8}Ala^{15}Leu^{27}Ser^{28}Hse^{30}]-bGHRF(1-30)NHEt; ■ [desNH$_2$Tyr^{1}D-Ala^2-Ala^{15}Leu^{27}]hGHRF(1-29)NH$_2$; ▲ [D-Ala^{2}Ala^{8,9,15,27}D-Arg^{29}]hGHRF(1-29)NH$_2$; ★ [N-MeTyr^{1}Ala^{8,9,15,22} Nle^{27}Ala^{28}]hGHRF(1-29)NH$_2$; □ water.

GHRF Analogues with Improved Pharmaceutical Properties

Drug potency, chemical stability, and resistance to various enzymes are important characteristics especially when there is a need for long-term delivery of a drug to be used as a growth enhancer. Although all these factors are important, as shown in this report, the practicality of producing a potent, long-lasting analogue in a cost-effective manner may be the key for successfully developing a GHRF analogue as a drug for human, and especially for animal, use.

Although the current synthetic methods (especially solid-phase technique) have enabled the synthesis of hundreds of analogues for SAR studies, scaling up to multikilogram quantities presents major difficulties and high production costs. With the current state of the art of chemical peptide synthesis, recombinant DNA synthesis might be the production method of choice for the preparation of large quantities of peptidic drugs, especially where economics are highly important. In this case, highly potent, chemically and metabolically stabilized GHRF analogues should be composed entirely of naturally occurring, DNA-coded amino acids. From this point of view, the production of many superior GHRF analogues identified over the years, which requires either chemical or a combination of biosynthetic and chemical/or enzymatic methods, might be hampered by economics. The pegylated GHRFs, desNH$_2$Tyr^1- and/or D-Ala^2-containing GHRF analogues, discussed earlier, fall into this category, unless better synthetic or enzymatic methods are found.

The Eli Lilly [Ala^0]pro-GHRF (54), the Hoffmann-LaRoche [His^{1}Val^{2}Gln^{8}Ala^{15}Leu^{27}]-hGHRF(1-32)OH (17, 40, 41) and our analogues from the [X^{2}Ser^{8}Ala^{15}Leu^{27}Ser^{28}Hse^{30}]bGHRF(1-30)NHY (X = Thr, Val or Ile; Y = H or alkyl) series are among the GHRF analogues with superior characteristics for which effective biosynthetic production methods have been found and reported.

In contrast to [Ala^0]pro-GHRF, which can be directly expressed in host cells (54), smaller peptides are usually destined for destruction by endogenous host system proteases. To overcome this problem, peptides are usually expressed as fusion proteins with larger carriers followed either by selective chemical or enzymatic release of the desired product as reported for [His^{1}Val^{2}Gln^{8}Ala^{15}Leu^{27}]hGHRF(1-32)OH (17). The combination of His^1 with Val^2 in this analogue resulted in resistance to plasma DPP-IV and high GH release in vivo. These features along with improved chemical stability, due to the Gln^8 for Asn^8 substitution, make this analogue very attractive for future practical applications (17, 40, 41).

Large scale biosynthesis of analogues from the [X^{2}Ser^{8}Ala^{15}Leu^{27}Ser^{28}Hse^{30}]bGHRF(1-30)NHY series can be accomplished by expressing tandem repeats of the precursor GHRF sequence having a Met residue at position 30, separated by linker peptides amenable for subsequent CNBr cleavage as described by Kirschner et al. (34). The CNBr-liberated monomeric GHRF, in its C-terminal homoserine^{30} (Hse^{30}) lactone form, can then be subjected to aminolysis or a reaction with alkylamines to generate analogues with the C-terminal Hse^{30}-NH$_2$ or Hse^{30}-N-alkylated amides, respectively.

Selected analogs from the [X^{2}Ser^{8}Ala^{15}Leu^{27}Ser^{28}Hse^{30}]bGHRF(1-30)-NHY series have been found to be about fivefold as potent as the native bGHRF in vivo in releasing serum GH in steers upon iv administration (unpublished data from our lab). Both [Ile^{2}Ser^{8}Ala^{15}Leu^{27}Ser^{28}Hse^{30}]-bGHRF(1-30)NH$_2$ (61) and [Ile^{2}Ser^{8}Ala^{15}Leu^{27}Ser^{28}Hse^{30}]bGHRF(1-30)NHEt (62) have been shown to be very effective in increasing milk production in dairy cows. In a recent trial, continuous iv infusion of 2.7 and 8.1 mg/day of [Ile^{2}Ser^{8}Ala^{15}Leu^{27}Ser^{28} Hse^{30}]bGHRF(1-30)NHEt and 12 mg/day of [Leu^{27}Hse^{45}]bGHRF(1-45)lactone produced similar increases in serum GH and insulin-like growth factor (IGF)-1 concentrations with no evidence of pituitary refractoriness for at least 92 days in dairy cows (62). For the same period, the yield of solids-corrected milk increased by about 8 kg/day (27%) in the GHRF-treated versus control animals (62). The next challenge is to design appropriate drug formulations that effectively deliver bioactive material for extended periods of times in a fashion that will sustain elevated serum GH concentrations and produce the desired biologic effects in target species.

In summary, the recombinant production of highly potent, metabolically and chemically stabilized GHRFs is currently most promising and offers a good potential for future commercialization of GHRF analogues for human and veterinary applications.

Acknowledgments. The authors are grateful to Drs. Jean Rivier (Salk Institute) and David Coy (Tulane University School of Medicine) for providing samples of their highly potent GHRF analogues for in vivo testing. Dr. Arthur Felix (Hoffmann-La Roche) is greatly appreciated for providing us with the preprints of the manuscripts listed as refs. 17 and 55. Dr. James Caputo (Upjohn) is acknowledged for adapting Wagner's model to computer simulated GHRF-induced GH release, and Mrs. Diane Cleary (Upjohn) is thanked for helping with graphical data representation and data analyses.

References

1. Guillemin R, Brazeau P, Böhlen P, Esch F, Ling N, Wehrenberg WB. Growth hormone-releasing factor from a human pancreatic tumor that causes acromegaly. Science 1982;218:585–7.

2. Rivier J, Spiess J, Thorner M, Vale W. Characterization of a growth hormone-releasing factor from a human pancreatic islet tumor. Nature 1982;300:276–8.
3. Ling N, Esch F, Böhlen P, Brazeau P, Wehrenberg WB, Guillemin R. Isolation, primary structure, and synthesis of human hypothalamic somatocrinin: growth hormone-releasing factor. Proc Natl Acad Sci USA 1984;81:4302–6.
4. Böhlen P, Esch F, Brazeau P, Ling N, Guillemin R. Isolation and characterization of the porcine growth hormone-releasing factor. Biochem Biophys Res Commun 1983;116:726–34.
5. Esch F, Böhlen P, Ling N, Brazeau P, Guillemin R. Isolation and characterization of the bovine hypothalamic growth hormone releasing factor. Biochem Biophys Res Commun 1983;117:772–9.
6. Brazeau P, Böhlen P, Esch F, Ling N, Wehrenberg WB, Guillemin R. Growth hormone-releasing factor from ovine and caprine hypothalamus: isolation, sequence analysis and total synthesis. Biochem Biophys Res Commun 1984;125:606–14.
7. Spiess J, Rivier JE, Vale W. Characterization of rat hypothalamic growth hormone-releasing factor. Nature 1983;303:532–5.
8. Frohman MA, Downs TR, Chomczynski P, Frohman LA. Cloning and characterization of mouse growth hormone-releasing hormone (GRH) complementary DNA: increased GRH messenger RNA levels in the growth hormone-deficient lit/lit mouse. Mol Endocrinol 1989;3:1529–36.
9. Suhr ST, Rahal JO, Mayo KE. Mouse growth hormone-releasing hormone: precursor structure and expression in brain and placenta. Mol Endocrinol 1989;3:1693–700.
10. Vance ML, Kaiser DL, Evans WS, Furlanetto R, Vale W, Rivier J, et al. Pulsatile growth hormone secretion in normal man during a continuous 24-hour infusion of human growth hormone-releasing factor (1-40): evidence for intermittent somatostatin secretion. J Clin Invest 1985;75:1584–90.
11. Wehrenberg WB. Continuous infusion of growth hormone-releasing factor: effects on pulsatile growth hormone secretion in normal rats. Neuroendocrinol 1986;43:391–4.
12. Brain CE, Hindmarsh PC, Brook CGD. Continuous subcutaneous GHRH(1-29)NH$_2$ promotes growth over one year in short, slowly growing children. Clin Endocrinol 1990;32:153–63.
13. Moseley WM, Krabill LF, Friedman AR, Olsen R. Administration of synthetic human pancreatic growth hormone-releasing factor for five days sustains raised serum concentrations of growth hormone in steers. J Endocrinol 1985; 104:433–9.
14. Moseley WM, Huisman J, VanWeerden EJ. Serum growth hormone and nitrogen metabolism responses in young bull calves infused with growth hormone-releasing factor for 20 days. Domestic Animal Endocrinol 1987;4:51–9.
15. Frohman LA, Jansson J. Growth hormone-releasing hormone. J Endocr Rev 1986;7:223–53.
16. Ling N, Zeytin F, Böhlen P, Esch F, Brazeau P, Wehrenberg WB, et al. Growth hormone-releasing factor. Annu Rev Biochem 1985;54:403–23.
17. Campbell RM, Bongers J, Felix AM. Rational design, synthesis, and biological evaluation of novel growth hormone-releasing factor (GRF) analogs. Biopolymers 1995;37:67–88.

18. Ling N, Baird A, Wehrenberg WB, Ueno N, Munegumi T, Brazeau P. Synthesis and in vitro activity of C-terminal deleted analogs of human growth hormone-releasing factor. Biochem Biophys Res Commun 1984;123:854–61.
19. Campbell RM, Lee Y, Rivier J, Heimer EP, Felix AM, Mowles TF. GRF analogs and fragments: correlation between receptor binding, activity and structure. Peptides 1991;12:569–74.
20. Kubiak TM, Martin RA, Leone JW, Cleary DL. Metabolism of mouse growth hormone-releasing factor, mGRF(1-42)OH, and selected analogs from the bovine GRF series in mouse and bovine plasma. Peptide Res 1994;7:153–61.
21. Frohman LA, Downs TR, Williams TC, Heimer EP, Pan YC-E, Felix AM. Rapid enzymatic degradation of growth hormone-releasing hormone by plasma in vitro and in vivo to a biologically inactive product cleaved at the NH_2 terminus. J Clin Invest 1986;78:906–13.
22. Frohman LA, Downs TR, Heimer EP, Felix AM. Dipeptidylpeptidase IV and trypsin-like enzymatic degradation of human growth hormone-releasing hormone in plasma. J Clin Invest 1989;83:1533–40.
23. Kubiak TM, Kelly CR, Krabill LF. In vitro metabolic stability of a bovine growth hormone-releasing factor analog Leu[27]-bGRF(1-29)NH_2 in bovine and porcine plasma. Correlation with plasma dipeptidylpeptidase activity. Drug Metab Dispos 1989;17:393–7.
24. Boulanger L, Roughly P, Gaudreau P. Catabolism of rat growth hormone-releasing factor (1-29) amide in rat serum and liver. Peptides 1992;13:681–9.
25. Su C-M, Jensen LR, Heimer EP, Felix AM, Pan Y-CE, Mowles TF. In vitro stability of growth hormone-releasing factor (GRF) analogs in porcine plasma. Horm Metab Res 1991;23:15–21.
26. Kubiak TM, Martin RA, Hillman RM, Kelly CR, Caputo JF, Alaniz GR, et al. Implications of improved metabolic stability of peptides on their performance in vivo. Resistance to DPP-IV-mediated cleavage of GRF analogs greatly enhances their potency in vivo. In: Smith JA, Rivier JE, eds. Peptides. Chemistry and biology. Proceedings of the Twelfth American Peptide Symposium, 1991, June 16–21, Cambridge, MA. Leiden: ESCOM Science Publishers, 1992:23–5.
27. Kubiak TM, Friedman AR, Martin RA, Ichhpurani AK, Alaniz GR, Claflin WH, et al. High in vivo bioactivities of position 2/Ala[15]-substituted analogs of bovine growth hormone-releasing factor (bGRF) with improved metabolic stability. In: Smith JA, Rivier JE, eds. Peptides. Chemistry and biology. Proceedings of the Twelfth American Peptide Symposium, 1991, June 16–21, Cambridge, MA. Leiden: ESCOM Science Publishers, 1992:858–7.
28. Kubiak TM, Friedman AR, Martin RA, Ichhpurani AK, Alaniz GR, Claflin WH, et al. Position 2 and position 2/Ala[15]-substituted analogs of bovine growth hormone-releasing factor (bGRF) with enhanced metabolic stability and improved in vivo bioactivity. J Med Chem 1993;36:888–97.
29. Walter R, Simmons WH, Yoshimoto T. Proline specific endo- and exopeptidases. Mol Cell Biochem 1980;30:111–27.
30. Yaron A, Naider F. Proline-dependent structural and biological properties of peptides and proteins. Clin Rev Biochem Mol Biol 1993;28:31–81.
31. Felix AM, Heimer EP, Wang C, Lambros TJ, Fournier A, Mowles TF, et al. Synthesis, biological activity and conformational analysis of cyclic GRF analogs. Int J Peptide Protein Res 1988;32:441–54.

32. Kubiak TM, Cleary DL, Krabill LF. N-terminally extended GRF analogs as pro-drugs for the liberation of the core GRF by bovine plasma *in vitro*. In: Schneider CH, Eberle AN, eds. Peptides 1992. Proceedings of the Twenty-Second European Peptide Symposium, 1992, Sept 13–19, Interlaken, Switzerland. Leiden: ESCOM Science Publishers, 1993:739–40.
33. Kubiak TM, Martin RA, Hillman RM, Caputo JF, Alaniz GR, Claflin WH, et al. N-terminally extended analogs of bGRF with a general formula $[X^{-1},Y^0,Leu^{27}]$bGRF(1-29)NH_2 as pro-drugs and potential targets for processing by plasma dipeptidylpeptidase IV (DPP-IV). In: Hodges RS, Smith JA, eds. Peptides. Chemistry and biology. Proceedings of the Thirteenth American Peptide Symposium, 1993, June 20–25, Edmonton, Alberta, Canada. Leiden: ESCOM Science Publishers, 1994:859–61.
34. Kirschner RJ, Hatzenbuhler NT, Moseley WM, Tomich CSC. Gene synthesis, *E. coli* expression and purification of the bovine growth hormone releasing factor analog, (Leu^{27},Hse^{45})bGRF. J Biotechnol 1989;12:247–60.
35. Felix AM, Heimer EP, Mowles TF, Eisenbeis H, Leung P, Lambros T, et al. Synthesis and biological activity of novel growth hormone releasing factor analogs. In: Theodoropoulis D, ed. Peptides 1986. Berlin-New York: Walter de Gruyter, 1987:481–4.
36. Felix AM, Wang CT, Heimer EP, Fournier A, Bolin D, Ahmad M, et al. Synthesis and biological activity of novel linear and cyclic GRF analogs. In: Marshall R, ed. Peptides. Chemistry and biology. Proceedings of the Tenth American Peptide Symposium, 1987, May 23–28. Leiden: ESCOM Science Publishers, 1988:465–7.
37. Martin RA, Cleary DL, Guido DM, Zurcher-Neely HA, Kubiak TM. Dipeptidyl peptidase IV (DPP-IV) from pig kidney cleaves analogs of bovine growth hormone-releasing factor (bGRF) modified at position 2 with Ser, Thr, or Val. Extended DPP-IV substrate specificity? Biochim Biophys Acta 1993;1164:252–60.
38. Bongers J, Lambros T, Ahmad M, Heimer EP. Kinetics of dipeptidyl peptidase IV proteolysis of growth hormone-releasing factor and analogs. Biochim Biophys Acta 1992;1122:147–53.
39. Friedman AR, Ichhpurani AK, Moseley WM, Alaniz GR, Claflin WH, Cleary DL, et al. Growth hormone releasing factor analogs with hydrophobic residues at position 19. Effects on growth hormone releasing activity *in vitro* and *in vivo*, stability in blood plasma *in vitro*, and secondary structure. J Med Chem 1992;35:3928–33.
40. Heimer EP, Bongers J, Ahmad M, Lambros T, Campbell RM, Felix AM. Synthesis and biological evaluation of growth hormone-releasing factor analogs resistant to degradation by dipeptidylpeptidase IV. In: Smith JA, Rivier JE, eds. Peptides. Chemistry and biology. Proceedings of the Twelfth American Peptide Symposium, 1991, June 16–21, Cambridge, MA. Leiden: ESCOM Science Publishers, 1992:80–1.
41. Campbell RM, Stricker P, Miller R, Bongers J, Heimer EP, Felix AM. Hybrid rodent-human growth hormone-releasing factor (GRF) analogs with enhanced stability and biological activity. In: Hodges RS, Smith JA, eds. Peptides. Chemistry and biology. Proceedings of the Thirteenth American Peptide Symposium, 1993, June 20–25, Edmonton, Alberta, Canada. Leiden: ESCOM Science Publishers, 1994:378–9.

42. Clore GM, Martin SR, Gronenborn AM. Solution structure of human growth hormone-releasing factor. Combined use of circular dichroism and nuclear magnetic resonance spectroscopy. J Mol Biol 1986;191:553–61.
43. Fry DC, Madison VS, Greeley DN, Felix AM, Heimer EP, Frohman LA, et al. Solution structures of cyclic and dicyclic analogs of growth hormone-releasing factor as determined by two-dimensional NMR and CD spectroscopies and constrained molecular dynamics. Biopolymers 1992;32:649–66.
44. Kloosterman DA, Scahill TA, Hillman RM, Cleary DL, Kubiak TM. ^{1}H NMR analysis and *in vitro* bioactivity of bovine growth hormone-releasing factor (bGRF) analogs: Leu^{27}-bGRF(1-29)NH$_2$, and its D-Ala^2, and des-(Tyr^1-Ala^2) analogs. Peptide Res 1991;4:72–8.
45. Kubiak TM, Kloosterman DA, Martin RA, Hillman RM, Cleary DL, Bannow CA, et al. Analogs of bovine growth hormone-releasing factor (bGRF) with various combinations of Arg and/or Lys at positions 11/12 and 20/21. Similar solution conformations are not sufficient for analogs' effective bioactivities *in vitro*. In: Giralt E, Andreu D, eds. Peptides 1990. Proceedings of the Twenty-First European Peptide Symposium, 1990, September 2–8; Platja d'Aro, Spain. Leiden: ESCOM Science Publishers, 1991:533–4.
46. Wagner JG. Kinetics of pharmacologic response. I. Proposed relationship between response and drug concentration in the intact animal and man. J Theor Biol 1968;20:173–201.
47. Friedman AR, Ichhpurani AK, Brown DM, Hillman RM, Krabill LF, Martin RA, et al. Degradation of growth hormone-releasing factor analogs in neutral solution is related to deamidation of asparagine residues. Int J Pept Protein Res 1991;37:14–20.
48. Lance VA, Murphy WA, Sueiras-Diaz J, Coy DH. Super-active analogs of growth hormone-releasing factor (1-29)-amide. Biochem Biophys Res Commun 1984;119:265–72.
49. Coy DH, Murphy WA, Hockart SJ, Taylor J. Pharmacology of GRF and somatostatin: positive and negative control of GH release. In: Bercu BB, Walker R, eds. Growth hormone II: basic and clinical aspects. Proceedings of the Symposium on Growth Hormone II, 1992, Dec 3–6, Tarpon Springs (FL). New York: Springer-Verlag, 1993:3–16.
50. Mowles T, Stricker P, Eisenbis H, Heimer E, Felix A, Brazeau P. Effects of novel GRF analog on GH secretion *in vitro* and *in vivo*. Endocrinol Jpn 1987;34(suppl):148.
51. Petitclerc D, Lapierre H, Pelletier G, Dubreuil P, Gaudreau P, Mowles T, et al. Effect of a potent analog of human growth hormone-releasing factor on growth hormone release and milk production of dairy cows. J Diary Sci 1987;(suppl 1):178(abstr).
52. Moseley WM, Alaniz GR, Claflin WH, Krabill LF. Food intake alters the serum growth hormone response to bovine growth hormone-releasing factor in meal-fed Holstein steers. J Endocrinol 1988;117:253–9.
53. Yoshimoto T, Ogita K, Walter R, Koida M, Tsuru D. Post-proline cleaving enzyme. Synthesis of a new fluorogenic substrate and distribution of endopeptidases in rat tissues and body fluids. Biochim Biophys Acta 1979;569:184–92.
54. Smith DP, Heiman LM, Wagner JF, Jackson RL, Bimm RA, Hsiung HM. Production and biological activity of hybrid growth hormone-releasing hormone propeptides. Biotechnology 1992;10:315–9.

55. Felix AM, Lu YA, Campbell RM. Pegylated Peptides. IV. Enhanced biological activity of site-directed pegylated GRF analogs. Int J Peptide Protein Res 1995;46:253–264.
56. Wright HT. Nonenzymatic deamidation of asparaginyl and glutaminyl residues in proteins. Crit Rev Biochem Mol Biol 1991;26:1–52.
57. Bongers J, Heimer, EP, Lambros T, Pan Y-CE, Campbell RM, Felix AM. Degradation of aspartic acid and asparagine residues in growth hormone-releasing factor. Int J Pept Protein Res 1992;39:364–74.
58. Cervini L, Galyean R, Donaldson CJ, Yamamoto G, Koerber SC, Vale W, Rivier JE. A SAR study of the complete Ala and partial Aib scans of the growth hormone-releasing factor [Nle^{27}]hGRF(1-29)NH$_2$. In: Smith JA, Rivier JE, eds. Peptides. Chemistry and biology. Proceedings of the Twelfth American Peptide Symposium, 1991, June 16–21, Cambridge, MA. Leiden: ESCOM Science Publishers, 1992:437–8.
59. Coy DH, Hocart SJ, Murphy WA. Human growth hormone-releasing hormone analogs with much improved in vitro growth hormone-releasing potencies in rat pituitary cells. Eur J Pharmacol 1991;204:179–85.
60. Rivier JE, et al. Patent application, USSN 701,414, filed 5/15/91.
61. Dahl GE, Chapin LT, Moseley WM, Kamdar MB, Tucker HA. Galactopoietic effects of a (1-30)NH$_2$ analog of growth hormone-releasing factor in dairy cows. J Dairy Sci 1994;77:2518–25.
62. Tucker HA, Terhune ALF, Chapin LT, Vanderhaar MJ, Sharma BK, Moseley WM. Long-term somatotropic and galactopoietic effects of a (1-30) NHEt analog of growth hormone-releasing factor. J Dairy Sci 1995;78:1489–97.

4

Structure, Function, and Regulation of the Pituitary Receptor for Growth Hormone Releasing Hormone

KELLY E. MAYO, PAUL A. GODFREY, VENITA DEALMEIDA, AND TERESA L. MILLER

Growth hormone releasing hormone (GHRH) is a peptide hormone synthesized and released from neurosecretory cells of the hypothalamic arcuate nuclei that stimulates the secretion of growth hormone from pituitary somatotroph cells (1–5). In addition to its expression in the brain, GHRH is also synthesized in the rodent placenta, where it may have paracrine functions or contribute to fetal growth (6, 7), and in the gonads, where it may be an autocrine or paracrine regulator of ovarian and testicular cell function (8–10). An important role for GHRH in body growth is suggested both by clinical studies with GHRH-producing tumors (3–5), and by animal studies with GHRH-expressing transgenes (11); in both cases growth hormone hypersecretion, pituitary somatotroph hyperplasia, and inappropriate growth are observed. GHRH belongs to a superfamily of structurally related peptides, often referred to as the "brain-gut" family that includes glucagon, glucagon-like peptide-1 (GLP-1), vasoactive-intestinal peptide (VIP), secretin, gastric inhibitory peptide (GIP), peptide with histidine as N-terminus and isoleucine as C-terminus (PHI), and pituitary adenylate cyclase activating peptide (PACAP) (12).

While a great deal is known about the biology of this important peptide hormone (3–5), there is somewhat less information pertaining to the molecular mechanisms by which the hormone exerts its actions on target cells. Specific, high-affinity binding sites for GHRH have been demonstrated on membranes of pituitary cells using several different GHRH analogues (13, 14). GHRH stimulates adenylate cyclase, resulting in increased adenosine 3′,5′-cyclic monophosphate (cAMP) production, indicating that the Gs protein is an intermediate in GHRH action (15). In agreement with the finding that cAMP is an important second messenger for GHRH signaling, both

GHRH and cAMP stimulate pituitary growth hormone secretion, augment growth hormone gene expression, and increase the proliferation of cultured pituitary somatotroph cells (16, 17). Although little is known about the signaling pathways through which cAMP exerts its effect on proliferation, it is probable that the pathway by which a GHRH-stimulated cAMP signal leads to increased growth hormone synthesis involves an activation of the transcription factor cAMP response element binding (CREB) by protein kinase A (18). Consistent with this, targeted expression of a dominant-negative form of CREB in somatotrophs results in pituitary hypoplasia and a dwarf phenotype (19). CREB activation could result in a subsequent CREB-induced transcriptional stimulation of the pituitary-specific regulatory factor Pit-1 or growth hormone factor (GHF)-1, which is known to activate transcription of the growth hormone gene (20).

The examination of genetic mutations in mice that result in aberrant growth has provided significant insight into the regulation of growth. For example, the Snell dwarf mouse (*dw*) was found to harbor a mutation in the gene encoding the Pit-1 transcription factor, resulting in a failure of Pit-1 expressing cells, including pituitary somatotrophs, to differentiate and express their hormone products (21). This led to the subsequent identification of Pit-1 mutations involved in pituitary disease in the human (22). The *little* mouse has been intensively studied as a model for human isolated growth hormone deficiency type 1B (23). Pituitary and growth hormone levels and growth hormone messenger RNA (mRNA) are substantially reduced in these mice, and while pituitary somatotrophs can be identified, they are significantly reduced in number and are sparsely granulated (24, 25). Pituitary somatotroph cells from *little* mice do not release growth hormone when treated with GHRH, but they do release growth hormone when treated with cAMP or with agents that elevate cellular cAMP levels, suggesting that the defect in the *little* mouse is related either to the binding of GHRH by its receptor or to the subsequent function of the hormone-receptor complex (26). Recent studies from our laboratory, and from others, demonstrate that the *little* mouse does indeed carry a mutation within the GHRH receptor gene, although the biochemical mechanisms by which this mutation leads to the observed phenotype have not been elaborated (27, 28).

The goals of the studies that are reviewed in this chapter are to investigate further the mechanisms by which GHRH transmits a biologic signal in target cells, contributes to the regulation of physiologically important growth responses, and participates in clinically important endocrine disorders or diseases. To further these goals, several laboratories have recently identified a specific receptor for GHRH in the pituitary gland (29–32). In this chapter, we will review the identification and structural characterization of the pituitary GHRH receptor, describe several recent studies regarding the structure, expression, and functional properties of the GHRH receptor and its gene, and discuss an ongoing analysis of the involvement of the GHRH receptor in diseases of growth, using the *little* mouse as a model system.

Results

Structure of the GHRH Receptor

The strategy used by several laboratories to identify cDNA clones encoding the GHRH receptor was based on the conviction that the receptor would signal through a G protein, as discussed above, and would therefore have the conserved structural features of other G protein–coupled receptors. This extremely large superfamily of receptors has as its hallmark the presence of seven hydrophobic membrane-spanning domains, and these domains are conserved in sequence among receptors within many subfamilies (33). Using degenerate oligonucleotide primers to these membrane-spanning domains and reverse transcription–polymerase chain reaction (RT-PCR), it has been possible to identify cDNAs encoding novel G protein–coupled receptors (34). Oligonucleotide primers corresponding to transmembrane domains 6 and 7 were designed based on the sequences of the receptors for calcitonin (35), parathyroid hormone (36), and secretin (37). Following RT-PCR amplification of rat pituitary cDNA with these primers, a clone that had many of the expected features of a GHRH receptor was identified and was used to screen rat and human pituitary cDNA libraries by conventional hybridization approaches (29). Conceptual translation of these cDNAs revealed a 423 amino acid protein in both species. Using very similar cloning strategies, human, rat, mouse, and porcine GHRH receptor cDNA clones have also been identified in several other laboratories (30–32).

The structure of the rat GHRH receptor protein is shown schematically in Figure 4.1. The GHRH receptor has many of the conserved structural features of other characterized G protein–coupled receptors (38). In addition to the seven transmembrane domains, these features include cysteine residues in the second and third extracellular loops that are believed to form a disulfide bond, a cysteine in the cytoplasmic tail that may be palmitoylated, one or more amino-terminal sites for N-linked glycosylation, potential phosphorylation sites in the third cytoplasmic loop, and numerous highly conserved residues within the membrane spanning domains. The GHRH receptor and related receptors have an unusually large amino-terminal domain that includes six conserved cysteine residues (Fig. 4.1); this domain is likely to be extracellular and to participate in ligand binding. As is illustrated in Figure 4.1, the GHRH receptor is homologous not only to the aforementioned receptors for calcitonin, parathyroid hormone (PTH),

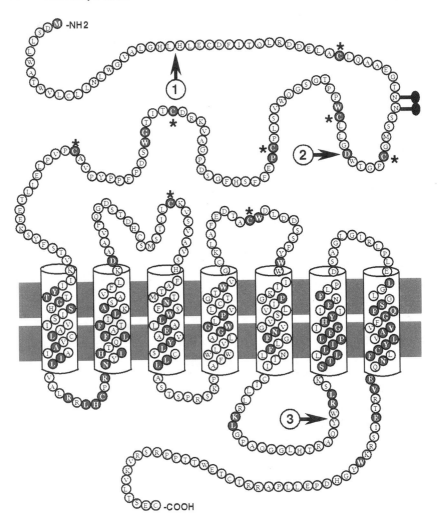

FIGURE 4.1. Structure of the pituitary GHRH receptor and homology to the brain-gut hormone receptor superfamily. The amino acid sequence shown in one-letter code is that of the rat GHRH receptor. The seven potential membrane-spanning domains are shown as cylinders crossing the cell membrane. The circles that are filled represent amino acids that are completely conserved in the related receptors for the hormones secretin, glucagon, glucagon-like peptide-1, gastric inhibitory polypeptide, vasoactive intestinal peptide, and pituitary adenylate cyclase activating polypeptide. The asterisks show eight conserved cysteine residues in the presumed extracellular domain. The arrow designated #1 indicates the approximate site of the signal sequence cleavage, arrow #2 indicates aspartic acid residue 60, which is mutated to glycine in the *little* mouse, and arrow #3 indicates the site of a 41 amino acid insertion generated by alternative RNA processing in a variant form of the GHRH receptor, as discussed in the text. The lollipop structures designate the location of two consensus sites for N-linked glycosylation in the presumed extracellular domain of the receptor protein.

and secretin, but also to a new emerging subfamily of receptors that includes the recently identified receptors for VIP (39), GLP-1 (40), glucagon (41), GIP (42), PACAP (43), and corticotropin-releasing hormone (44).

Hormone Binding and Cellular Signaling

To establish that the receptor we identified was a functional GHRH receptor, human cDNA clones were expressed in human 293S kidney cells using a cytomegalovirus promoter, and stably transfected cell lines were cloned and used for subsequent analysis. An example of this analysis is shown in Figure 4.2. Membranes from the GHRH receptor-expressing cell line (HPR9-3C) show high-affinity binding to human GHRH, as assessed by a competition binding experiment. In numerous experiments, the median effective concentration (EC_{50}) for competition is <1 nM, indicating high affinity of the receptor for this ligand. Other ligands including secretin and VIP do not appreciably compete for binding of the GHRH tracer (Fig. 4.2).

Figure 4.2 also demonstrates that the cloned receptor mediates GHRH-induced cellular signaling, as assessed by measuring cAMP accumulation. GHRH stimulates cAMP accumulation in the HPR9-3C cells that express the GHRH receptor, but not in control 293S cells. cAMP stimulation is also quite specific for GHRH. VIP does exhibit ~5% of the activity of GHRH in this system, indicating a potential interaction with the receptor. Others have demonstrated that the cloned GHRH receptor can mediate the transcriptional activation of cAMP-responsive reporter genes in transfected cells (30).

We have also investigated the ability of a growth hormone releasing peptide (GHRP)-6, to interact with the cloned human pituitary GHRH receptor (45). These data are shown in Figure 4.3. GHRP-6 at 1 μM concentration does not compete with labeled GHRH for binding to the cloned GHRH receptor. In addition, GHRP-6 is not able to independently activate the cloned GHRH receptor, as assessed by measuring cAMP accumulation in the HPR9-3C cells. These results are consistent with data indicating that GHRP-6 signals growth hormone release through a receptor that is distinct from the GHRH receptor (46). Similar findings regarding the inability of GHRP-6 to compete with GHRH for binding to the receptor have been reported by others (31).

Expression of the GHRH Receptor

To determine whether there is a tissue-specific pattern of GHRH receptor mRNA expression, RNA blot analysis was used to examine the levels of GHRH receptor mRNA in various tissues (29–32). An example of this is shown in Figure 4.4. Transcripts of approximately 2.5 and 4 kilobases in size are observed in rat pituitary, but not in other rat tissues including liver, gut,

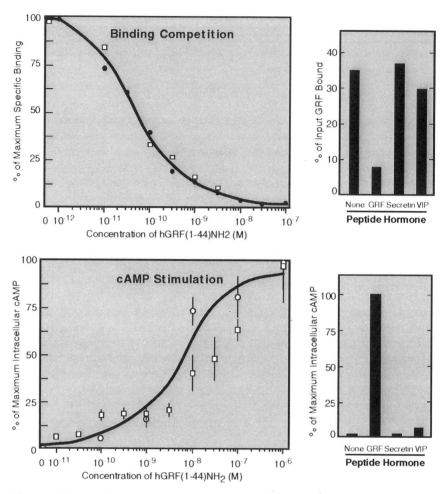

FIGURE 4.2. Functional analysis of the cloned and expressed human pituitary GHRH receptor. The human receptor cDNA was expressed in human kidney 293S cells to produce the clonal cell line HPR9-3C used in these studies. The top-left panel is a competition binding experiment using the tracer ligand [^{125}I-Tyr^{10}] human GHRH (1–44)-amide and competing with various doses of hGHRH. The bottom left panel is a cAMP stimulation experiment in which HPR9-3C cells were treated with various doses of hGHRH. The two right panels represent similar experiments using additional hormones (all present at 1 μM) to acertain the specificity of binding competition or cAMP stimulation. The data are adapted from ref. 29.

brain, gonads, or placenta. Similar findings have been made in other species. The pattern of cell-specific expression of the GHRH receptor gene within the pituitary gland remains to be established, although expression is limited to the anterior pituitary (29).

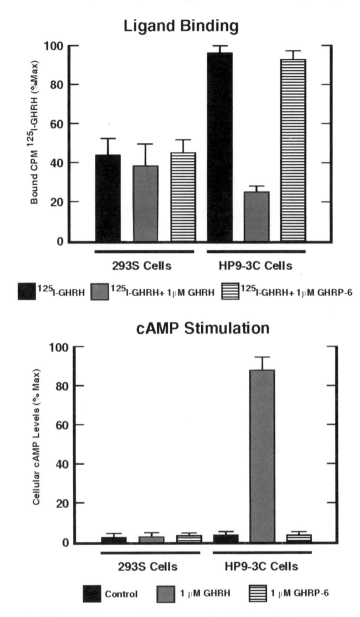

FIGURE 4.3. Effects of the growth hormone releasing peptide GHRP-6 on GHRH binding to the cloned GHRH receptor and on cAMP stimulation by the cloned GHRH receptor. The left panel is a competition binding experiment using control 293S cells or the HPR9-3C cells expressing the human GHRH receptor. GHRP-6 does not compete for binding to the receptor. The right panel is a cAMP stimulation test using the same cell groups. GHRH activates cAMP accumulation in the HPR-3C cells, whereas GHRP-6 does not.

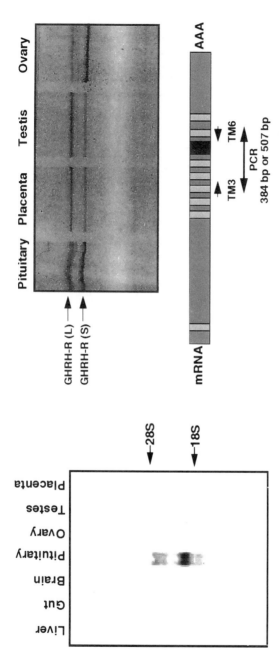

FIGURE 4.4. Tissue-specific expression of the GHRH receptor mRNA in the rat. The left panel is a Northern RNA blot containing 10μg of RNA from the indicated tissues that was hybridized to a radiolabeled rat GHRH receptor cDNA clone. The positions of the ribosomal RNAs are indicated. Two transcripts are detected in the pituitary gland. The right panel is a reverse transcription–polymerase chain reaction (RT-PCR) experiment (35 PCR cycles) to detect low-level expression of the GHRH mRNA. Two PCR-generated species are observed in the pituitary, placenta, ovary, and testis. These correspond to two splice variants of the GHRH receptor mRNA, as indicated in the schematic diagram and discussed in the text. No RT-PCR products were generated from liver RNA using these conditions (not shown).

As discussed briefly in the introduction, GHRH is synthesized in sites outside the brain, including the placenta and gonads, where it is expected to have paracrine or autocrine roles. It was therefore somewhat surprising to us that the GHRH receptor mRNA was not detected in these tissues. To determine whether this was due to the relative insensitivity of RNA blot analysis, an RT-PCR procedure was used to attempt to amplify GHRH receptor transcripts from these tissues. As shown in Figure 4.4, primers specific for the GHRH receptor within transmembrane domains 3 and 6 amplify two predominant species that are of the size expected based on analysis of the pituitary GHRH receptor and of a variant form of the receptor generated by alternative RNA processing, which will be discussed in the following section. To verify that these PCR products correspond to the GHRH receptor, the PCR products were cloned and sequenced. All of the clones from placenta, ovary, and testis were identical in sequence to the two previously cloned rat pituitary GHRH receptor mRNAs. No PCR products could be obtained when using RNA from the liver as a control tissue. These results indicate that the GHRH receptor mRNA is indeed expressed in reproductive tissues, albeit at low levels compared with the pituitary gland.

The GHRH Receptor Gene and Receptor Isoforms

An eventual goal of our studies is to understand the appropriate regulation of GHRH receptor gene expression within the somatotroph cells of the pituitary gland. One approach to this goal is to characterize the gene encoding the receptor to provide reagents for subsequent gene regulation studies. As shown in Figure 4.5, the GHRH receptor gene is fairly complex, particularly for one encoding a G protein–coupled receptor, and the gene includes 14 identified exons. There is no apparent segregation of functional domains of the protein into the exons of the gene, and nearly all of the exons are small and of uniform size. Sequencing of regions upstream of the first exon of this gene has revealed the presence of numerous consensus or near-consensus binding sites for transcriptional regulatory factors with the potential to modulate GHRH gene expression. Functional testing of the relevance of these sites is now beginning in our laboratory.

During the cloning of the GHRH receptor cDNA from the rat pituitary, two cDNA forms were identified that differed by the presence or absence of 123 basepairs encoding 41 amino acids within the coding region (29, 30). The shorter form of the receptor (423 amino acids) is identical in size to the cloned human and mouse receptors, while the longer form of the receptor (464 amino acids) does not yet have a counterpart in those species, although a 451 amino acid porcine receptor has been described (32). The characterization of the GHRH receptor gene revealed that the insertion in the long form of the receptor occurred at an intron-exon junction, suggesting that

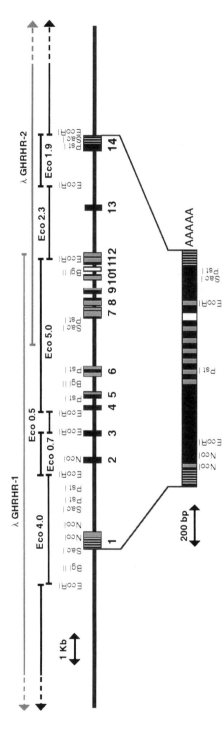

FIGURE 4.5. Schematic structure of the rat gene encoding the pituitary GHRH receptor. Clones from a rat lambda genomic library were used to identify the indicated contiguous Eco R1 subclones that define the complete gene. Exons are numbered and indicated as boxes, and are drawn larger than scale for convenience. The lighter-shaded regions represent the seven potential membrane-spanning domains in both the gene and the cDNA clone. The white exon is alternatively spliced to generate two cDNA forms as described in the text. Representative restriction enzyme sites used in gene mapping are indicated for both the gene and the cDNA clone.

alternative RNA processing was the mechanism used to generate the two receptor forms. This is schematically shown in Figure 4.6. An alternative exon is either retained during RNA processing to generate the long form of the receptor, or is spliced out to generate the short form. In the pituitary gland, semiquantitative RT-PCR analysis indicates that the short form of the receptor mRNA is predominant. This might result from inefficient utilization of a nonconsensus splice donor site following the alternative

Insert Sequence

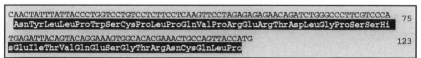

Generation via Splicing

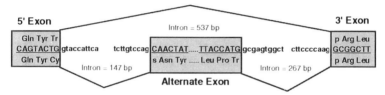

Structural Isoforms

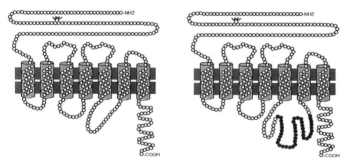

FIGURE 4.6. Alternative RNA processing has the potential to generate two predominant GHRH receptor structural isoforms. The top panel shows the sequence of the 123 bp insertion found in some rat pituitary cDNA clones. The middle panel indicates the generation of this insertion by alternative RNA processing to include an alternate exon (see also Figure 4.5). The alternate exon is followed by a nonconsensus splice donor sequence (gc rather than gt). The lower panel indicates the location of the 41 amino acid insertion in the third cytoplasmic domain of the protein. Note that while the variant mRNA has been described, the generation of this larger protein isoform has not yet been demonstrated. Sequence data are adapted from ref. 29.

exon of the loner receptor isoform. Both forms of the GHRH receptor mRNA do exist in tissues other than the pituitary, as can be seen in Figure 4.4. It has not been established if these mRNAs encode structurally or functionally distinct GHRH receptor proteins.

Interestingly, the insertion in the long form of the GHRH receptor occurs within the third cytoplasmic loop of the receptor, just upstream of trans-membrane domain 6. In some receptors, this region has been implicated as important for interaction with the G protein (47). Consistent with this, multiple isoforms of the related PACAP receptor that are generated by alternative RNA processing at the same intron-exon junction as that de-scribed for the GHRH receptor have recently been identified (48). These receptors exhibit differential coupling to downstream effector pathways when they are expressed in a frog oocyte system. These findings suggest RNA processing as a mechanism for generating functional diversity within this receptor subfamily.

A Mutation of the GHRH Receptor in the little Mouse

As discussed earlier, the dwarf mouse strain *little* is deficient in some aspect of GHRH signal transduction. Following the identification of the mouse GHRH cDNA (49) we determined that the GHRH gene is on mouse chromosome 2 (27), and thus is not a candidate for the *little* mutation, which resides on mouse chromosome 6 (23). To determine if a defect in the GHRH receptor might be responsible for the GHRH-resistant phenotype of the *little* mouse, the chromosomal location of the GHRH receptor gene was established. The gene was localized to the middle region of mouse chromosome 6, near the *little* mutation (26). This is shown schematically in Figure 4.7. The GHRH receptor locus is within a region syntenic with human chromosome 7, and the human gene has recently been mapped to human chromosome 7p14 (50).

The GHRH receptor cDNA was cloned from the pituitary of +/+, +/lit, and lit/lit mice and sequenced, both in our laboratory and that of Rosenfeld and coworkers (27, 28). This resulted in the identification of a point muta-tion within the receptor coding region that changes an aspartic acid as position 60 to a glycine. This mutation was found in both cDNA and genomic DNA clones. As is shown in Figure 4.7, aspartic acid 60, which is in the presumed extracellular domain of the receptor (see Fig. 4.1), is absolutely conserved in all members of the receptor subfamily character-ized to date, suggesting an important function. Consistent with this, directed mutation of the analogous aspartic acid residue within the related glucagon receptor has been found to completely abolish glucagon binding (51).

When the *little* GHRH receptor is transiently expressed in COS monkey kidney cells, it is unable to mediate GHRH-stimulation of intracellular

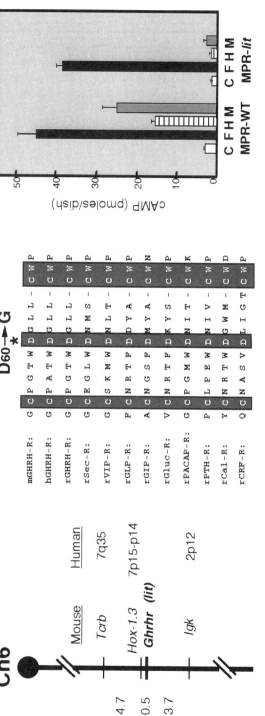

FIGURE 4.7. A mutation in the GHRH receptor in the dwarf *little* mouse. The left panel is a schematic diagram of mouse chromosome 6, indicating the relative position of the GHRH receptor gene as determined using interspecific backcross analysis, as described in ref. 27. *Ghrhr* maps to the same region of mouse chromosome 6 as *little*. Syntenic regions on human chromosomes 7 and 2 are indicated on the right, recombination distances (in centimorgans) to nearby markers used for the analysis are indicated on the left. The middle panel shows the location of the mutation in the *little* mouse, in which aspartic acid 60 (*) is changed to a glycine, as well as the conservation of amino acids sequence in the vicinity of the *little* mutation in the family of receptors related to the GHRH receptor. Shaded positions are absolutely conserved in sequence. The designations m, h, and r indicate the mouse, human, and rat GHRH receptors. Sec, secretin; VIP, vasoactive intestinal peptide; GLP, glucagon-like peptide-1; GIP, gastric inhibitory polypeptide; Gluc, glucagon; PACAP, pituitary adenylate cyclase activating polypeptide; PTH, parathyroid hormone; Cal, calcitonin; CRF, corticotropin releasing factor. The right panel demonstrates the loss of signaling activity of the mutant GHRH receptor from the *little* mouse, determined by the ability of the cloned and expressed receptor to stimulate cAMP production. C, control; F, forskolin (10^{-5} M); H, hGHRH (10^{-7} M); M, mGHRH (10^{-7} M). MPR-WT are COS monkey kidney cells transiently expressing the wild-type receptor; MPR-*lit* cells are expressing the mutant receptor. Data are adapted from ref. 27.

cAMP accumulation in comparison to the normal mouse GHRH receptor, suggesting a defect in either ligand binding or signal transduction. An example of this is shown in Figure 4.7, in which cells expressing the mutant receptor respond to the direct activation of adenylate cyclase using forskolin, but not to the regulated activation of adenylate cyclase using GHRH. Recent studies with clonal 293S cell lines that stably express the wild-type or mutant receptor forms have confirmed this loss of hormone binding and/or signal transduction potential by the *little* mouse GHRH receptor.

Discussion

These studies have led to the identification of a new protein that has the features expected of the receptor that mediates the effects of GHRH on growth hormone secretion. These features include the following: (1) the conceptually translated receptor protein has the seven membrane-spanning domains of a G protein–coupled receptor; (2) the predicted receptor protein is homologous to the receptors for related peptide hormones within the brain-gut family; (3) the conceptually translated receptor protein is the size expected from GHRH photoaffinity cross-linking studies (52); (4) the cloned receptor, when expressed in mammalian cells, confers specific and high-affinity GHRH binding ability to these cell lines; (5) the cloned receptor, when expressed in mammalian cells, confers a physiologic response, GHRH-induced cAMP accumulation, on these cells; (6) the receptor mRNA is predominantly expressed in the anterior pituitary, the major site of GHRH action; and (7) the receptor mRNA is also found in nonpituitary tissues where GHRH is reported to have biologic effects.

A schematic model illustrating the potential signaling pathways through which GHRH stimulates growth hormone synthesis and secretion by the pituitary somatotroph is presented in Figure 4.8. Characterization of the GHRH receptor as a G protein alpha subunit (G_α)-coupled receptor supports the notion that the early signaling events leading to growth hormone secretion and generation of a cAMP second messenger are G protein mediated. The neuropeptide hormone somatostatin exerts opposing actions to inhibit adenylate cyclase and growth hormone release, and a family of G protein–coupled somatostatin receptors has recently been identified (53). The subsequent signaling pathways leading to increased growth hormone gene transcription are less clear. As discussed earlier, many of the effects of cAMP on gene expression are mediated by the transcription factor CREB, and CREB is likely to be important either through Pit-1 mediated pathways, or through direct actions on the GHRH receptor gene. Recent studies indicate that Pit-1 can directly activate expression of the GHRH receptor as indicated in Figure 4.8 (28), potentially amplifying the initial GHRH signal by increasing receptor numbers. The pathways by which GHRH and

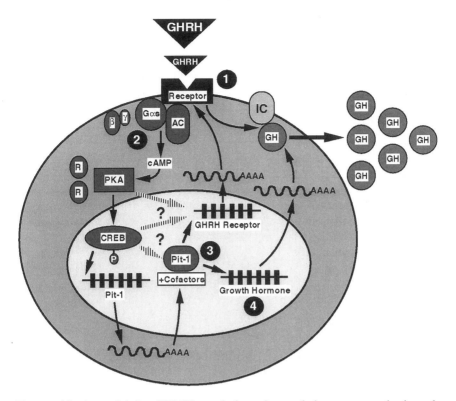

FIGURE 4.8. A model for GHRH regulation of growth hormone synthesis and secretion by the pituitary somatotroph cell, illustrating points affected in diseases of growth. GHRH interaction with the receptor leads rapidly to growth hormone (GH) secretion, perhaps through the receptor-induced interaction of a G protein with ion channels (IC) or through ion channel phosphorylation. Stimulation of adenylate cyclase (AC) by the Gs protein leads to increased intracellular cAMP accumulation and activation of the catalytic subunit of protein kinase A (PKA). PKA can phosphorylate the cAMP response element binding protein (CREB), leading to CREB activation and enhanced transcription of the gene encoding the pituitary-specific transcription factor Pit-1. Pit-1 is also subject to phosphorylation, although the functional role of this is unknown. Pit-1, together with cofactors, activates transcription of the growth hormone gene, leading to increased growth hormone mRNA and protein and replenishing cellular stores of growth hormone. Pit-1 also stimulates transcription of the GHRH receptor gene, perhaps leading to increased GHRH receptors on the somatotroph cell. Upstream components of the cAMP signaling pathway might have direct effects on GHRH receptor gene expression (indicated by the question marks). The numbered circles indicate sites at which the pathway is disrupted in human or animal diseases of GH secretion: (1) Inactivating mutation of the GHRH receptor in the dwarf mouse strain *little*. (2) Activating mutations in the $G_{\alpha s}$ protein in human growth hormone-secreting pituitary adenoma. (3) Inactivating mutations of the transcription factor Pit-1 in the Snell dwarf mouse strain (*dw*) and in human combined pituitary hormone deficiency. (4) Inactivating mutations if the GH gene in the dwarf rat (SDR) and in human isolated growth hormone deficiency.

cAMP regulate proliferation of the somatotroph cell remain essentially uncharacterized.

Numerous points in this cellular signaling cascade are affected in diseases of growth hormone secretion in man or in animal models. Four of these control points are indicated by the filled numbers in Figure 4.8. Mutation of the GHRH receptor in the *little* mouse has already been discussed, and it is important to establish if similar inactivating mutations in this receptor are found in individuals with isolated growth hormone deficiency. Mutations in the stimulatory G_α protein, which the GHRH receptor is believed to couple with, have been identified in patients with acromegaly resulting from growth hormone secreting pituitary tumors (54). This suggests that activating mutations in the GHRH receptor have the potential to be causative in pituitary tumorigenesis, and it will be important to establish if this is the case. Mutation of the transcription factor Pit-1 has been described in the Snell dwarf mouse and in human combined pituitary hormone deficiency, as discussed previously (21, 22). Lastly, disruption of the growth hormone gene itself is common in human isolated growth hormone deficiency (55) and has been observed in the spontaneous dwarf rat (56).

The identification of the pituitary receptor for GHRH represents an important step toward establishing the role of GHRH and its receptor in endocrine signaling in both health and in disease. Further investigation of the signaling pathways modulated by this receptor promises to lend insight into the basic biology of the somatotroph cells and to have direct clinical applicability toward the study of pituitary tumorigenesis and disorders of growth hormone secretion and growth.

References

1. Guillemin R, Brazeau P, Bohlen P, Esch F, Ling N, Wehrenberg WB. Growth hormone-releasing factor from a human pancreatic tumor that caused acromegaly. Science 1982;218:585–7.
2. Rivier J, Spiess J, Thorner M, Vale W. Characterization of a growth hormone-releasing factor from a human pancreatic islet tumour. Nature 1982;300:276–8.
3. Frohman L, Jansson J-O. Growth hormone-releasing hormone. Endocr Rev 1986;7:223–53.
4. Gelato MC, Merriam CR. Growth hormone releasing hormone. Annu Rev Physiol 1986;48:569–91.
5. Cronin MJ, Thorner MO. Growth hormone-releasing hormone: basic physiology and clinical implications. In: DeGroot LJ, ed. Endocrinology. Philadelphia: WB Saunders, 1994:280–302.
6. Margioris AN, Brockman G, Bohler HLC, Grino M, Vamvakopoulos N, Chrousos GP. Expression and localization of growth hormone-releasing hormone messenger ribonucleic acid in the rat placenta: in vitro secretion and regulation of its peptide product. Endocrinology 1990;126:151–8.
7. Endo H, Yamaguchi M, Farnsworth R, Thordarson G, Ogren L, Alonso FJ, Sakata M, Hirota K, Talamantes F. Mouse placental cells secrete immunoreactive growth hormone-releasing factor. Biol Reprod 1994;51:1206–12.

8. Bagnato A, Moretti C, Ohnishi J, Frajese G, Catt K. Expression of the growth hormone-releasing hormone gene and its peptide product in the rat ovary. Endocrinology 1992;130:1097–102.

9. Chiampani T, Fabbri A, Isidori A, Dufau ML. Growth hormone-releasing hormone is produced by rat Leydig cell in culture and acts as a positive regulator of Leydig cell function. Endocrinology 1992;131:2785–92.

10. Srivastava CH, Breyer PR, Rothrock JK, Peredo MJ, Pescovitz OH. A new target for growth hormone releasing-hormone in the rat: the Sertoli cell. Endocrinology 1993;133:1478–81.

11. Mayo KE, Hammer RE, Swanson LW, Brinster RL, Rosenfeld MG, Evans RM. Dramatic pituitary hyperplasia in transgenic mice expressing a human growth hormone-releasing factor gene. Mol Endocrinol 1988;2:606–12.

12. Campbell RM, Scanes CG. Evolution of the growth hormone-releasing factor (GRF) family of peptides. Growth Regulation 1992;2:175–91.

13. Seifert H, Perrin M, Rivier J, Vale W. Binding sites for growth hormone releasing factor on rat anterior pituitary cells. Nature 1985;313:487–9.

14. Velicelebi G, Santacroce T, Harpold M. Specific binding of human pancreatic growth hormone releasing factor (1-40-OH) to bovine anterior pituitaries. Biochem Biophys Res Commun 1985;126:33–9.

15. Labrie F, Gagne B, Lefevre G. Growth hormone-releasing factor stimulates adenylate cyclase activity in the anterior pituitary gland. Life Sci 1983;33:2229–33.

16. Barinaga M, Yamamoto G, Rivier C, Vale W, Evans R, Rosenfeld MG. Transcriptional regulation of growth hormone gene expression by growth hormone-releasing factor. Nature 1983;306:84–5.

17. Billestrup N, Swanson LW, Vale W. Growth hormone-releasing factor stimulates proliferation of somatotrophs in vitro. Proc Natl Acad Sci USA 1986; 83:6854–7.

18. Gonzalez GA, Montminy MR. Cyclic AMP stimulates somatostatin gene transcription by phosphorylation of CREB at serine 133. Cell 1989;59:675–89.

19. Struthers RS, Vale WW, Arias C, Sawchenko PE, Montminy MR. Somatotroph hypoplasia and dwarfism in transgenic mice expressing a non-phosphorylatable CREB mutant. Nature 1991;350:622–4.

20. McCormick A, Brady H, Theill LE, Karin M. Regulation of the pituitary-specific homeobox gene GHF1 by cell-autonomous and environmental cues. Nature 1990;345:829–32.

21. Li S, Crenshaw EB III, Rawson EJ, Simmons DM, Swanson LW, Rosenfeld MG. Dwarf locus mutants lacking three pituitary cell types result from mutations in the POU-domain gene pit-1. Nature 1990;347:528–33.

22. Radovick S, Nations M, Du Y, Berg LA, Weintraub BD, Wondisford FE. A mutation in the POU-homeodomain of pit-1 responsible for combined pituitary hormone deficiency. Science 1992;257:1115–8.

23. Eicher EM, Beamer WG. Inherited ateliotic dwarfism in mice: characterization of the mutation, little, on chromosome 6. J Hered 1976;67:87–91.

24. Cheng TC, Beamer WG, Phillips JA III, Bartke A, Mallonee RL, Dowling C. Etiology of growth hormone deficiency in little, Ames, and Snell dwarf mice. Endocrinology 1983;133:1669–78.

25. Wilson DB, Wyatt DP, Gadler RM, Baker CA. Quantitative aspects of growth hormone cell maturation in the normal and little mutant mouse. Acta Anat 1988;131:150–5.

26. Jansson J-O, Downs TR, Beamer WG, Frohman LA. Receptor-associated resistance to growth hormone-releasing factor in dwarf "little" mice. Science 1986;232:511–2.

27. Godfrey PG, Rahal JO, Beamer WG, Copeland NG, Jenkins NA, Mayo KE. GHRH receptor of *little* mice contains a missense mutation in the extracellular domain that disrupts receptor function. Nature Genet 1993;4:227–32.

28. Lin S-C, Lin CR, Gukovsky I, Lusis AJ, Sawchenko PE, Rosenfeld MG. Molecular basis of the *little* mouse phenotype and implications for cell-type specific growth. Nature 1993;364:208–13.

29. Mayo KE. Molecular cloning and expression of a pituitary-specific receptor for growth hormone-releasing hormone. Mol Endocrinol 1992;6:1734–44.

30. Lin C, Lin S-C, Chang C-P, Rosenfeld MG. Pit-1 dependent expression of the receptor for growth hormone releasing factor mediates pituitary cell growth. Nature 1992;360:765–8.

31. Gaylinn BD, Harrison JK, Zysk JR, Lyons CE, Lynch KR, Thorner MO. Molecular cloning and expression of a human anterior pituitary receptor for growth hormone-releasing hormone. Mol Endocrinol 1993;7:77–84.

32. Hsiung HM, Smith DP, Zhang XY, Bennet T, Rosteck Jr PR, Lai MH. Structure and functional expression of a complementary DNA for porcine growth hormone-releasing hormone receptor. Neuropeptides 1993;25:1–10.

33. Larhammar D, Blomqvist AG, Wahlestedt C. The receptor revolution: multiplicity of G-protein-coupled receptors. Drug Design Discovery 1993;9:179–88.

34. Libert F, Parmentier M, Lefort A, Dinsart C, Van Sande J, Maenhaut C, Simons M-J, Dumont JE, Vassart G. Selective amplification and cloning of four new members of the G protein-coupled receptor family. Science 1989;244:569–72.

35. Lin H, Tonja H, Flannery M, Aruffo A, Kaji E, Gorn A, Kolakowski LJ, Lodish HF, Goldring S. Expression cloning of an adenylate cyclase-coupled calcitonin receptor. Science 1991;254:1022–4.

36. Juppner H, Abdul-Badi A-S, Freeman M, Kong X, Schipani E, Richards J, Kolakowski LJ, Hock J, Potts JT, Kronenberg H, Segre G. A G-protein-linked receptor for parathyroid hormone and parathyroid hormone-related peptide. Science 1991;254:1024–6.

37. Ishihara T, Nakamura S, Toro Y, Takahashi T, Takahashi K, Nagata S. Molecular cloning and expression of a cDNA encoding the secretin receptor. EMBO J 1991;10:1635–41.

38. Probst WC, Snyder LA, Schuster DJ, Borsius J, Sealfon S. Sequence alignment of the G protein coupled receptor superfamily. DNA Cell Biol 1992;11:1–20.

39. Ishihara T, Shigemoto R, Mori K, Takahashi K, Nagata S. Functional expression and tissue distribution of a novel receptor for vasoactive intestinal peptide. Neuron 1992;8:811–9.

40. Thorens B. Expression cloning of the pancreatic β cell receptor for the glucoincretin hormone glucagon-like peptide 1. Proc Natl Acad Sci USA 1992;89:8641–5.

41. Jelinek L, Lok S, Rosenberg GB, Smith RA, Grant FJ, Biggs S, Bensch PA, Kuijer JL, Sheppard PO, Sprecher CA, O'Hara PJ, Foster D, Walker KM, Chen LHJ, McKernan PA, Kindsvogel W. Expression cloning and signaling properties of the rat glucagon receptor. Science 1993;259:1614–6.

42. Usdan TB, Mezey E, Button DC, Brownstein MJ, Bonner TI. Gastric inhibitory polypeptide receptor, a member of the secretin-vasoactive intestinal peptide

receptor family, is widely distributed in peripheral organs and the brain. J Biol Chem 1993;133:2861–70.
43. Pisegna JP, Wank SA. Molecular cloning and functional expression of the pituitary adenylate cyclase activating polypeptide type I receptor. Proc Natl Acad Sci USA 1993;90:6345–9.
44. Perrin MH, Donaldson CJ, Chen R, Lewis KA, Vale WW. Cloning and functional expression of a rat brain corticotropin releasing factor (CRF) receptor. Endocrinology 1993;133:3058–61.
45. Bowers CY, Momany FA, Reynolds GA, Hong A. On the *in vitro* and *in vivo* activity of a new synthetic hexapeptide that acts on the pituitary to specifically release growth hormone. Endocrinology 1984;114:1537–45.
46. Blake AD, Smith RG. Desensitization studies using perifused pituitary cells show that growth hormone-releasing hormone and His-D-Trp-Ala-Trp-D-Phe-Lys-NH$_2$ stimulate growth hormone release through distinct receptor sites. J Endocrinol 1991;129:11–19.
47. Strader C, Sigal I, Dixon R. Structural basis of β-adrenergic receptor function. FASEB J 1989;3:1825–32.
48. Spengler D, Waeber C, Pantaloni C, Holsboer F, Bockert J, Seeburg PH, Journot L. Differential signal transduction by five splice variants of the PACAP receptor. Nature 1993;365:170–5.
49. Suhr ST, Rahal JO, Mayo KE. Mouse growth hormone-releasing hormone: precursor structure and expression in brain and placenta. Mol Endocrinol 1989;3:1693–700.
50. Gaylinn BD, von Kap-Herr C, Golden WL, Thorner MO. Assignment of the human growth hormone-releasing hormone receptor gene (GHRHR) to 7p14 by in situ hybridization. Genomics 1994;19:193–5.
51. Carruthers CJL, Unson CG, Kim HN, Sakmar TP. Synthesis and expression of a gene for the rat glucagon receptor: replacement of an aspartic acid in the extracellular domain prevents glucagon binding. J Biol Chem 1994;269:29321–8.
52. Gaylinn BD, Lyons CE, Zysk JR, Clarke IJ, Thorner MO. Photoaffinity cross-linking to the pituitary receptor for growth hormone-releasing factor. Endocrinology 1994;135:950–5.
53. Bell GI, Reisine T. Molecular biology of somatostatin receptors. Trends Neurosci 1993;16:34–8.
54. Landis C, Masters S, Spada A, Pace A, Bourne H, Vallar L. GTPase inhibiting mutations activate the α chain of Gs and stimulate adenyl cyclase in human pituitary tumours. Nature 1989;340:692–6.
55. Phillips JA, Cogan JD. Molecular basis of familial human growth hormone deficiency. J Clin Endocrinol Metab 1994;78:11–16.
56. Takeuchi T, Suzuki H, Sakurai S, Nogami H, Okuma S, Ishikawa H. Molecular mechanism of growth hormone (GH) deficiency in the spontaneous dwarf rat: detection of abnormal splicing of GH mRNA by the polymerase chain reaction. Endocrinology 1990;126:31–8.

5

Computer-Assisted Modeling of Xenobiotic Growth Hormone Secretagogues

FRANK A. MOMANY AND CYRIL Y. BOWERS

Growth hormone releasing peptides (GHRPs) are chemically synthesized and designed entities with no sequence relationship to any known peptide hormone (1–4). Unusual in the realm of drug design, they are peptides that have been invented prior to the discovery of the endogenous peptide hormone. In fact after more than 13 years of study, the putative GHRP-like hormone and its receptor have yet to be isolated. Further, GHRPs act in fundamentally different ways than GHRH, as described by Bowers et al. (5), and synergistically release GH when administered together with GHRH. The endocrinology and structure/activity relationships of GHRPs pose challenging problems to scientists, and answers are being actively pursued.

We report here our progress in conformational analysis designed to determine the bioactive structure of GHRP-6 and enhance our understanding of structure/function relationships. This is part of an ongoing series of studies on GHRP-6 (His-DTrp-Ala-Trp-DPhe-Lys-NH$_2$) and other bioactive growth hormone releasing ligands (Table 5.1) using molecular mechanic and dynamic computational methods, and experimental results from nuclear magnetic resonance (NMR) solution studies. Conformational analysis of sterically constrained analogues, including disulfide-linked cystine-containing peptides, N-methylated and α-methylated residues, and proline substitution, has made it possible to reduce the very large number of low-energy conformational states of GHRP-6 to a reasonable and exclusive set. Bathing the peptides in water droplets and carrying out extensive dynamic simulations at room temperature provides information about the stability of various conformations in solution. These results in turn are compared to J-coupling constants obtained from the solution ^{1}H NMR (Tennessee Eastman Animal Products Division of Eastman Kodak). From the ^{1}H NMR and the calculation data GHRP-6 appears to have a unique and stable conformation in solution (6). The working hypothesis is, if the solution conformation is similar in its three-dimensional form to the

TABLE 5.1. Growth hormone releasing peptides with steric and other conformational constraints.

Compound name[a]	Residues								
	−1	−0	1	2	3	4	5	6	7
GHRP-6			His	DTrp	Ala	Trp	DPhe	Lys	
Cys[1,9]-Ala-GHRP-6	Cys	Ala	His	DTrp	Ala	Trp	DPhe	Lys	Cys
Cys[1,9]-NMe-DPhe[7]-Ala-GHRP-6	Cys	Ala	His	DTrp	Ala	Trp	NMeDPhe	Lys	Cys
NMeDPhe[5]-GHRP-6			His	DTrp	Ala	Trp	NMeDPhe	Lys	
αMeDPhe[6]-Ala-CHRP-6		Ala	His	DTrp	Ala	Trp	αMeDPhe	Lys	
αMeTrp[5]-Ala-GHRP-6		Ala	His	DTrp	Ala	αMeTrp	DPhe	Lys	
Pro[3]-GHRP-6			His	DTrp	Pro	Trp	DPhe	Lys	

[a] All peptides have C-terminal amide.

bioactive conformation when bound to the receptor, then knowing the solution conformation implies knowing the bioactive conformation. The application of conformationally restricted analogues at different sites in the sequence limits the local structure to specific regions of conformational space. Upon testing for biologic activity, these constrained analogues further help to confirm the predicted bioactive conformation. Future development of new and more potent analogues of the GHRPs depends on just such an understanding of the bioactive structure.

Computational Methods

All calculations were carried out on a IBM RS/6000 workstation using the InsightII (version 2.3.0, Molecular Modeling Program from Biosym, San Diego, CA) graphics program for setup and viewing, and the Discover (version 3.1, Molecular Mechanics Program from Biosym) program for energy and dynamics calculations. Conformational states of GHRP-6 were searched by using dynamic simulations and traditional search algorithms in an attempt to find conformations of low energy. Results of conformational dynamics simulations in vacuo at 300 K revealed that several stable conformations persist over large periods of time (Table 5.2). Water molecules, using TIP3P potentials, were added in a shell of space surrounding each of the conformations found from the simulations carried out in vacuo and each complex of water plus peptide was energy minimized to a gradient less than 0.02 kcal/mol-Å. Water layers were added so that the volume was filled with solvent 4.5 to 5.0 Å beyond the atomic van der Waals radius surface. After the complexes of peptide and water were energy minimized in order to relieve "hot-spots" (high-energy interactions that may result in unrealistic motion upon initiating dynamics), 1 to 3 psec ($1–3 \times 10^{-12}$ sec) of dynamics were carried out with time steps of 10^{-15} sec (1 fsec) and atomic velocities (i.e., kinetic energy) consistent with heating to a temperature of 300°K. No

TABLE 5.2. Starting structure and final conformation of
GHRP-6 after dynamic simulation in vacuo.

Residue	Starting[a] (ϕ,Ψ) (deg)	Final[b] (ϕ,Ψ)	χ^1 (deg)
His	—, 104	—, −59	179
DTrp	144, −152	134, −156	167
Ala	−138, 147	−85, −33	—
Trp	−147, 141	−102, 76	−58
DPhe	146, −145	123, −130	−178
Lys-NH2	−92, 82	−73, 92	−43

[a] Starting conformation after energy minimization of the fully
beta-sheet conformer.
[b] Conformation after 200 psec dynamics, not energy minimized.

constraints were applied to the water molecules or peptide, leaving the
water free to move around the molecule and to be restricted only by the
energetics of the system. After the preliminary heating, the water complexes were again energy minimized to a gradient less than 0.02 kcal/mol-Å.
This completed the preparation phase. The resulting solvent and peptide
complex was finally heated to 300°K and then several hundred picoseconds
of molecular dynamics were carried out. Several in vacuo conformations
were unstable in the solvated state, since they moved to other conformations during the dynamics simulation. Solvated conformations that were
stable over long simulation times (i.e., >100 psec) were considered to be
possible solution structures.

GHRP analogues studied computationally are listed in Table 5.1. These
peptides were synthesized and tested for bioactivity over the past 13 years
at Tulane and Eastman Animal Products laboratories. They represent a
significantly varied group in terms of growth hormone releasing activity and
conformational constraint. Each analogue of Table 5.1 (with the exception
of GHRP-6 itself) is conformationally restricted as a result of a chemical
modification that adds constraints on the dihedral angles of the peptide that
otherwise would be allowed. For example, molecular constraints were
added to GHRP by formation of a disulfide bridge as in Cys[1,9]-Ala-GHRP-
6, by substitution of an amide hydrogen or α-carbon hydrogen by a methyl
group as in NMeDPhe[5]-GHRP-6 and αMeDPhe[5]-Ala-GHRP, or by substitution of proline for alanine as in Pro[3]-GHRP-6. In each instance the
structural restrictions placed on the particular residues backbone and/or
side chain are different, resulting in differences in the energetics and shape
of the low-energy regions. Different low-energy conformational profiles
affect the preferred molecular conformations, which affects the bioactivity.

The relative growth hormone releasing activities (3, 4) of the GHRP
analogues listed in Table 5.1 taking GHRP-6 = 100%, are Cys[1,9]-Ala-
GHRP-6 (30%), Cys[1,9]-NMeDPhe[7]-Ala-GHRP-6 (25%), NMeDPhe[5]-

GHRP-6 (130%), αMeDPhe[6]-Ala-GHRP (0%), αMeTrp[5]-Ala-GHRP (0%), and Pro[3]-GHRP-6 (<1%).

^{1}H NMR Study

In a ^{1}H NMR study on an earlier GHRP analogue, Tyr-DTrp-Ala-Trp-DPhe-NH$_2$ (6), a relatively stable conformation was deduced from several nuclear Overhauser contacts, HN-HC alpha coupling constants, and temperature coefficients. The NMR-derived dihedral angles (6) are given in Table 5.3. Because of the flexibility of the linear peptide above, it was thought that a better model to study experimentally was the conformationally restricted bioactive analogue, Cys[1,9]-GHRP-6. ^{1}H NMR coupling constants for side chain and backbone protons were determined for most of the residues except the L-Trp residue, and preliminary φ and χ[1]

TABLE 5.3. NMR-derived and calculated dihedral angles of several low energy conformations of Cys[1,9]-Ala-GHRP-6 and Tyr-DTrp-Ala-Trp-DPhe-NH$_2$.

Residues	Dihedral angles φ,χ[1]	NMR derived	NMR Tyr[1b]	C-1	C-1 dyn[a] 100ps	C-2	C-3	C-4 dyn[a] 100ps	Tyr[1] solv mini
Cys[1]	φ	—		—	—	—	—	—	
	Ψ	—		−112	148	138	−161	145	
	χ[1]	43		37	70	59	67	57	
Ala[2]	φ	−163		−70	−70	−74	−77	−54	
	Ψ	—		94	138	134	−75	−28	
His[3](Tyr[1])	φ	−77	—	−118	−128	−87	−69	−107	—
	Ψ	—	−160	135	115	149	−55	−60	129
	χ[1]	−50	−60	−96	−81	−74	−73	−72	−67
DTrp[4]	φ	157	85	158	142	77	159	121	113
	Ψ		−80	−130	−120	−144	−104	−131	−87
	χ[1]	148	60	170	69	179	179	158	74
Ala[5]	φ	−77	−80	−71	−75	−75	−80	−86	−74
	Ψ	—	170	−74	−54	−43	−61	−29	174
Trp[6]	φ	—	−80	−89	−88	−102	147	−101	−94
	Ψ	—	110	91	92	80	−95	139	74
	χ[1]	—	−60	−69	−72	−74	−68	−74	−81
DPhe[7]	φ	153	85	140	103	149	94	97	130
	Ψ	—	−110	107	114	−108	−138	−155	−83
	χ[1]	138	60	170	165	169	169	−160	60
Lys[8]	φ	−84		−111	−124	−155	−93	−68	
	Ψ	—		80	95	−45	85	124	
	χ[1]	—		−86	165	−168	−173	−167	
Cys[9]	φ	−86		−89	−80	−82	−77	−149	
	Ψ	—		86	105	69	−72	−42	
	χ[1]	−130		−175	−179	−66	177	−172	
Average deviation vs NMR				28	39	34	28	39	21

[a] Average of last 5 psec of solvated (4.0 Å H$_2$O layer) 300 K dynamics.
[b] (Tyr[1]) is the first residue of the analogue studied by Biosym, San Diego, CA.

dihedral angles were calculated. In Table 5.3, only the major side chain population for the χ^1 values are reported.

Results and Discussion

The dynamics simulation of "uncharged" GHRP-6 was performed in vacuo for 200 psec at 300 K, starting from an extended beta-sheet conformation [all backbone dihedral angles, $\phi(\text{N-C}^\alpha) = -150°$, $\psi(\text{C}^\alpha\text{-C}') = 150°$ and side chains in either the trans ($\chi = 180°$) or gauche ($\chi = +60°$ for D or $-60°$ for L-residues] form. After 130 psec a compact, folded conformation was found that was retained for the next 70 psec with moderately large librational motion occurring along the lysine side chain. The aromatic rings oscillated modestly about the side chain dihedral angles but retained their configuration once the stable folded conformation was reached. The starting and final conformations (after 200 psec) are given in terms of their dihedral angles in Table 5.2 and shown schematically in Figure 5.1. Since it

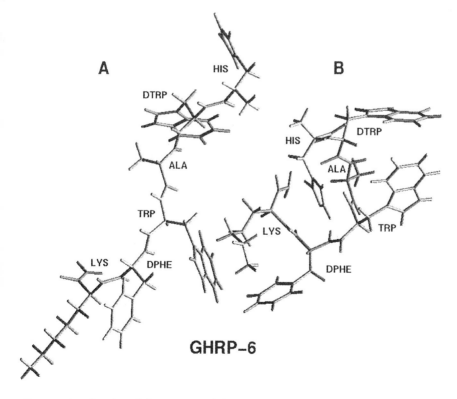

FIGURE 5.1. Starting (A) and final (B) structures of GHRP-6 before and after dynamics simulation in vacuo. The simulation was carried out for 200 psec at a temperature of 300 K. The amino acids are labeled.

was difficult to find a way to damp the repulsive charge-charge terms such that they didn't dominate the interactions, the fully charged molecule was not studied in vacuo.

From Tables 5.2 and 5.3 and Figures 5.1–5.3 it is apparent that the L-D combination of amino acids at both the 1-2 and 4-5 positions in GHRP-6 and its analogues creates an energetically favorable folded structure. These bends allow the head and tail of the molecule to come reasonably close to one another, as for example in the Cys[1,9]-Ala-GHRP-6 peptide where it is necessary that they be close in order to create the disulfide S-S bond.

The fully positively (+3) charged hydrated beta-sheet extended GHRP-6 + water complex after being subjected to more than 400 psec of dynamics at 300 K, did not fold into a compact conformation during that time. When fully charged the extended conformation was retained throughout the time course of the simulation. Neutralization of the histidine side

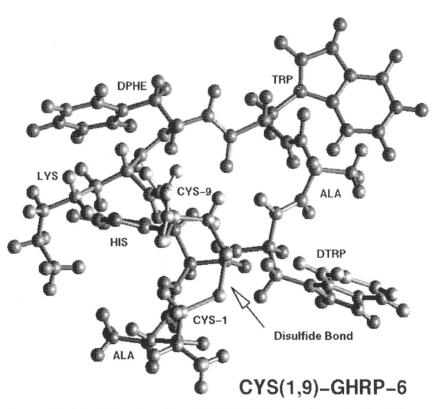

CYS(1,9)–GHRP-6

Figure 5.2. Solvated configuration of GHRP-6 (His-neutral) in water after 200 psec of dynamics simulation performed at 300 K. The small structures are water molecules. Some water molecules have been removed in front to better view the GHRP-6 molecule.

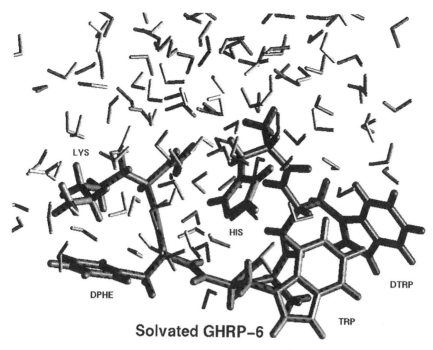

FIGURE 5.3. Conformer C-4 of Cys[1,9]-Ala-GHRP-6. The disulfide bridge is facing the viewer.

chain (protonated in the 1 position) and retaining the charges on the N-terminal amine and lysine residue, resulted in a folded stable conformation after about 200 psec of dynamics. The result is interesting and argues for a neutral charge on the histidine residue as might be expected at neutral pH. Simulations in which all the possible charge groups were taken to be neutral (no net charge) failed to hold the solvent around the polar regions of the molecule and this approach was not pursued further. The folded conformation found after 200 psec of dynamics of the doubly charged solvated complex is shown in Figure 5.2. The water has migrated away from the aromatic side chains during the dynamics simulation and can be seen collecting on the upper or polar side of the molecule in Figure 5.2. The implication of this is that the aromatic side chains do not interact strongly with water which is consistent with our understanding of hydrophobic surfaces of molecules. Perhaps in the receptor these residues are interacting with nonpolar regions leaving the more polar side of the molecule facing a water/polar environment where the N-terminal amine, His, and Lys side chains could interact with other polar groups in the receptor. This is consistent with the data from other analogues where the histidine is removed, the N-terminus is extended and modified, and lysine is replaced. Activity is not necessarily lost when these changes are made.

From NMR studies of Cys^{1,9}-Ala-GHRP-6 (Tennessee Eastman Animal Products) a set of J-coupling constants were found. These results were consistent with a stable conformation in solution, and when compared with calculated forms of the structure, it was clear that the conformation was a compact folded form, not an extended form. Because of these results an attempt was made to find a folded conformation that was consistent with the calculated linear GHRP-6 conformers and the bioactive disulfide bridged compounds, Cys^{1,9}-Ala-GHRP-6 and Cys^{1,9}-NMeDPhe^{7}-Ala-GHRP-6. These analogues were the only active disulfide bridged forms found from 23 attempts to covalently bridge different positions while still maintaining the GHRP-6 like framework. Clearly, since these disulfide-bridged analogues are active, it precludes a fully extended conformation from being the bioactive form since in that case the two ends cannot come close enough together to make the disulfide bond. Utilizing the different folded conformations found for GHRP-6, the additional residues (CysH^{1}-Ala^{2}, CysH^{9}) were added and folded into energetically allowed forms until the N- and C-terminal Cys-SH groups came near one another. The thiol hydrogens were then deleted and the S-S bond formed. Upon gentle minimization (steepest descents) the disulfide bridge was closed and the high-energy contacts were relieved. The conformations were then fully energy minimized, solvated with water molecules, and dynamic simulations run for 100 to 200 psec or more. It was found that a number of different conformers were possible that closed the disulfide bridge, and they were nearly the same energy in vacuo. Each of these conformations was relatively stable during the time course of the solvated dynamics simulations. It was not possible to distinguish energetically between several of the most stable forms. The NMR dihedral angles, which were derived from a variety of J-coupling constants, and the conformations of four low-energy structures (C-1 to C-4) for Cys^{1,9}-Ala-GHRP-6 are given in Table 5.3. The agreement with the NMR results is only modest. Dihedral angles as a general rule, tend to pucker and close up upon energy minimization relative to the dynamic state at 300 K, but that is not the cause for the disagreement with the Ala^{2}-His^{3} residues of the C-1 conformation. In this case the averaged dynamics dihedrals are in worse agreement with the NMR than are the minimized values.

Results from the dynamics simulation of conformers C-1 and C-4 with explicit solvent, carried out for 100 psec at 300 K, are also presented in Table 5.3. Conformer C-4 is recorded in Figure 5.3. The increase in average deviation of the C-1 conformer after the dynamics simulation is largely a result of the side chain movement of DTrp^{4}; however, it is clear that the Ala^{2}-His^{3} φ-dihedral angles are not in agreement with the NMR results. This suggests that either the solution structure has not been sufficiently sampled during the dynamics simulations or that the NMR may have some intrinsic problem of which we are not aware. Also, conformer C-3 has a relatively low average deviation and has altogether different positioning of the N- and C-terminal ends of the molecule relative to the core structure, residues 3 to 8. The vacuum energy for C-3 is only a few kcal/mol above that of C-1, and nearly the same as the other conformations recorded. From this analysis it is difficult to choose which conformer is the bioactive one. One favorable result from the disulfide bridged analogue is the observation that the central core residues (i.e., GHRP-6) retain conformations very close to those found from unrestrained calculations for linear GHRP-6.

Comparison of the NMR data for the Tyr^1-GHRP analogue to the NMR result on the Cys^{1,9}-Ala-GHRP-6 analogue is interesting. As deduced from the two studies, there are significant variances in the backbone and side chain dihedral angles. Our calculations suggest that the Tyr^1-GHRP conformation is of low energy, but this conformation does not persist over time during the dynamics solvation studies. In the last column of Table 5.3 is given the minimized solvated NMR conformation of the Tyr^1-GHRP with the cvff force field used above (Biosym). The agreement with the NMR results is very good and the conformation is of low energy. However, the conformation changes during 100 psec of dynamic simulation of this solvated structure and results in a conformational transition with the molecule moving closer to that described above for GHRP-6. Since the NMR studies were carried out in deuterated dimethylsulfoxide (DMSO), it is possible that the conformation would be different had the NMR been carried out in water. At this time it cannot be determined for what reason(s) the results differ. Simulations using DMSO as the solvent may help resolve this difference.

The analogue NMeDPhe^5-GHRP-6 is of interest since it retained high activity even after a methyl group had replaced the hydrogen of the amide preceding the DPhe^4 residue. Not only does N-methylation affect the conformation of the DPhe^5 residue [eliminating α-helical states at that residue (7)], it also helps to confirm the conformation of the preceding Trp^4 residue. In Table 5.3 we see that the Trp^4 residue is generally in an opened C$_7^{eq}$ conformation $(\phi,\psi) = (-90°,+90°)$. This open form is acceptable for substituting a methyl group on the $i + 1$ residue amide nitrogen, while if the conformation were a closed hydrogen bonded seven-membered ring ($-70°$, $+80°$), methylation would cause hydrogen bond disruption and open the seven-membered ring much wider. In contrast, the DPhe^5 residue is in a conformation that accepts the N-methyl without distortion. Another effect of N-methylation of DPhe^5 is that it causes the Trp^4 side chain indole ring to flip from χ^2 of $\sim +90°$ to $\sim -90°$. Additionally, this analogue argues against the amide hydrogen of DPhe^4 taking part in hydrogen bonding to the receptor. The analogue with α-methylation at the Trp^5 position of Ala^1-GHRP-6 is inactive, and is also in agreement with our expectations from the backbone conformations shown in Table 5.3. The reason for this conclusion is the established preference of α-methylated amino-acid residues (such as α-isobutyric acid, Aib) to energetically allow the α-helical (usually a 3/10 helix) backbone peptide conformation for α-methyl substituted residues. α-methylation at the DPhe^5 position in GHRP-6 would be expected to destroy the activity, and it did. Thus this analogue helps one eliminate the possibility of α-helical dihedral angles at this position, and is in agreement with the previous discussion.

Finally we should consider the Pro^3 analogue, which is relatively inactive. Clearly, many analogues of GHRP have been synthesized and tested and found to have little or no activity. Since in almost all conformations found for GHRP-6 and its analogues the Ala^3 backbone conformation will easily allow the formation of the pyrolidine ring, it is not readily apparent why the Pro^3 analogue is not fully active. That is, the φ value of $-70°$ to $-80°$ found for alanine is also acceptable for closing the five-membered pyrolidine ring. However, when other analogues with different substitutions at the 3-position are considered, it is clear that any enlargement of the side chain greater than the size of a methyl group adversely affects the activity. For example, Abu^3-GHRP-6 (Abu is 2-amino-butyryl) has no significant activity, NMeAla^3 analogues are also inactive, and Ala^1, Ser^3-GHRP-6 has only 20% of the activity of GHRP-6 even though the additional N-terminal Ala preceding His in GHRP-6 generally enhanced the activity. From these and other analogues, we are led to believe that the proline ring sterically interferes in the receptor binding and for this reason has no significant activity. However, a second possibility that these data do not totally rule out is that the NH hydrogen of Ala^3 is a hydrogen-bonding site interacting with the receptor, and we have removed this interaction.

Conclusion

When histidine is uncharged, GHRP-6 and analogues have been found by computational analysis to have stable conformations in water. This result is in agreement with ^{1}H NMR data on GHRP-6. The core GHRP conformation has been found to be retained for different analogues. Two specific disulfide linked structures that retained the conformation of the basic GHRP molecules were found to be biologically active. Those analogues that perturb the preferred GHRP conformation, were generally inactive or have much reduced activity. Position 3 of the GHRP-6 analogues was found to be very sensitive to steric bulk and receptor contact. Thus, even though the conformation was not perturbed by proline at the 3-position, the activity is dramatically reduced. N-methylation at selected sites either enhances or lowers activity and becomes analogues, which helps to define the bioactive conformation. Further studies on much more potent analogues are in progress.

References

1. Momany FA, Bowers CY, Reynolds GA, Chang D, Hong A, Newlander K. Design, synthesis, and biological activity of peptides which release growth hormone in vitro. Endocrinology 1981;108:31–9.

2. Momany FA, Bowers CY, Reynolds GA, Hong A, Newlander K. Conformational energy studies and in vitro activity data on active GH releasing peptides. Endocrinology 1984;114:1531–6.
3. Bowers CY, Momany FA, Reynolds GA, Hong A. On the in vitro and in vivo activity of a new synthetic hexapeptide that acts on the pituitary to specifically release growth hormone. Endocrinology 1984;14:1537–45.
4. Bowers CY, Momany FA, Chang CH, Cody W, Hubbs JC, Foster CH. Combinations having synergistic growth hormone releasing activity and methods for the use thereof. US Patent #4880778, Nov. 14, 1989.
5. Bowers CY, Veeraragavan K, Sethumadhavan K. Atypical growth hormone releasing peptides. In: Bercu BB, Walker RF, eds. Growth hormone II. New York: Springer-Verlag, 1994;15:203–22.
6. Xing Q, Chubiao X, Chongxi L, Huitong T. Conformational study of the growth hormone releasing peptide in DMSO-d_6 solution by NMR spectroscopy. J Mol Sci (Wuhan, China) 1987;5:27–36.
7. Momany FA. Conformational analysis and polypeptide drug design. In: Metzger RM, ed. Topics in current physics. Berlin: Springer-Verlag, 1981;26:41–61.

6

Growth Hormone Releasing Peptides

Romano Deghenghi

Growth hormone releasing peptides (GHRP) are small peptides, having no sequence homology with growth hormone releasing hormone (GHRH) and characterized by high potency and selectivity in releasing growth hormone (GH). The best known series have been prepared and studied by Momany and Bowers (1). Their pioneering and outstanding research over several years has allowed the clinical development of promising compounds such as GHRP-6, GHRP-1, and GHRP-2 has spurred intense investigations on their mode of action and on analogues, including nonpeptides (Table 6.1).

Structure-Activity Relationship

Historically the starting point of the Momany-Bowers GHRPs was the pentapeptide MetEnkephalin Tyr-Gly-Gly-Phe-Met and the observation that the analogue Tyr-D-Trp-Gly-Phe-Met-NH$_2$ possessed GH releasing properties (2, 3). Subsequent modifications, such as Tyr-D-Trp-Ala-Trp-DPheNH$_2$ (4) led eventually to GHRP-6 (5), which was found to be active in vivo, in numerous animal species including man, and effective when administered orally, albeit at a large multiple dosage compared with its parenteral potency. These peptides have totally lost the opioid properties of their progenitors.

I became interested in GHRP-6 through my involvement with tryptophan, the largest aromatic amino acid and one of the most reactive, being known to undergo photodegradation, oxidation, alkylation, acylation, dimerization, and cyclization, and to be prone to electrophilic and free radical attack. This high chemical reactivity does pose a problem during the synthesis and the analysis of Trp-containing peptides and it is a challenge to the formulation chemist who is concerned with the purity and stability of peptide drugs.

TABLE 6.1. Milestones of GHRP's development.

1977	Tyr-D-Trp-Gly-Phe-Met-NH$_2$	(D-Trp^2)MetEKNH$_2$
	↓	
1981	Tyr-D-Trp-Ala-Trp-D-Phe-NH$_2$	
1984	His-D-Trp-Ala-Trp-D-Phe-NH$_2$	
	His-D-Trp-Ala-Trp-D-PheLys-NH$_2$	GHRP-6
	↓	
1991	Ala-His-DβNal-Ala-Trp-D-Phe-Lys-NH$_2$	GHRP-1
1992	His-D2MeTrp-Ala-Trp-D-Phe-Lys-NH$_2$	Hexarelin

(L-692,429) nonpeptide R = H

| 1993 | D-Ala-DβNal-Ala-Trp-DPhe-Lys-NH$_2$ | GHRP-2 |
| 1994 | | |

OH
|
(L-692,585) R=—CH$_2$CHCH$_3$

| | D-Thr-D-2MeTrp-Ala-Trp-DPheLys-NH$_2$ | T-relin |

Since a preferred point of attack is the position 2 of the indole nucleus (Fig. 6.1) it was reasonable to assume that a suitable substituent in that position, such as a methyl group, could lessen or prevent some of these problems. 2-Methyl-tryptophan (Mrp) has been known since 1914 (6), its D-isomer was isolated in 1936 (7), and its L-isomer in 1979. Yabe et al. (8) substituted it for Trp in the natural luteinizing hormone releasing hormone (LHRH) sequence, and observed a drastic (90%) loss of biologic activity of the methylated analogue, a finding that perhaps discouraged other researchers from employing it as a replacement for Trp in biologically active peptides.

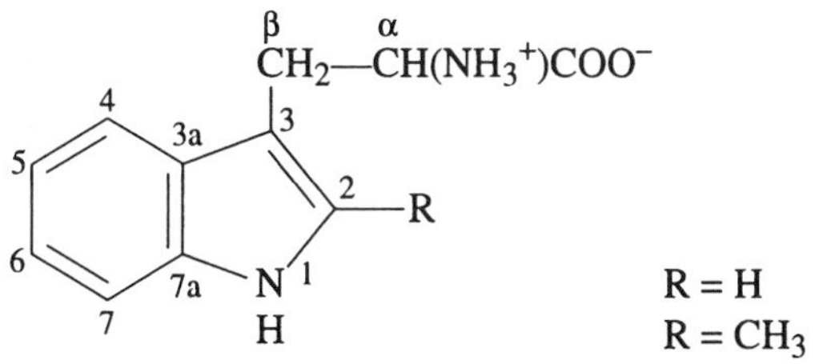

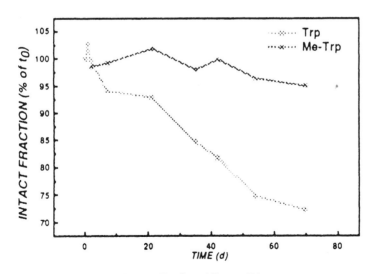

FIGURE 6.1 Tryptophan (Trp) and its 2-methyl derivative (Mrp).

Encouraged by a successful substitution of D-Mrp instead of D-Trp in position 6 of an LHRH analogue (9) and having verified the greatly enhanced chemical stability or Mrp, when compared with Trp, as shown in Figures 6.2 and 6.3, we substituted either one of the two Trp present in GHRP-6 with the corresponding L or D-Mrp derivative and obtained His-D-Trp-Ala-Mrp-D-Phe-LysNH$_2$ and His-D-Mrp-Ala-Trp-D-Phe-LysNH$_2$.

Similarly to what was observed in the LHRH case, the Mrp substitution for Trp resulted in great loss of GH releasing activity, but an enhancement of potency, respective to GHRP-6, with the D-Mrp isomer replacing D-Trp (Tables 6.2 and 6.3) (10, 11).

As expected, the newer analogues were chemically more stable than the parent compound (even if only one Trp was "stabilized" through its methyl derivative (Fig. 6.4). What we did not expect, however, was a somewhat

FIGURE 6.2 Stability of Me-Trp vs. Trp in acidic conditions.

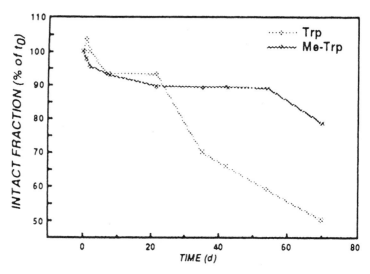

FIGURE 6.3 Stability of Me-Trp vs. Trp in oxidative conditions.

lesser toxicity (at higher doses in the rat) of our methylated derivative (hexarelin) when carefully compared with GHRP-6 (Table 6.4), nor its important stimulation of in vivo adenosine 3',5'-cyclic monophosphate (cAMP) accumulation in the infant rat pituitary (vide infra). Other better known pairs of homologous compounds (diazepam and nordiazepam, simvastatin and lovastatin) do not exhibit such differences between themselves, although there are well-documented large differences between methyl-desmethyl pairs exhibited, for instance, between some N-furylmethyl–substituted benzomorphans with agonist or antagonist action on opiate receptors (12).

TABLE 6.2. Effect of GHRP-6 and hexarelin on GH release on pentobarbital-anesthetized adult rats ($n = 5$).

	Plasma GH (ng/ml)	
	0 min	15 min
Saline sc	32 ± 15	43 ± 21
GHRP-6 sc		
25 μg/kg	41 ± 16	222 ± 95
50 μg/kg	57 ± 39	262 ± 58
Hexarelin sc		
25 μg/kg	56 ± 21	$388 \pm 99*$
50 μg/kg	32 ± 16	$439 \pm 69**$

Student's t-test: $*.01 \leq p \leq .05$; $** .001 \leq p \leq .01$.

TABLE 6.3. GH releasing activity of 2MeTrp (Mrp) containing peptides in groups of infant rats (sc; assay at 15 minutes).

Code	Dose (µg/kg)	Structure	Plasma GH (ng/ml)
		Group I ($n = 8$)	
Controls		Saline	15.86 ± 0.66
GHRP-6	300	His-D-Trp-Ala-Trp-D-Phe-Lys-NH$_2$	94.70 ± 6.84
Hexarelin	300	His-D-Mrp-Ala-Trp-D-Phe-Lys-NH$_2$	101.86 ± 8.60
EP92632	300	Ala-His-D-Mrp-Ala-Trp-D-Phe-Lys-NH$_2$	111.00 ± 5.23
EP92439	300	His-D-Mrp-D-Lys-Trp-D-Phe-Lys-NH$_2$	7.78 ± 1.28
EP92440	300	His-Ala-D-Trp-D-Lys-Mrp-D-Phe-Lys-NH$_2$	6.28 ± 0.61
EP92441	300	His-D-Mrp-D-Lys-Mrp-D-Phe-Lys-NH$_2$	9.94 ± 1.82
EP9399 (cyclic hexarelin)	300	His-D-Mrp-Ala-Trp-D-Phe-Lys Artimsp154-B	25.87 ± 3.75
EP9399	1200		23.87 ± 5.12
		Group II ($n = 8$)	
Controls		Saline	36.9 ± 22.3
GHRP-6	300	His-D-Trp-Ala-Trp-D-Phe-Lys-NH$_2$	127.0 ± 18.6
Hexarelin	300	His-D-Mrp-Ala-Trp-D-Phe-Lys-NH$_2$	163.5 ± 21.5
EP7459	300	His-Ala-D-Trp-Ala-Mrp-D-Phe-Lys-NH$_2$	9.0 ± 2.3
EP7458	300	His-D-Trp-Ala-Mrp-D-Phe-Lys-NH$_2$	60.8 ± 21.6
EP7460	300	His-D-Trp-Ala-Mrp-D-Phe-Ala-Lys-NH$_2$	21.6 ± 9.5
		Group III ($n = 5$)	
Controls		Saline	20.00 ± 3.2
EP251	300	4-Aminobenzoyl-D-Mrp-Ala-Trp-D-Phe-Lys-NH$_2$	18.8 ± 4.0
EP252	300	2-Aminobenzoyl-D-Mrp-Ala-Trp-D-Phe-Lys-NH$_2$	10.4 ± 0.8
EP253	300	3-Aminobenzoyl-D-Mrp-Ala-Trp-D-Phe-Lys-NH$_2$	10.4 ± 2.0
EP254	300	2-N-Methylaminobenzoyl-D-Mrp-Ala-Trp-D-Phe-Lys-NH$_2$	17.6 ± 2.0
EP255	300	Pipecolyl-D-Mrp-Ala-Trp-D-Phe-Lys-NH$_2$	18.0 ± 3.2
EP256	300	D-Pipecolyl-D-Mrp-Ala-Trp-D-Phe-Lys-NH$_2$	12.8 ± 1.6
EP257	300	4-Amino-Phe-D-Mrp-Ala-Trp-D-Phe-Lys-NH$_2$	26.0 ± 5.2
EP258	300	D-4-Amino-Phe-D-Mrp-Ala-Trp-D-Phe-Lys-NH$_2$	29.2 ± 8.0
EP259	300	Imidazolacetyl-D-Mrp-Ala-Trp-D-Phe-Lys-NH$_2$	158.8 ± 26.8
Hexarelin	300	His-D-Mrp-Ala-Trp-D-Phe-Lys-NH$_2$	250.4 ± 38.3
		Group IV ($n = 10$)	
Controls		Saline	69.2 ± 14.0
EP40473	5000	Trp-D-Phe-Lys-NH$_2$	11.8 ± 1.4
EP40474	5000	Ac-Trp-D-Phe-Lys-NH$_2$	22.1 ± 3.5
EP40475	5000	His-D-Mrp-Ala-Trp-NH$_2$	123.3 ± 37.2
L-692,429	5000	(see above)	131.0 ± 18.1
		Group V ($n = 9$)	
Controls		Saline	31 ± 8
EP40735	300	His-D-Mrp-Ala-Trp-D-Phe-Lys-Thr-NH$_2$	176 ± 20
EP40736	300	His-D-Mrp-Ala-Trp-D-Phe-Lys-D-Thr-NH$_2$	169 ± 27
EP40737	300	D-Thr-D-Mrp-Ala-Trp-D-Phe-Lys-NH$_2$	266 ± 20
EP40738	300	D-Thr-His-D-Mrp-Ala-Trp-D-Phe-Lys-NH$_2$	86 ± 19
Hexarelin	300	His-D-Mrp-Ala-Trp-D-Phe-Lys-NH$_2$	235 ± 23
		Group VI ($n = 8$)	
Controls	300	Saline	31 ± 5
Hexarelin	300	His-D-Mrp-Ala-Trp-D-Phe-Lys-NH$_2$	299 ± 14
EP92111	300	His-D-Mrp-Ala-Trp-D-Phe-Lys-OH	167 ± 19
EP40904	300	Thr-D-Mrp-Ala-Trp-D-Phe-Lys-NH$_2$	56 ± 10
EP40804	300	His-D-Mrp-Ala-Phe-D-Trp-Lys-NH$_2$	22 ± 2
EP40894	5000	D-Thr-D-Mrp-Ala-Trp-NH$_2$	60 ± 11
EP40954	5000	Ala-Trp-D-Phe-Lys-NH$_2$	36 ± 12

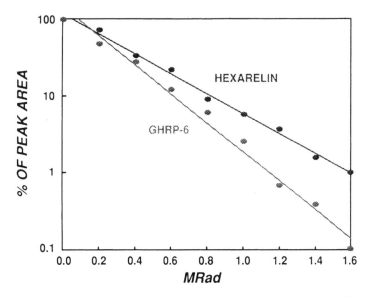

FIGURE 6.4 Effect of gamma irradition on GHRP-6 and Hexarelin stability in acetate buffer pH 5.4.

An interesting finding in our Mrp substituted series is the total inactivity of the cyclic analogue of hexarelin, His-DMrp-Ala-Trp-DPhe-Lys, which incidentally does not cross-react with a sensitive hexarelin antibody, pointing out the importance of having a linear structure and/or the presence of NH_2 terminal functions.

Returning to the amino acid sequences depicted in Table 6.1, one cannot be impressed by the relative importance of the C-terminal tetrapeptide Ala-Trp-D-Phe-Lys-NH_2, which one is tempted to consider a "pharmacophore" responsible for at least part of the GH releasing properties of GHRPs. Indeed Momany et al. (4) have shown that practically any attempted substitution in the C-terminal tetrapeptide sequence was deleterious to activity, the Trp and D-Phe moieties being particularly sensitive to changes.

TABLE 6.4. Number of treated and dead animals after single intravenous doses of GHRP-6 and hexarelin to male rats.

Dose (mg/kg)	GHRP-6		Hexarelin	
	Treated	Dead	Treated	Dead
2.5	6	0		
5.0	6	1	6	0
7.5	6	2	6	1
10.0			6	2

One change that was not reported was the inversion of Trp^4 with Phe^5. We wanted to explore this inversion by preparing the hexapeptide His-D-Mrp-Ala-Phe-D-Trp-Lys-NH$_2$ and found that it liberates less GH than saline in the control rats (vide infra). At the N-terminal end, it was known (13) that His^1 could be favorably substituted with an imidazoleacetate, a finding confirmed in our Mrp series (Table 6.2), and later it was found that His^1 could be favorably replaced by a D-Ala (as in GHRP-2) or by other groups.

Mechanisms of Action

In the GHRP series, GHRP-6 has been the most studied analogue so far. A word of caution, suggested by recent findings, is that generalization of results to other members of the series even structurally closely related is not justified.

GHRP-6 and GHRH

There is reasonable consensus that GHRH is needed for GHRP-6 activity, since passive immunization with GHRH antibodies blunts the GH response to GHRP. Receptors to GHRP-6, different from GHRH, have been shown on rat pituitary cells (14) and have been postulated since the intracellular pathways are different for the two agents (cAMP dependent for GHRH, Ca^{2+} dependent for GHRP-6 and GHRP-1) (15), but it is likely that a common receptor domain exists, activated by both agents. GHRP-6 does not appear to release GHRH (1) but hexarelin does, at least in sheep (16). The in vivo synergism between GHRP-6 and GHRH (1) is postulated at the hypothalamic level and direct evidence of hypothalamic action has been shown (17). GHRP-6 increases the *number* of pituitary GH-secreting cells, whereas GHRH increases the *amount* of GH secreted per cell (18).

The existence of at least three functionally distinct subpopulations of somatotroph cells (rat pituitary) responsive to GHRP, GHRH, and to both secretagogues was described indicating a partial overlapping of at least two distinct receptors (19).

The Unknown Factor (U-Factor) Hypothesis

The U-factor hypothesis was proposed to explain the synergistic GH response that is induced when GHRP(s) + GHRH are coadministered in vivo (20). The unknown GHRP-like endogenous putative ligand released by GHRP(s) in the hypothalamus would inhibit somatotropin release inhibiting factor (SRIF) action at the hypothalamic and/or pituitary level with enhanced GH response relative to the action observed when GHRP(s) or GHRH are given separately.

GHRPs and Somatostatin (SRIF)

GHRP-6 is inhibited by SRIF in vitro and in vivo (5, 18). Bowers was unable to show any significant action of SRIF on binding of GHRP-6 to pituitary and hypothalamic membranes (21, 22) but displacement of binding at 50 nM of SRIF was described by Codd et al. (23).

What about the reverse effect of GHRPs on SRIF? According to Bowers (1), Tannenbaum and Bowers (24), and Locatelli et al. (25) (in the infant rat), GHRP-6 does not disrupt or inhibit SRIF action. Wehrenberg's team (26) advocates an effect (adult rats), and Clark et al. (27) have suggested that GHRP-6 disrupts cyclic somatostatin release (conscious male rats) and noted their structural similarity.

If one concurs with the suggestion (28) that the pharmacophore of SRIF comprises at least the contiguous Phe^7-Trp^8-Lys^9-(Thr^{10}) of the sequence present in SRIF and analogues such as Octreotide (DTrp instead of Trp^8), then the C-terminal tripeptide of GHRP-6, Trp-D-Phe-Lys-NH$_2$, may conceivably sit on SRIF receptors. In a 10-day rat model this tripeptide (5 mg/kg sc), appears to decrease GH release respective to controls (11.8 ± 1.4 ng/ml vs. 69.3 ± 14.1), suggesting an SRIF-like action. The rest of the whole sequence then would provide the SRIF-antagonistic effect that releases GH. The C-terminal tetrapeptide of GHRP(s), Ala-Trp-DPhe-Lys-NH$_2$, is inactive in vivo at 5 mg/kg in the infant rat.

In vitro binding studies on SRIF human receptors have shown that neither hexarelin nor its putative receptor antagonist, His-Ala-D-Trp-D-Lys-D-Mrp-D-Phe-Lys-NH$_2$, nor the C-terminal tripeptide of GHRPs Trp-D-Phe-Lys-NH$_2$, bind to the receptors. Additional studies with radioligands are needed to determine whether they bind to different receptor subpopulations interfering with SRIF action.

In contrast to the effect in the infant/adult rat model, GHRPs are not as potent in prepubertal children as they are in adult humans, where the somatostatin tone is higher and therefore more susceptible to disruption, through a mechanism as yet unclear.

GHRP-6 and α-MSH

α-Melanotropin [α-melanocyte stimulating hormone (α-MSH)] is a tridecapeptide structurally homologous to adrenocorticotropic hormone (ACTH), which, in addition to its pigmentary effects, is of interest for a number of neurochemical-neurotrophic activities.

α-MSH: Ac-Ser^1-Tyr-Ser-Met^4-Glu-*His^6*-Phe-Arg-*Trp^9*-Gly-*Lys^{11}*-ProValNH$_2$
GHRP-6:   *His*-DTrp-Ala-*Trp*-DPhe*Lys*-NH$_2$

Sawyer et al. (29) have found that GHRP-6 is an inhibitor of the melanotropic activity of α-MSH in the frog (*R. pipians*) in vitro bioassay. This competitive antagonism has been attributed to structural similarity of GHRP-6 with the sequence 6–11 of α-MSH.

GHRP-6 and Substance P

Substance P (Arg^1-Pro^2-Lys-Pro-Glu^5-Glu-Phe^7-Phe-Gly^9-Leu-Met^{11}NH$_2$) has no effect on GHRP-6 binding on rat and porcine pituitary and hypothalamic membranes, but its antagonist (D-Arg^1, D-Phe^5, DTrp^{7,9}, Leu^{11}, substance P) is a potent inhibitor (21, 22). This inhibition was also seen in vivo (anesthetized rats) (30).

From Peptides to Nonpeptides . . . and Back!

GHRP-6 fulfills the requirement of a potentially useful drug: reasonable selectivity, high potency when administered parenterally, and an acceptable toxicity profile at therapeutic dose. Its oral availability however is low (0.3%) when compared with parenteral absorption, and this fact has prompted a team from Merck to look for nonpeptidyl analogues with a better oral availability.

Several years of brilliant computer-assisted research were crowned by the discovery of the series of benzazepinones, related to angiotensin II antagonists, possessing GH releasing properties, high selectivity, and, for some, the sought oral activity. The prototype (L-692,429) had rather modest potency, but a hydroxypropyl derivative (L692,585) (Table 6.1) was 20 times more potent in a rat pituitary cell assay (31). Clinical data of the orally active secretagogue L-163,191 (MK-677), are expected in due time.

In their medicinal chemistry paper the Merck team (32) presented some molecular overlays to show three-dimensional similarity between the model GHRP-6 and L-692,429. The authors readily admit that several other overlays or mappings are possible.

Even with the (maximum biased manual) Dreiding models one can readily obtain convincing overlays between L-692,429 (MW 509) and either the C-terminal tripeptide of GHRP-6 (Trp-D-Phe-Lys-NH$_2$, MW 478) (Fig. 6.5) or the N-terminal tetrapeptide (His-D-Trp-Ala-Trp-NH$_2$, MW 597) (Fig. 6.6). More objective computer-assisted overlays can also be obtained.

As we have seen, the C-terminal peptide has SRIF-like activity (in rats in vivo) but the N-terminal tetrapeptide (as its D-Mrp^2 analogue) has modest

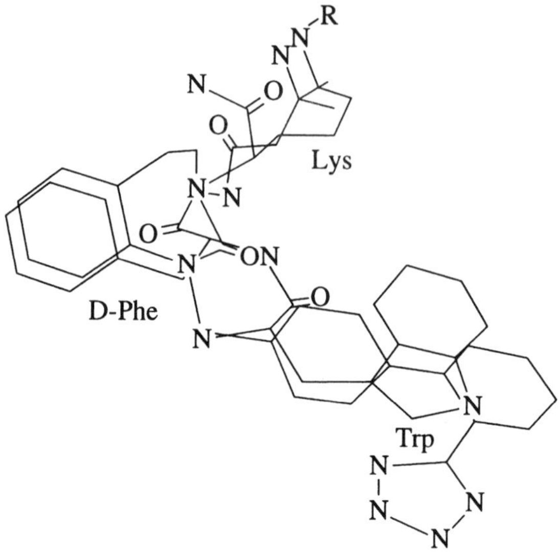

FIGURE 6.5 Overlay L692, 585 and Trp-D-Phe-Lys-NH$_2$.

in vivo GH releasing activity comparable to L-692,429 (GH 123 ± 37 ng/ml vs. 131 ± 18 ng/ml, both at 5 mg/kg sc). Tetrapeptides have better oral absorption than hexapeptides, thus opening the way to explore and compare their respective oral activity/potency ratio.

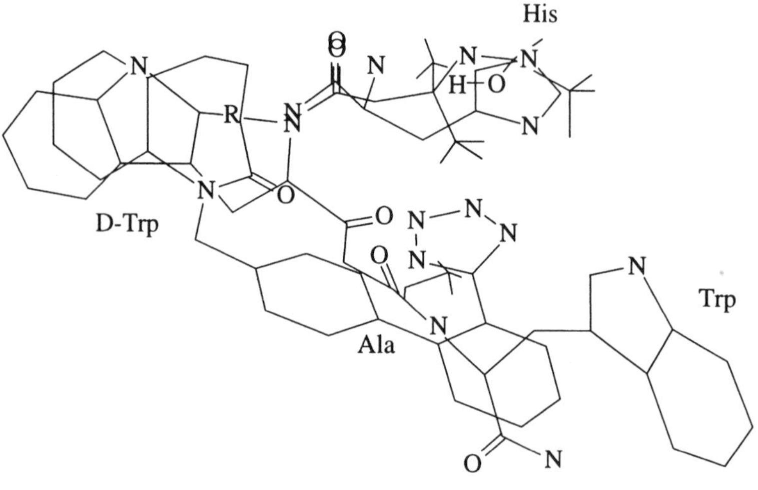

FIGURE 6.6 Overlay L692, 585 and His-D-Trp-Ala-Trp-NH$_2$.

Having thus determined that biologic activity points to the overlay L-692,429 with the N-terminal tetrapeptide of GHRP-6 (or of hexarelin) the modification of the side chain that resulted in the potent analogue L-692,585 was adapted to the hexapeptide structure of hexarelin, the hydroxylpropyl residue being incorporated as Thr, both D and L, instead of His[1]. Thus the hexapeptides

EP40904: Thr-D-2MeTrp-Ala-Trp-D-Phe-Lys-NH$_2$
EP40737: DThr-D-2MeTrp-Ala-Trp-D-Phe-Lys-NH$_2$ (T-relin)

were tested in our infant rat model (10) (Table 6.2).
    As expected, the D-Thr[1] analogue is more potent than the L-Thr[1] derivative.

Hexarelin: Animal Studies and Mechanism of Action

A time-response study of hexarelin on GH release was performed in conscious adult male and female rats, pretreated with GHRH and somatostatin antiserum (to suppress the confounding effects of endogenous hormones). In this model hexarelin is active when given iv, sc, and po. The dose range was from 1 µg to 400 µg/kg iv with peak response at 7 ± 1 min posttreatment. Subcutaneously, the peak plasma GH occurred at 13 ± 2 min (100–1000 µg/kg range), whereas it occurred at 80 ± 11 min when the secretagogue was given po (dose range 100–5000 µg/kg) (11).
    In another study, hexarelin and GHRP-6 were compared when given iv in conscious and freely moving nonstressed adult rats (pretreated with GHRH and SRIF antisera). Hexarelin was more effective in eliciting GH release than was GHRP-6 at the dose tested (25 µg/kg); GHRH inhibited the GH response to both hexarelin and GHRP-6, and SRIF antiserum pretreatment did not affect the GH response to the GH releasing peptides. Therefore, it was concluded that these two GH releasing peptides elicit GH release via a GHRH-dependent pathway. In addition, because both hexarelin and GHRP-6 are bioactive regardless of circulating somatostatin tone, these GH releasing peptides must also act to suppress somatostatin release from the hypothalamus or somatostatin effectiveness at the pituitary (26).
    In the infant rat (10-day old pups) hexarelin dose-dependently stimulated GH secretion (ED$_{50}$ = 40 µg/kg sc) with peak effect at 15 min. Passive immunization with GHRH and/or SRIF antisera did not prevent the plasma GH rise evoked by 300 µg/kg sc of hexarelin, indicating that in the infant rat its action is not mediated by endogenous GHRH and/or SRIF. In addition, hexarelin was shown to stimulate GH mRNA expression (GH synthesis) in this animal model (25).
    In conscious, freely moving adult rats, hexarelin given iv (indwelling jugular catheter) at repeated doses at 2-hour intervals or as a continuous infusion for 6, 30, or 174 hours elicits significant GH responses when admin-

istered repeatedly (three times). The plasma GH pulse pattern was unaltered and no downregulation in response was noted with the infusions of the secretagogue (100μg/h) (33).

Long-term studies with hexarelin in old dogs (E.E. Müller, unpublished observations) have shown an increased GH response to daily sc injections of 500μg/kg repeated up to 4 weeks and a decreased response to pretreatment value at 6 weeks. However, after a washout period of 2 weeks, the GH response to hexarelin again increased for the next 3 weeks then declined, to resume again after a second 2-week washout.

A number of previous studies (15, 18, 34, 35) have indicated that the intracellular mechanism by which GHRP-6 releases GH from somatotrophs is Ca^{2+} dependent and cAMP independent. More recently (36) it was shown that in an in vitro perfusion system of primary cultured ovine anterior pituitary cell GHRP-2, while releasing GH through an increase in Ca^{2+} influx, can be blocked by [Ac-Tyr^1, D-Arg^2] GHRH 1-29, an antagonist to GHRH. It was shown moreover (Chen Chen, unpublished observations) that GHRP-2, like GHRH, increases cAMP levels through an activation of adenylate cyclase.

Independently in studies in pituitary preparations from infant and adult rats it was shown (V. Locatelli, unpublished observations) that hexarelin and GHRP-6 both elevate adenylate cyclase activity in vitro in 10-day-old rat anterior pituitary glands, but are devoid of activity in 6-week-old glands, whereas GHRH is fully active in both preparations.

Clinical Pharmacology of Hexarelin

The GH releasing activity of hexarelin in humans (a dose-response study) was recently reported (37). The doses were 0.5, 1.0, and 2.0μg/kg iv. There was no substantial difference in the release of GH after the 1- and 2-μg/kg dose, thus indicating that the effect after these doses is close to the maximal response (Fig. 6.7).

As with other GHRPs, the action of hexarelin is not completely specific for GH release. There was a slight elevation of both prolactin and cortisol levels, although they remained within normal limits after all hexarelin treatments. In a different study (38) the effects of hexarelin after the intravenous, subcutaneous, intranasal, and oral administration in male and female volunteers were measured. The results are given in Table 6.5.

Hexarelin induces GH release in thalassemic patients (39) and in women with polycystic ovarian syndrome (40). Additional studies were done in short, normal, GH-deficient, and obese children (41) and given alone or in combination with GHRH or arginine, in normal elderly subjects (42). In normal males, as expected, a synergistic effect on GH secretion was seen after coadministration of hexarelin and GHRH. In the same study it was

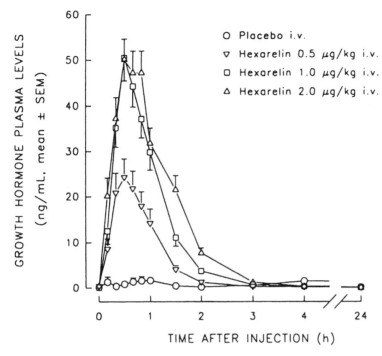

FIGURE 6.7 Time-course of growth hormone plasma concentrations after different intravenous doses of Hexarelin in 12 healthy young male volunteers.

shown that somatostatin abolished the GH response to GHRH, while it only blunted that to hexarelin. In addition it was shown that pyridostigmine and arginine did not modify the GH response to hexarelin, whereas pyrenzepine blunted it. Glucose and lipid heparin only blunted the

TABLE 6.5. GH-releasing activity of hexarelin (HEX) after intravenous, subcutaneous, intranasal, and oral administration in 12 healthy volunteers (6 men and 6 women, age: 26.2 ± 4.4 and 25.7 ± 3.6 years; weight 70.0 ± 11.0 and 56.0 ± 7.4 kg).

Treatment/dose	C_{max} (µg/L)		T_{max} (min)		AUC (µg/L/h)	
	Men	Women	Men	Women	Men	Women
Saline iv	0.7 ± 0.3	4.4 ± 0.4	93 ± 27	78 ± 24	17.4 ± 6.9	156.0 ± 16.6
GHRH 1.0 µg/kg iv	20.0 ± 2.6	25.7 ± 2.9	30 ± 3	22 ± 5	466.8 ± 76.2	562.4 ± 77.3
HEX 1.0 µg/kg iv	53.7 ± 10.6	49.4 ± 8.6	32 ± 4	22 ± 3	1227.8 ± 302.3	889.0 ± 149.2
HEX 2.0 µg/kg iv	59.0 ± 6.6	57.7 ± 14.9	25 ± 3	20 ± 3	1620.7 ± 221.6	1232.0 ± 315.5
HEX 1.5 µg/kg sc	57.3 ± 4.7	40.5 ± 11	50 ± 3	35 ± 6	1310 ± 107.8	809.5 ± 196.3
HEX 3.0 µg/kg sc	63.5 ± 7.6	52.4 ± 7.7	55 ± 3	37 ± 3	1820.5 ± 280.0	1152.1 ± 183.7
HEX 20.0 µg/kg in	39.2 ± 6.6	41.4 ± 13.5	45 ± 3	25 ± 3	871.3 ± 141.4	889.9 ± 282.8
HEX 20.0 mg os	23.9 ± 4.3	39.6 ± 10.1	72 ± 9	55 ± 15	554.4 ± 138.0	964.5 ± 244.2
HEX 40.0 mg os	43.4 ± 7.5	66.1 ± 9.2	65 ± 5	58 ± 3	1092.0 ± 187.0	1627.2 ± 255.9

C_{max}, the maximum GH serum concentration observed; T_{max}, the time of its occurrence; AVC, area under curve.

hexarelin-induced GH release while they strongly inhibited the GHRH-induced GH rise (43).

In normal children hexarelin releases more GH than GHRH, the effect being greater in pubertal age and more marked in girls than in boys. Intranasal administration to children with familial short stature was effective and devoid of side effects (44). Hexarelin was shown to counteract the inhibitory effect of glucocorticoids on GH secretion in humans (45). Naloxone did not inhibit or blunt the GH response of hexarelin in normal volunteers (M. Korbonits, unpublished observations). Hexarelin effects on circulating prolactin (PRL) levels in normal subjects and acromegalic patients (46) and on GH, PRL, thyroid stimulating hormone (TSH) and gonadotropins in the latter patients were studied (47). PRL is not significantly released by hexarelin in normal subjects, but an increment was seen in acromegalic patients (47) and in hyperprolactinemic women (48).

Conclusion

GHRPs are a recently discovered, intensely studied class of man-made secretagogues. As their discoverer (1) has said, they are important at the theoretical, physiologic, and, potentially, at the therapeutic level. Although structurally related, it is now apparent that even minor structural variations may result in unexpected differences in their mechanism of action. They are not entirely specific in their action and their "minor" activities merit careful study.

Would they lead to the discovery of a mysterious endogenous ligand? Or would they turn out to be a novel kind of somatostatin antagonist? Is their mechanism related to substance P? Would their orally active, low-cost nonpeptidyl analogues be the wonder drugs of the new millennium? Or would they be regarded as a transient pharmacologic curiosity? Shouldn't we here agree with Stephen Jay Gould (49) when he says that scientists reach conclusions for the damnedest of reasons?

Acknowledgments. I am deeply indebted to Eugenio Müller, Vittorio Locatelli, and their coworkers at the University of Milan for having performed the initial assays and most follow-up studies with the novel peptides described in this chapter. The enthusiastic and careful attention given by Bill Wehrenberg and his team to hexarelin and its mechanism of action is highly appreciated. I am grateful to Dr. Robert Collu for studies on human somatotroph preparations, to Drs. Y.C. Patel and C.B. Srikant for binding studies on somatostatin receptors, to Dr. Chen Chen for testing hexarelin in his ovine pituitary experiments, and to Drs. Robert Kirby and Suman Rakhit for computer assisted modeling. My colleagues in France and in Italy (Vincent Lenaerts, François Boutignon, Patrick Wüthrich, Paolo

Lucchelli, and Bruno Imbimbo) have given much of their time and ability to this project.

The list of clinical investigators who have used hexarelin is a long one, but I am particularly indebted to Dr. Ezio Ghigo, Prof. Franco Camanni, and their team for their competent and innovative contributions. George Tolis in Athens has explored new aspects, and, last but not least, Prof. Zvi Laron has advanced our knowledge of this series of compounds in pediatric applications.

References

1. Bowers CY. GH releasing peptides—structure and kinetics. J Pediar Endocrinol 1993;6:21–31.
2. Bowers CY, Chang J, Momany F, Folkers K. Effect of the enkephalins and enkephalin analogues on release of pituitary hormones in vitro. In: MacIntyre and Szelke, editors. Molecular Endocrinology. Elsevier/North Holland Biomedical Press, 1977;287–92.
3. Bowers CY, Momany F, Reynolds GA, Chang D, Hong A, Chang K. Structure-activity relationships of a synthetic pentapeptide that specifically releases growth hormone in vitro. Endocrinology 1980;106:663–7.
4. Momany FA, Bowers CY, Reynolds GA, Chang D, Hong A, Newlander K. Design, synthesis and biological activity of peptides which release growth hormone in vitro. Endocrinology 1981;108:31–9.
5. Bowers CY, Momany FA, Reynolds GA, Hong A. On the in vitro and in vivo activity of a new synthetic hexapeptide that acts on the pituitary to specifically release growth hormone. Endocrinology 1984;114:1537–45.
6. Ellinger A, Matsuoka. Preparation of [2]-methyltryptophan and its behaviour in the animal organism. Z Physiol Chem 1914;91:45–7.
7. Majima S. Ein biologisches Verfahren der d-Tryptophandarstellung. Hoppe Seylers Z Physiol Chem 1936;243:250–3.
8. Yabe Y, Miura C, Horikoshi H, Miyagawa H, Baba Y. Synthesis and biological activity of LH-RH analogues substituted by alkyltryptophans at position 3. Chem Pharm Bull 1979;27:1907–11.
9. Deghenghi R, Boutignon F, Wüthrich P, Lenaerts V. EP23904 (Meterelin)—a novel sustained release LHRH-agonist. Gynecol Endocrinol 1993;7:29(abstr).
10. Deghenghi R, Cananzi MM, Torsello A, Battisti C, Müller EE, Locatelli V. GH-releasing activity of hexarelin, a new growth hormone releasing peptide, in infant and adult rats. Life Sci 1994;18:1321–8.
11. Conley LK, Giustina A, Imbimbo BP, Stagg LC, Deghenghi R, Wehrenberg WB. Hexarelin (His-D-2-MethylTrp-Ala-Trp-D-Phe-Lys-NH$_2$): a new GH-releasing peptide, is biologically active in male and female rats. Endocr J 1994;2:691–5.
12. Bowman WC, Rand MJ. Textbook of pharmacology, 2nd ed. Oxford: Blackwell Scientific, 1980:1612.
13. Walker RF, Codd EE, Yellin T, Fishman B. Growth hormone (GH) releasing activity and binding correlations following structural modification of a GH-releasing hexapeptide His-D-Trp-Ala-Trp-D-Phe-Lys-NH$_2$ (GHRP-6; SK&F 110679). 71st Annual Meeting, The Endocrine Society, 1989; Abstract 777.

14. Blake AD, Smith RG. Desensitization studies using perfused rats pituitary cells show that growth hormone-releasing hormone and His-D-Trp-Ala-Trp-D-Phe-Lys-NH$_2$ stimulate growth hormone release through distinct receptor sites. J Endocrinol 1991;129:11–19.

15. Chick CL, Akman M, Girard M, Ho AK. Growth hormone releasing peptide stimulates GH via a calcium-dependent c-AMP-independent mechanism. 74th Annual Meeting, The Endocrine Society 1992; Abstract 521.

16. Guillaume V, Magnan E, Catald M, Dutour A, Sauze N, Renard M, Razafindraibe H, Conte-Devolx B, Lenaerts V, Deghenghi R, Oliver C. GHRH secretion is stimulated by a new GHRP hexapeptide in sheep. Endocrinology 1994;135:1073–6.

17. Dickson SL, Leng G, Robinson ICAF. Systemic administration of growth hormone-releasing peptide activates hypothalamic arcuate neurons. Neuroscience 1993;53:303–6.

18. Goth MI, Lyons CE, Canny BJ, Thorner MO. Pituitary adenyl cyclase activating polypeptide, GH-releasing peptide and GH-releasing hormone stimulate GH release through distinct pituitary receptors. Endocrinology 1992;130:939–44.

19. Chihara K, Kaji H, Hayashi S, Yagi H, Takeshima Y, Mitami M, Ohashi S, Abe H. Growth hormone releasing hexapeptide: basic research and clinical application. In: Bercu BB, Walker RF, eds. Growth hormone II, basic and clinical aspects. New York: Springer-Verlag, 1994:223–30.

20. Bowers CY, Sartor AO, Reynolds GA, Badger TM. On the actions of growth hormone-releasing hexapeptide, GHRP. Endocrinology 1991;128:2027–35.

21. Sethumadhavan K, Veeraragavan K, Bowers CY. Demonstration and characterization of the specific binding of growth hormone-releasing peptide to rat anterior pituitary and hypothalamic membranes. Biochem Biophys Res Commun 1991;178:31–7.

22. Veeraragavan K, Sethumadhavan K, Bowers CY. Growth hormone releasing peptide (GHRP) binding to porcine anterior pituitary and hypothalamic membranes. Life Sci 1992;50:1149–55.

23. Codd EE, Shu AYL, Walker RF. Binding of a growth hormone releasing hexapeptide to specific hypothalamic and pituitary binding sites. Neuropharmacology 1989;28:1139–44.

24. Tannenbaum GS, Bowers CY. Growth hormone-releasing hexapeptide GHRP stimulates GH release via central GH-releasing factor (GRF) pathways. Society of Neuroscience Annual Meeting 1991; Abstract.

25. Locatelli V, Torsello A, Grilli R, Ghigo MC, Cella SG, Deghenghi R, Wehrenberg WB, Müller EE. GHRP-6 stimulates GH secretion and synthesis independently from endogenous GHRH and SRIF in the infant rat. J Endocrinol Invest 1993;16:OP27(abstr).

26. Conley LK, Teik JA, Deghenghi R, Imbimbo BP, Giustina A, Locatelli V, Wehrenberg WB. The mechanism of action of hexarelin and GHRP-6: analysis of the involvement of GHRH and somatostatin in the rat. Neuroendocrinology 1995; 61:44–50.

27. Clark RG, Carlsson LMS, Trojnar J, Robinson ICAF. The effects of a growth hormone-releasing peptide and GH-releasing factor in conscious and anaesthetized rats. J Endocrinol 1989;1:249–55.

28. Papageorgiou C, Haltiner R, Bruns C, Petcher TJ. Design, synthesis and binding affinity of a nonpeptide mimic of somatostatin. Biorg Med Chem Lett 1992;2:135–40.
29. Sawyer TK, Staples DJ, De Lauro Castrucci AM, Hadley ME. Discovery and structure-activity relationships of novel α-melanocyte-stimulating hormone inhibitors. Peptide Res 1989;2:140–6.
30. Cheng K, Wei L, Chaung LY, Chan WWS, Butler B, Pong SS, Smith RG. Substance P antagonist inhibits L-692,429 and GHRP-6 stimulated rat growth hormone release both in vitro and in vivo. 76th Annual Meeting, The Endocrine Society 1994; Abstract 662.
31. Schoen WR, Ok D, DeVita RJ, Pisano JM, Hodges P, Cheng K, Chan WWS, Butler BS, Smith RG, Wyvratt MJ, Fischer MH. Structure-activity relationships in the amino acid sidechain of L-692,429. Bioorg Med Chem Lett 1994;9:1117–22.
32. Schoen WR, Pisano JM, Prendergast K, Wyvratt MJ, Fischer MH, Cheng K, Chan WWS, Butler B, Smith RG, Ball RG. A novel 3-substituted benzazepinone growth hormone secretagogue (L-692,429). J Med Chem 1994;37:897–906.
33. Conley LK, Stagg LC, Giustina A, Imbimbo BP, Deghenghi R, Wehrenberg WB. Plasma GH responses are maintained following repeated doses and continuous infusions of hexarelin. 76th Annual Meeting, The Endocrine Society 1994; Abstract 659.
34. Chen K, Chan WWS, Barreto A, Convey EM, Smith RG. The synergistic effects of His-D-Trp-Ala-Trp-D-Phe-Lys-NH$_2$ on GHRF-stimulated GH release and intracellular adenosine 3′,5′-monophosphate accumulation in rat primary pituitary cell culture. Endocrinology 1989;124:2791–8.
35. Akman MS, Girard M, O'Brien LF, Ho AK, Chik CL. Mechanisms of action of a second generation growth hormone-releasing peptide (Ala-His-D-βNal-Ala-Trp-D-Phe-Lys-NH$_2$) in rat pituitary cells. Endocrinology 1993;132:1286–91.
36. Wu D, Chen C, Katoh K, Zhang J, Clarke IJ. The effect of GH-releasing peptide-2 (GHRP-2 or KP102) on GH secretion from primary cultured ovine pituitary cells can be abolished by a specific GH-releasing factor (GRF) receptor antagonist. J Endocrinol 1994;140:R9–R13.
37. Imbimbo BP, Mant T, Edwards M, Amin D, Dalton N, Boutignon F, Lenaerts V, Wüthrich P, Deghenghi R. Growth hormone-releasing activity of hexarelin in humans. Eur J Clin Pharmacol 1994;46:421–5.
38. Ghigo E, Arvat E, Giannotti L, Imbimbo BP, Lenaerts V, Deghenghi R, Camanni F. Growth hormone-releasing activity of hexarelin, a new synthetic hexapeptide, after intravenous, subcutaneous, intranasal and oral administration in man. J Clin Endocrinol Metab 1994;78:693–8.
39. Tolis G, Mesimeris T, Deghenghi R, Boutignon F, Wüthrich P, Lenaerts V. Growth hormone release in thalassemic patients by a new GH-releasing peptide administered intravenous or orally. 75th Annual Meeting, The Endocrine Society 1993; Abstract 331.
40. Tolis G, Markusis V, Krassas G, Skaltsas T, Ponticidis J, Moschoyannis H, Boutignon F, Lenaerts V, Deghenghi R. Enhancement of hexarelin induced GH release by an inhibitor of lipolysis in women with polycystic ovarian syndrome. 76th Annual Meeting, The Endocrine Society, 1994; Abstract.

41. Loche S, Cambiaso P, Carta D, Setzu S, Casini MR, Imbimbo BP, Borrelli P, Pintor C, Cappa M. GH-releasing effect of hexarelin a new synthetic peptide, in short normal, GH-deficient and obese children. 76th Annual Meeting, The Endocrine Society 1994; Abstract.

42. Ghigo E, Arvat E, Gianotti L, DiVito L, Imbimbo B, Lenaerts V, Deghenghi R, Camanni F. Effects of hexarelin, a synthetic hexapeptide, alone and combined with GHRH or arginine on GH secretion in normal elderly subjects. 76th Annual Meeting, The Endocrine Society 1994; Abstract.

43. Arvat E, Ghigo E, Gianotti L, Maccario M, Procopio M, Grottoli S, Imbimbo BP, Lenaerts V, Deghenghi R, Camanni F. Neuroendocrinological and metabolic modulation of the growth hormone-releasing activity of hexarelin in man. 76th Annual Meeting, The Endocrine Society 1994; Abstract.

44. Bellone J, Ghigo E, Aimaretti G, Benso L, Bartolotta E, Imbimbo BP, Deghenghi R, Camanni F. GH-releasing activity of hexarelin, a new synthetic hexapeptide, after intravenous administration in normal children. J Endocrinol Invest 1993;16(suppl 1–8).

45. Giustina A, Bussi AR, Deghenghi R, Imbimbo B, Licini H, Poiesi C, et al. Comparison of the effects of growth hormone-releasing hormone and hexarelin, a novel growth hormone-releasing peptide-6 analog, on growth hormone secretion in humans with or without glucocorticoid excess. J Endocrinol 1995;146:227–32.

46. Casati G, Palmieri E, Biella O, Guglielmino L, Arosio M, Faglia G. Effects of hexarelin (EP23905-MF6003) on circulating GH, PRL, TSH gonadotropins and α-submit of glycoprotein hormone levels in acromegalic patients. 3rd European Congress of Endocrinology, Amsterdam 1994; Abstract.

47. Palmieri E, Casati G, Guglielmino L, Biella O, Gambino G, Arosio M, Faglia G. Effects of hexarelin on circulating PRL levels in normal subjects and in acromegalic patients. 3rd European Congress of Endocrinology, Amsterdam 1994; Abstract.

48. Giusti M, Porcella E, Imbimbo BP, Cuttica MC, Falivene MR, Sessarego P, Giordano G. PRL response to hexarelin in hyperprolactinaemic women: dopaminergic and opioid control on hexapeptide induced PRL release. 3rd International Congress of Neuroendocrinology. Budapest, 1994; Abstract.

49. Gould SJ. In the mind of the beholder. Natural History 1994;2:14–23.

7

Nonpeptidyl Growth Hormone Secretagogues[1]

MATTHEW J. WYVRATT

The availability of recombinant human growth hormone (rhGH) in the mid-1980s has fostered a renewed interest in potential clinical applications of growth hormone (GH). In addition to the treatment of GH-deficient children and adults, rhGH may have beneficial effects in the treatment of patients with burns, bone fractures, or Turner's syndrome, in improving the exercise capacity of elderly individuals, and in reversing the catabolic effects of glucocorticoids, chemotherapy, and AIDS (1–5). While GH is synthesized and stored in the pituitary, its release is regulated by two hypo-thalamic peptides: growth hormone releasing hormone/factor (GHRH/GHRF) and the inhibitory hormone somatostatin (Fig. 7.1). Most cases of idiopathic GH deficiency are due to a hypothalamic defect and not a pituitary deficiency in GH, and therefore an alternate approach to rhGH treatment of GH-deficient patients would be to administer a GH secretagogue to release endogenous GH (6). For this reason, GRF(1-44)-NH$_2$ and its trun-cated analogues, e.g., GRF(1-29)-NH$_2$, have been studied over the past decade as an alternate approach in the treatment of GH deficiency (7, 8). Although rhGH and GHRF analogues are clinically effective, their high cost and lack of oral bioavailability have restricted their clinical applications.

In the late 1970s, Bowers and coworkers (9) reported the discovery of a series of peptides derived from Leu and Met enkephalins that specifically released growth hormone from the pituitary. These growth hormone releas-ing peptides (GHRPs) act directly on the pituitary via a mechanism distinct from GHRF (Fig. 7.1). A hypothalamic component, which is partially re-sponsible for the GHRPs' and the nonpeptidyl secretagogues' GH releasing activity in vivo, has recently been identified (10–12; also see Chapters 11 and 16, this volume). Assisted by molecular modeling, Bowers and Momany (13–16) discovered the hexapeptide GHRP-6 (Fig. 7.2), which is

---

[1] Numbers in boldface in this chapter refer to compound numbers (see figures and tables).

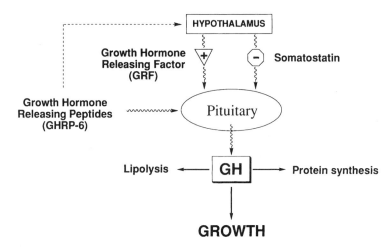

FIGURE 7.1. Growth hormone regulation.

an extremely potent and specific GH secretagogue in animals and in humans. While GHRP-6 is a much smaller peptide and therefore, possibly better absorbed than either rhGH, GHRF, or its analogues, it has low oral bioavailability in humans (0.3%). However, the demonstrated efficacy of GHRP-6 in releasing endogenous GH in vivo validated the hypothesis that a relatively small molecule, with its many possible positive attributes of cost and oral bioavailability, may be a viable alternative to treatment with rhGH.

Discovery of a Nonpeptidyl Growth Hormone Secretagogue

Merck researchers in 1988 embarked on a research program to discover a nonpeptidyl mimic of GHRP-6. At the time this research was initiated, nonpeptidyl mimics of peptide agonists were extremely rare, e.g., non-

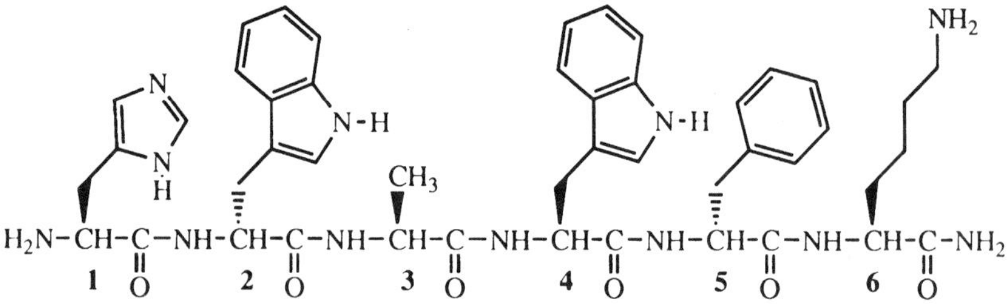

FIGURE 7.2. Structure of GHRP-6: His-D-Trp-Ala-Trp-D-Phe-Lys-NH$_2$.

peptidyl kappa opioid receptor agonists (17) were known. Extensive structure-activity relationships (SAR) had already been published for the hexapeptide GHRP-6 (13–15). It was clear from the published SAR that aromatic amino acids were favored at positions 2, 4, and 5. In particular, a Trp residue at position 2 was critical for biologic activity and the unnatural D-configuration was dramatically favored over the L-configuration. Since modifications at the N-terminus (e.g., desamino analogue of GHRP-6) of this hexapeptide reduced GH releasing activity, a basic amine appears to be a critical determinant for significant GH releasing activity. With these attributes in mind, potential nonpeptidyl templates were selected from the Merck sample collection and screened in a rat pituitary cell assay for release of GH. In early 1989 the lead compound L-158,077 (**1**) (Fig. 7.3), which was prepared for an angiotensin II antagonist program at Merck, was found to stimulate GH release in a dose-dependent [median effective dose (ED$_{50}$) = 3 µM; GHRP-6, ED$_{50}$ = 10 nM] and specific manner (18). Replacement of

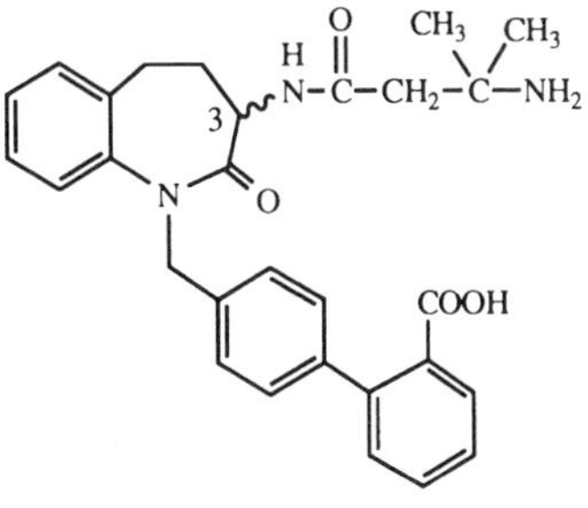

1
(L-158,077)

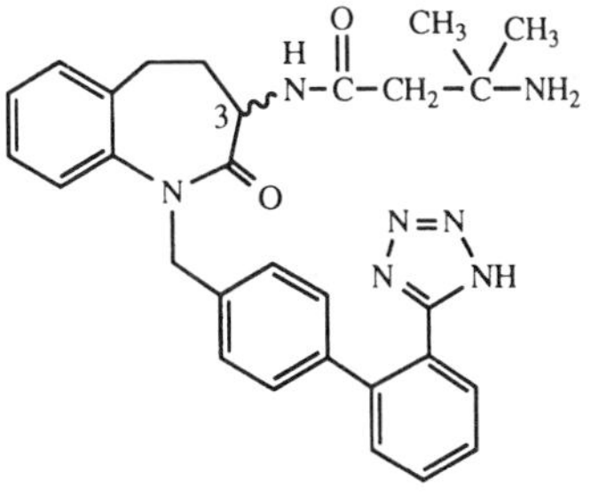

2

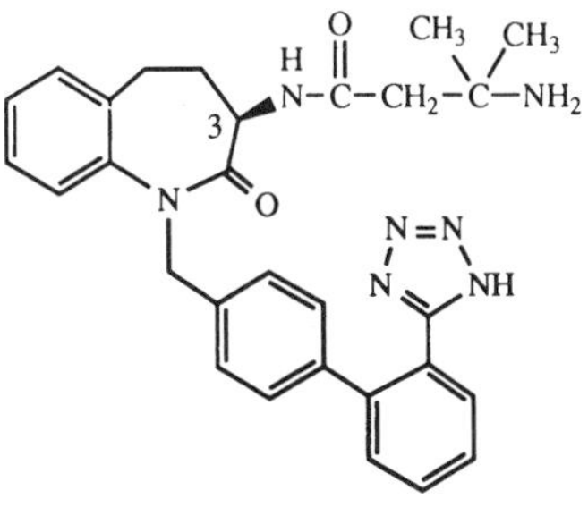

3
(L-692,429)

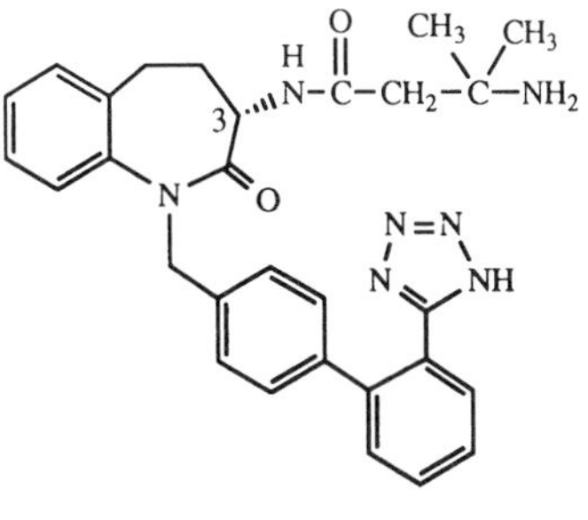

4

FIGURE 7.3. Nonpeptidyl growth hormone secretagogue lead structures.

the 2'-carboxylic acid function in L-158,077 with a tetrazole ring, a known carboxylic acid bioisostere, afforded the more potent (ED_{50} = 120 nM) racemic analogue **2**. Resolution of the chiral center at C-3 of compound **2** produced the individual enantiomers **3** and **4**. Growth hormone releasing activity was found to reside entirely in the R-isomer **3** (L-692,429; ED_{50} = 60 nM). L-692,429 was evaluated in over 50 receptor binding assays and was inactive at <10 μM in all assays except for weak angiotensin II receptor binding (IC_{50} = 6 μM) (18, 19). Mechanistically L-692,429 was found to be identical to GHRP-6 in vitro and to release GH in animal models (20–22). Recently, L-692,429 (MK-0751) has been shown to be a well-tolerated and specific nonpeptidyl GH secretagogue in humans when administered intravenously (23–25).

Molecular modeling techniques were used to examine the structural similarity between GHRP-6 and L-692,429 (3). SEAL (26) was employed to automatically generate superpositions of the two secretagogues. This approach resulted in the structural overlay for GHRP-6 and L-692,429 as displayed in Figure 7.4 (19). The amino group of L-692,429 is in the same vicinity as the N-terminus amino group of GHRP-6 and the benzolactam nucleus is aligned with the D-Trp residue of GHRP-6. The asymmetric center at C-3 in L-692,429 is overlaid onto the α-carbon of the D-Trp residue. The bridging phenyl group of the biphenyl appendage in L-692,429 positions the second phenyl group in the same region of space as that occupied by the D-Phe side chain of GHRP-6. In addition the tetrazole moiety can occupy the same space as the imidazole ring of the His residue

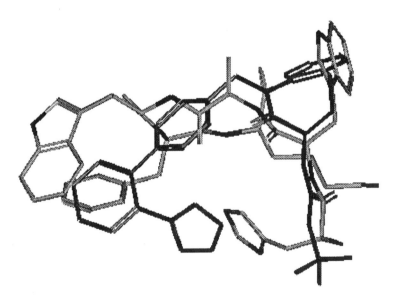

FIGURE 7.4. Structural overlay of hexapeptide GHRP-6 (gray) and benzolactam L-692,429 (black).

in GHRP-6, although it can also map onto the indole of the L-Trp residue or the amide backbone of the D-Phe residue in GHRP-6. Since the C-terminal lysine in GHRP-6 does not contribute significantly to its intrinsic in vitro activity, it is not surprising that there is no obvious alignment of this residue with any fragment in L-692,429. This comparison has served as a working hypothesis of how the benzolactam secretagogue L-692,429 may mimic GHRP-6.

Structure-Activity Relationships for L-692,429

Although the clinical results with L-692,429 were very promising and established the validity of our nonpeptidyl approach, its low oral bio-availability in animal models and modest potency in humans did not warrant its full development as an orally active GH secretagogue. There-fore, a detailed investigation of the structure-activity relationships for this benzolactam lead was initiated. All data presented here, unless indi-cated otherwise, are from the rat pituitary cell culture assay that has been previously described (27).

The benzazepin-2 one-ring system in L-692,429, which is also found in orally active angiotensin-converting enzyme (ACE) inhibitors (28, 29) and cholecystokinin (CCK) (30, 31) antagonists, was modified to evaluate the effects of ring size on GH releasing activity (Fig. 7.5). Both the six- and eight-membered lactams were found to be 10- to 50-fold less active (5, ED_{50} = $2\mu M$; 6, ED_{50} = $12\mu M$) than the seven-membered analogue 2, thus suggesting that the benzazepin-2 one-ring system in 2 and L-692,429 posi-tions the amino acid side chain and biphenyltetrazole moiety optimally for interaction with its receptor (19). The importance of the aromatic compo-nent of the benzolactam nucleus in L-692,429 is illustrated by the significant decrease in GH releasing activity for caprolactam analogue 7 (EC_{50} = $7\mu M$), although this decrease in activity may be attributed, in part, to a conformational effect.

L-692,429 possesses both a basic amine (pK_a 9.2) and an acidic tetrazole (pK_a 4.6), and is therefore zwitterionic (isoelectric point 6.9) at physiologic pH. Our initial structure-activity studies focused on the role of these charged functionalities in releasing GH and because they are a major deter-minant of the physicochemical properties of this class of GH secretagogues. Table 7.1 depicts several analogues (8 and 9) and indicates that the basic amine group in L-692,429, similar to that observed for the N-terminal amine in GHRP-6, is required for significant GH releasing activity (32). In addi-tion, the position of this protonated amine relative to the benzolactam nucleus is also important. Complete removal of the amino acid side chain to afford truncated derivative 10 resulted in a 150-fold loss in GH releasing activity, while an increase in the chain length by an additional methylene unit (11) or realignment of the gem-dimethyl substituents (12) caused a

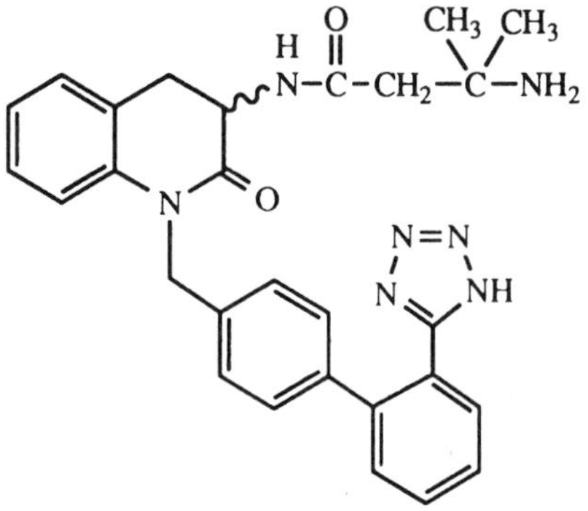

5

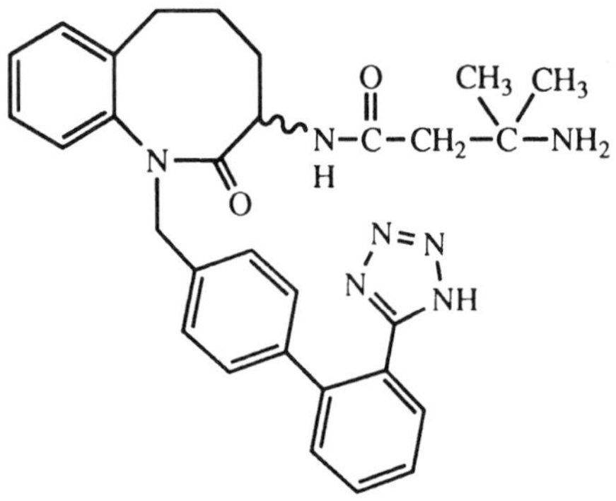

6

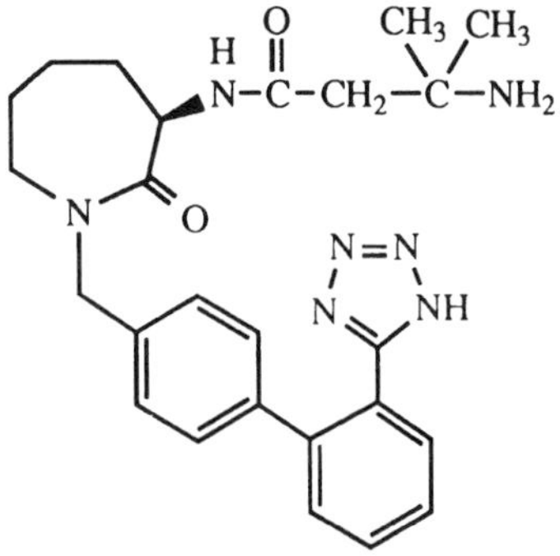

7

FIGURE 7.5. Alternative ring systems to the benzazepin-2 one-ring nucleus in L-692,429.

TABLE 7.1. L-692,429 Structure-activity relationships: amino acid side chain.

No.	R	ED$_{50}$(µM)	No.	R	ED$_{50}$(µM)
3	—C-CH$_2$-C-NH$_2$ (with O, CH$_3$, CH$_3$)	0.06	15	—C-CH-NH$_2$ (with O, CH$_3$)	1.4
8	—C-CH$_2$-C-NHAc (with O, CH$_3$, CH$_3$)	Inactive	16	—C-CH-NH$_2$ (with O, CH$_3$)	0.1
9	—C-CH$_2$-C-CH$_3$ (with O, CH$_3$, CH$_3$)	Inactive	17	—C-CH-NH$_2$ (with O, benzyl)	2
10	—H	9	18	—C-CH-NH$_2$ (with O, imidazole)	0.8
11	—C-CH$_2$CH$_2$-C-NH$_2$ (with O, CH$_3$, CH$_3$)	3	19	—C-CH-NH$_2$ (with O, imidazole)	3
12	—C—C-CH$_2$-NH$_2$ (with O, CH$_3$, CH$_3$)	3	20	—C-CH-NH$_2$ (with O, indole)	2
13	—C—C-NH$_2$ (with O, CH$_3$, CH$_3$)	0.03			
14	—C-CH$_2$-NH$_2$ (with O)	9			

significant decrease in activity. However, the shortened α-methylalanyl side chain **13** is slightly more active than L-692,429, although it has a less basic amine (pK$_a$ 7.9) (32).

Systematic deletion of the methyl groups from **13** illustrates the importance of geminal methyl substitution adjacent to the amine, although the D-Ala analogue **16** is only threefold less active than **13** (Table 7.1). D-Phenylalanine analogue **17** is less active than **16** and, consequently, suggestive of a sterically limited binding pocket. Based on our premise that the amino group in L-692,429 is occupying the same position on the receptor as the N-terminal amine of GHRP-6, the D-His (**18**), L-His (**19**), and D-Trp (**20**) analogues were prepared and found to be less active than the D-Ala analogue **16** and similar in potency to the D-Phe derivative **17**. The potency-enhancing methyl groups in both the alanine (e.g., **13**) and β-alanine (e.g., **3**) side chains appear to exert a conformational influence on the amino acid side chain to optimally position the basic amine for receptor interaction.

Since the basic amine in the benzolactam class of secretagogues is critical for GH releasing activity, a series of amino substituents were designed to modulate basicity and lipophilicity. As illustrated in Table 7.2, the basic amine can be extensively modified with alkyl side chains (**32**, **33**). The n-propyl (**21**), isopropyl (**22**), benzyl (**23**), and acetate (**24**) analogues have comparable potency with the unsubstituted parent **3**, suggesting that con-

TABLE 7.2. L-692,429 Structure-activity relationships: amino substituents.

No.	R	$ED_{50}(\mu M)$	No.	R	$ED_{50}(\mu M)$
3	—H	0.06	**30**	—CH$_2$C(CH$_3$)$_2$ (OH)	0.01
21	—CH$_2$CH$_2$CH$_3$	0.05	**31**	—CH$_2$CHCH(CH$_3$)$_2$ (OH)	1
22	—CH(CH$_3$)$_2$	0.08	**32**	—CH$_2$CHCH(CH$_3$)$_2$ (OH)	0.2
23	—CH$_2$Ph	0.07	**33**	—CH$_2$CH$_2$CHCH$_3$ (OH)	0.01
24	—CH$_2$CO$_2$Et	0.06	**34**	—CH$_2$CH$_2$CHCH$_3$ (OH)	Weakly active[a]
25	—CH$_2$CO$_2$H	>1	**35**	—CH$_2$CHCH$_3$ (OCH$_3$)	0.02
26	—CH$_2$CH$_2$NH$_2$	0.23	**36**	—CH$_2$CHCH$_3$ (OAc)	0.2
27	—CH$_2$CH$_2$OH	0.03	**37**	—CH$_2$CHCH$_3$ (F)	0.09[b]
28	—CH$_2$CHCH$_3$ (OH)	0.007			
29	—CH$_2$CHCH$_3$ (OH)	0.003			

[a] At 1-μM drug concentration.
[b] Mixture of diastereomers.

siderable steric bulk and lipophilicity is well tolerated in this region. In contrast, this region does not accept additional charged functionality such as acetic acid derivative **25** and, to a somewhat lesser extent, aminoethyl analogue **26**. However, a hydroxy group is potency enhancing as illustrated by the 2-hydroxyethyl derivative **27** and, in particular, the 2-hydroxypropyl analogues **28** (S-isomer) and **29** (R-isomer). The absence of a strong stereochemical preference for one diastereomer over the other is supported by the near equivalency of the gem dimethyl analogue **30**. It is of significance that these 2-hydroxypropyl analogues (**28** and **29**) are more potent in the rat pituitary assay than the peptidyl secretagogue GHRP-6 (ED_{50} = 10 nM). In addition to this 20-fold increase in in vitro activity for **29** (L-692,585) over L-692,429, this analogue has been reported (34) to be very active in vivo in beagle dogs at doses as low as 5 μg/kg when intravenously administered. Larger hydroxyalkyl substituents (e.g., **31** and **32**) are less active in releasing GH. Homologation of the 2-hydroxypropyl side chain yields the diastereomeric 3-hydroxybutyl derivatives **33** and **34**, which, unlike the 2-hydroxypropyl series, show a surprising difference in GH releasing activity. The potency enhancing effects of the 2(R)-hydroxy propyl function in **29** may be due to a conformational effect or accessing an additional binding site on the receptor. The results for methyl ether **35**, acetate **36**, and fluoro derivative **37** are consistent with the formation of a hydrogen bond between the hydroxy group in **29** and the receptor, thus leading to greater GH releasing activity relative to **3** or the propyl analogues **21** and **22** (33).

Since the α-methylalanine analogue **13** is slightly more potent in releasing GH than **3**, the potency enhancing 2-hydroxypropyl derivatives of **13** were examined (Table 7.3) (33). Attachment of the 2-hydroxypropyl side chains to **13** yielded diastereomers **38** and **39**, which resulted in a surprising decrease in GH releasing activity. Since analogues **38** and **39** are now one methylene shorter than **28** and **29**, the 3-hydroxybutyl analogues **40** and **41** were prepared and found also not to be potency enhancing relative to unsubstituted derivative **13**. Subtle conformational and binding interactions are most likely evolved in the interaction of this class of GH secretagogues with its receptor. However, a receptor binding assay is not available at this time to more accurately examine these interesting structure-activity relationships.

While the basic amine in the benzolactam secretagogues is critical for GH releasing activity, it was hoped that the tetrazole moiety in L-692,429 could be replaced with non–charge bearing functionality, which would alter its physicochemical properties, i.e., it would no longer be zwitterionic. The earlier SAR established for GHRP-6 indicated that a negatively charged functionality was not required for potent GH releasing activity. In fact, the C-terminus of GHRPs must be aminated for significant GH releasing activity (13). A representative subset of the modifications that have been made at the acidic 2′-position of the biphenyl moiety in L-692,429 is depicted in Table 7.4 (35, 36). The tetrazole function in **3** is approximately 50-fold more

TABLE 7.3. L-692,429 Structure-activity relationships: α-methylalanine modifications.

No.	R	$ED_{50}(\mu M)$
13	—H	0.03
38	—CH₂CHCH₃ (OH)	0.30
39	—CH₂CHCH₃ (OH)	0.70
40	—CH₂CH₂CHCH₃ (OH)	0.10
41	—CH₂CH₂CHCH₃ (OH)	0.05

active than its carboxylic acid (**42**) or its acylsulfonamide (**44**) counterparts, despite their similar acidity. Since neutral ester **43** or the simple unsubstituted biphenyl **45** is equal in GH releasing activity to acidic **42** or **44**, the obvious conclusion was that the potency enhancement of the tetrazole in this class of GH secretagogues is probably due to other effects than its anionic nature. It is possible that the positive contributions of the tetrazole could be due to its aromatic character or its ability to form a hydrogen bond.

Simple methylation of the tetrazole yielded the N-1 and N-2 isomers **46** and **47**, which exhibited only a five- and eightfold loss in potency, respectively, relative to the parent tetrazole **3**. Comparison of these tetrazole derivatives to the simple phenyl derivative **48** and other heterocycles (not depicted), strongly suggested that the potency enhancing effects associated with the 2'-tetrazole was not attributable to its aromatic character, but was most likely due to its ability to form a hydrogen bond with its receptor. Consequently, other functionalities that could donate or accept a hydrogen bond with the receptor were investigated. While the sulfonamide derivative **49** was significantly less active than tetrazole derivative **3**, the simple carboxamide variant **50** was found to be approximately equipotent with **3** (35). Thus, the critical nature of the 2'-substituent for this class of GH

TABLE 7.4. L-692,429 Structure-activity relationships: tetrazole replacements.

No.	R	$ED_{50}(\mu M)$	No.	R	$ED50(\mu M)$
3	(tetrazole, N—N / N / H)	0.06	47	(N-methyl tetrazole, N=N / N—N—CH₃)	0.3
42	—CO₂H	3	48	(phenyl)	3
43	—CO₂CH₃	2[a]			
44	—SO₂NHCOPh	3	49	—SO₂NH₂	3
45	—H	3	50	—CONH₂	0.08
46	(N-methyl tetrazole, N—N / N / CH₃)	0.5	51	—CONHCH₂CH₃	0.09
			52	—CONHCH₂Ph	0.9
			53	—CON(CH₂CH₃)₂	2

[a] Racemic mixture.

secretagogues is not due to its anionic charge or aromatic character, but to its ability to form a hydrogen bond with its receptor. As illustrated in Table 7.4, the carboxamide can be mono-substituted (e.g., **51**) with small alkyl groups without loss of GH releasing activity; however, larger monoalkyl groups (**52**) or bis-alkyl substitution (**53**) of the carboxamide group results in a decrease in GH releasing activity (36). These results suggest that the carboxamide occupies a site on the receptor of limited steric volume. While the carboxamide function is capable of either donating or accepting a hydrogen bond, the results for tetrazole derivatives **46** and **47** suggest, assuming there is no change in the mode of binding between these two subclasses of GH secretagogues, that the 2′-substituents accept a hydrogen bond from its receptor.

With the discovery of a suitable neutral surrogate for the tetrazole in **3**, it was of interest to see whether or not the potency enhancing 2-hydroxypropyl substituent found in **29** would carry over to the carboxamide series (Fig. 7.6). It was quite gratifying to observe that the 2-hydroxypropyl analogue **54** (ED_{50} = 4 nM) also exhibited greater GH releasing activity (20-fold) than its unsubstituted parent **50** (ED_{50} = 80 nM) (36). The related diol **55** (L-700,653) is also very potent in vitro (ED_{50} = 3 nM) and has recently been reported to be very active in releasing GH in swine (37).

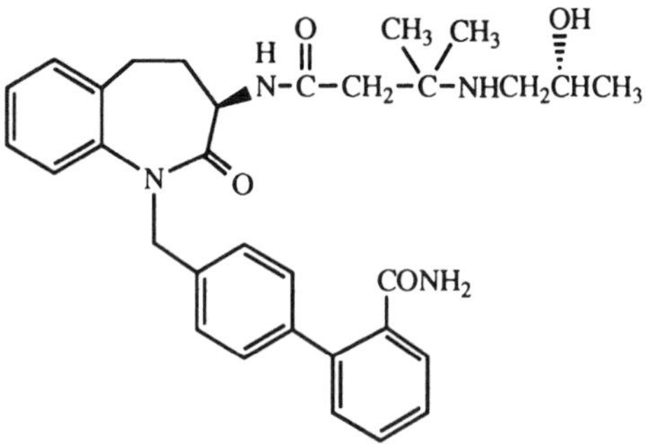

54

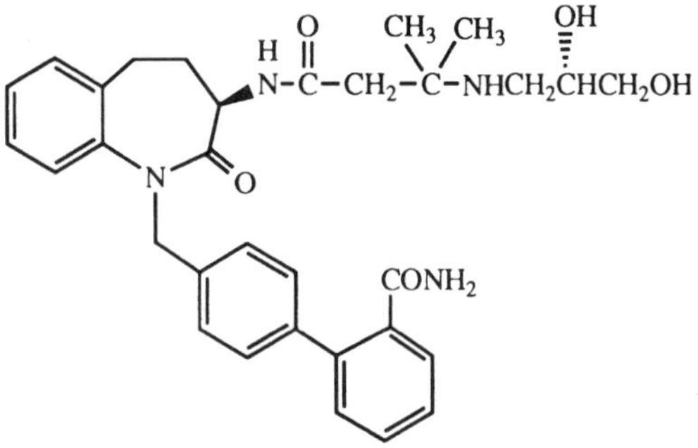

55

FIGURE 7.6. Carboxamide derivatives with potency-enhancing amino substituents.

Conclusion

The discovery of the novel nonpeptidyl class of benzolactam GH secreta-gogues has stimulated considerable scientific and clinical interest. L-692,429, the prototype nonpeptidyl GH secretagogue, has been shown to be a specific, well-tolerated GH secretagogue in humans and has demonstrated the potential of a nonpeptidyl secretagogue for use in GH therapy. The structure-activity relationships established for the benzolactam class of GH secretagogues have shown that potency can be significantly improved over L-692,429 in addition to major changes in its physicochemical properties.

Acknowledgments. In addition to their contributions in the many references cited in this chapter, special recognition for the work described here on the benzolactam GH secretagogues is due to William Schoen, Robert DeVita, Dong Ok, Kristine Prendergast, Michael Fisher, Kang Cheng, and Roy Smith.

References

1. Strobl JS, Thomas MJ. Human growth hormone. Pharmacol Rev 1994;46:1–34.
2. Chipman JJ. Recent advances in hGH clinical research. J Pediatr Endocrinol 1993;6:325–8.
3. Jorgensen JOL, Christiansen JS. Growth hormone therapy. Brave new senescence: GH in adults. Nature 1993;341:1247–8.
4. Sacca L, Cittadini A, Fazio S. Growth hormone and the heart. Endocr Rev 1994;15:555–73.
5. Rudman D, Feller AG, Nagraj HS, Gergans GA, Lalitha PY, Goldberg AF, et al. Effects of human growth hormone in men over 60 years old. N Engl J Med 1990;323:1–6.
6. Schoen WR, Wyvratt MJ, Smith RG. Growth hormone secretagogues. In: Bristol JA, ed. Annual reports in medicinal chemistry. New York: Academic Press, 1993;28:177–86.
7. Thorner MO. On the discovery of growth hormone-releasing hormone. Acta Pediatr (Suppl) 1993;388:2–7.
8. Low LCK. Growth hormone-releasing hormone: clinical studies and therapeutic aspects. Neuroendocrinology 1991;53(suppl 1):37–40.
9. Bowers CY, Chang J, Momany F, Folkers K. Effects of the enkephalins and enkephalin analogs on release of pituitary hormones in vitro. In: MacIntyne I, ed. Molecular endocrinology. Elsevier/North Holland Biomedical Press, 1977;287–92.
10. Codd EE, Shu AY, Walker RF. Binding of a growth hormone releasing hexapeptide to specific hypothalamic and pituitary binding sites. Neuropharmacology 1989;28:1139–44.
11. Bowers CY, Sartor AO, Reynolds GA, Badger TM. On the actions of the growth hormone-releasing hexapeptide, GHRP. Endocrinology 1991;128:2027–35.
12. Hickey GJ, Drisko J, Faidley T, Chang C, Anderson LL, Nicolich S, et al. Mediation by the central nervous system is critical to the in vivo activity of the GH secretagogue L-692,585. J Endocrinol 1996;148:371–80.
13. Bowers CY, Momany F, Reynolds GA, Chang D, Hong A, Chang K. Structure activity relationships of a synthetic pentapeptide that specifically releases growth hormone in vitro. Endocrinology 1980;106:663–7.
14. Momany FA, Bowers CY, Reynolds GA, Chang D, Hong A, Newlander K. Design, synthesis, and biological activity of peptides which release growth hormone in vitro. Endocrinology 1981;108:31–9.
15. Momany FA, Bowers CY, Reynolds GA, Hong A, Newlander K. Conformational energy studies and in vitro and in vivo activity data on growth hormone-releasing peptides. Endocrinology 1984;114:1531–6.
16. Bowers CY. GH releasing peptides—structure and kinetics. J Pediatr Endocrinol 1993;6:21–31.
17. Horwell DC. Kappa opioid analgesics. Drugs Future 1988;13:1061–71.
18. Smith RG, Cheng K, Schoen WR, Pong S-S, Hickey G, Jacks T, et al. A nonpeptidyl growth hormone secretagogue. Science 1993;260:1640–3.
19. Schoen WR, Pisano JM, Prendergast K, Wyvratt MJ, Fisher MH, Cheng K, et al. A novel 3-substituted benzazepinone growth hormone secretagogue (L-692,429). J Med Chem 1994;37:897–906.

20. Cheng K, Chan WW-S, Butler B, Wei L, Schoen WR, Wyvratt MJ, et al. Stimulation of growth hormone release from rat primary pituitary cells by L-692,429, a novel non-peptidyl growth hormone secretagogue. Endocrinology 1993;132:2729–31.

21. Hickey G, Jacks J, Judith F, Taylor J, Schoen WR, Krupa D, et al. Efficacy and specificity of L-692,429, a novel nonpeptide growth hormone secretagogue in beagles. Endocrinology 1994;134:695–701.

22. Bowers CY. Editorial: on a peptidomimetic growth hormone-releasing peptide. J Clin Endocrinol Metab 1994;79:940–2.

23. Gertz BJ, Barrett JS, Eisenhandler R, Krupa DA, Wittreich JM, Seibold JR, et al. Growth hormone response in man to L-692,429, a novel nonpeptide mimic of growth hormone releasing peptide. J Clin Endocrinol Metab 1993;77:1393–7.

24. Aloi JA, Gertz BJ, Hartman ML, Huhn WC, Pezzoli SS, Wittreich JM, et al. Neuroendocrine responses to a novel growth hormone secretagogue, L-692,429, in healthy older subjects. J Clin Endocrinol Metab 1994;79:943–9.

25. Gertz BJ, Sciberras DG, Yogendran L, Christie K, Bador K, Krupa D, et al. L-692,429, a nonpeptide growth hormone (GH) secretagogue, reverses glucocorticoid suppression of GH secretion. J Clin Endocrinol Metab 1994;79:745–9.

26. Kearsley S, Smith G. An alternative method for the alignment of molecular structures: maximizing electrostatic and steric overlap. Tetrahedron Comput Methodol 1992;3:615–33.

27. Cheng K, Chan WW-S, Barreto A, Convey EM, Smith RG. The synergistic effects of His-D-Trp-Ala-Trp-D-Phe-Lys-NH$_2$ on growth hormone(GH)-releasing factor stimulated GH release and intracellular adenosine-3',5'-monophosphate accumulation in rat primary pituitary cell culture. Endocrinology 1989;124:2791–8.

28. Parsons WH, Davidson JL, Taub D, Aster SD, Thorsett ED, Patchett AA, et al. Benzolactams. A new class of converting enzyme inhibitors. Biochem Biophys Res Commun 1983;117:108–13.

29. Watthey JWH, Stanton JL, Desai M, Babiarz JE, Finn BM. Synthesis and biological properties of (carboxyalkyl)amino-substituted bicyclic lactam inhibitors of angiotensin converting enzyme. J Med Chem 1985;28:1511–6.

30. Parsons WH, Patchett AA, Holloway MK, Smith GM, Davidson JL, Lotti VJ, et al. Cholecystokinin antagonists. Synthesis and biological evaluation of 3-substituted benzolactams. J Med Chem 1989;32:1681–5.

31. Bock MG, DiPardo RM, Veber DF, Chang RSL, Lotti VJ, Freedman SB, et al. Benzolactams as non-peptide cholecystokinin receptor ligands. BioMed Chem Lett 1993;3:871–4.

32. Schoen WR, Ok D, DeVita RJ, Pisano JM, Hodges P, Cheng K, et al. Structure activity relationships in the amino acid sidechain of L-692,429. BioMed Chem Lett 1994;4:1117–22.

33. Ok D, Schoen WR, Hodges P, DeVita RJ, Brown JE, Cheng K, et al. Structure activity relationships of the non-peptidyl growth hormone secretagogue L-692,429. BioMed Chem Lett 1994;4:2709–14.

34. Jacks T, Hickey G, Judith F, Taylor J, Chen H, Krupa D, et al. Effects of acute and repeated intravenous administration of L-692,585, a novel nonpeptidyl growth hormone secretagogue, on plasma growth hormone, IGF-1, ACTH, cortisol, prolactin, insulin and thyroxine (T4) levels in beagles. J Endocrinol 1994;143:399–406.

35. DeVita RJ, Schoen WR, Ok D, Barash L, Brown JE, Fisher MH, et al. Benzolactam growth hormone secretagogues: replacements for the 2'-tetrazole moiety of L-692,429. BioMed Chem Lett 1994;4:1807–12.
36. DeVita RJ, Schoen WR, Fisher MH, Frontier AJ, Pisano JM, Wyvratt MJ, et al. Benzolactam growth hormone secretagogues: carboxamides as replacements for the 2'-tetrazole moiety of L-692,429. BioMed Chem Lett 1994;4:2249–54.
37. Chang CH, Rickes EL, Marsilio F, McGuire L, Cosgrove S, Taylor J, et al. Activity of a novel non-peptidyl growth hormone secretagogue L-700,653 in swine. Endocrinology 1995;136:1065–71.

8

A Weak Substance P Antagonist Inhibits L-692,585-Stimulated GH Release in Swine

C. CHANG, E. RICKES, L. MCGUIRE, S. COSGROVE, E. FRAZIER,
H. CHEN, K. CHENG, R. SMITH, AND G. HICKEY

Growth hormone (GH) releasing hexapeptide, His-DTrp-Ala-Trp-DPhe-LysNH2 (GHRP-6), was first described as a potent GH-specific secretagogue by Bowers, et al. (2). The first described nonpeptidyl GH secretagogue, L-692,429 (3), has been shown to elicit GH release in many species including swine (4), rats (5), beagles (6), and man (7), while the more potent analogue L-692,585 has also been shown to elicit GH release in beagles (8) and swine (9). L-692,585, a novel nonpeptidyl GH secretagogue analogue of L-692,429, was found to be approximately 20-fold more potent than L-692,429 in rat pituitary cell assays (10) and in beagles in vivo (8). The GH secretagogue activity of L-692,585 has also been reported in swine (9). A weak substance P antagonist, H_2N-D-Arg-Pro-Lys-Pro-D-Phe-Gln-D-Trp-Phe-D-Trp-Leu-Leu-NH_2 (SPA), has been shown to inhibit GHRP-6 binding in membrane preparations of rat pituitary and hypothalamus (11) and to inhibit both GHRP-6- and L-692,429-stimulated GH responses in rat pituitary cell cultures and in rats in vivo (1). The objective of this study was to determine if SPA could inhibit the GH response elicited by L-692,585 in swine.

Materials and Methods

Animals

Cross-bred castrated male swine (barrows) weighing approximately 50 kg were used in this study. The animals were housed in individual pens for the entire study and were fed a commercial diet *ad libitum* containing 18% crude protein. All animals were exposed to 12 hours of light and 12 hours of dark during each 24-hour period. Each animal was instrumented with an indwelling jugular cannulae 2 to 3 days prior to the first blood sampling.

Test Compound and Method of Administration

H_2N-D-Arg-Pro-Lys-Pro-D-Phe-Gln-D-Trp-Phe-D-Trp-Leu-Leu-NH_2 (SPA, Bachem PSUB45), a weak substance P antagonist ($IC_{50} = 20\,\mu M$) was first dissolved in 10 mM acetic acid and adjusted to a 1–2% concentration in saline solution suitable for intravenous administration (10 ml dose volume). L-692,585, a substituted benzolactam (Fig. 8.1) (10), was dissolved in saline for intravenous administration (10 ml dose volume). CP-99,994, a potent nonpeptidyl substance P antagonist ($IC_{50} = 0.2\,nM$; 12, 13), was also prepared in 10 ml saline solution. All treatments were administered through an indwelling jugular catheter.

Blood Samples

Five milliliters of blood were collected at each time point by jugular catheter into tubes containing heparin. Blood was kept on ice until centrifuged and the plasma removed. Plasma samples were stored at −80°C until assayed.

Hormone Assays

Plasma concentration of GH was determined by using a double antibody radioimmunoassay. When possible, samples to be compared directly were run in a single assay to avoid interassay variation. A homologous assay system was developed using reagents supplied by Dr. A. F. Parlow, Pituitary Hormones and Antisera Center, Harbor-UCLA Medical Center, Torrance, California (pGH for iodination and standards, AFP10864B; anti-pGH, AFP10318545). Intra-assay and interassay variations were 7% and 11%, respectively.

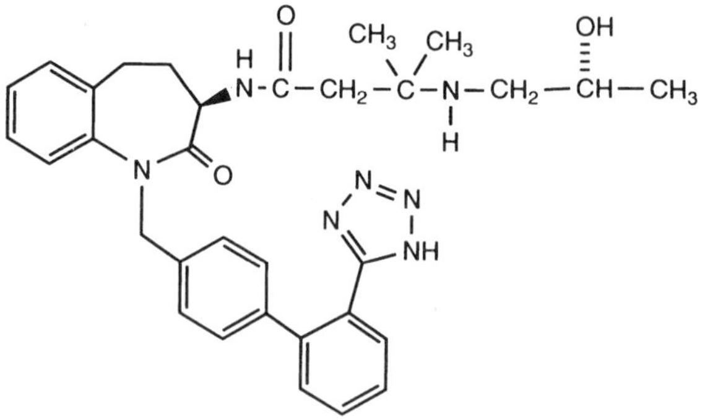

FIGURE 8.1. The structure of L-692,585.

Statistical Analysis

Several variables were used to characterize the hormone-response profile of individual pigs including peak hormone level, area under the hormone-time concentration curve and average hormone level at each postdose sampling interval. GH areas under the curves were calculated using the trapezoidal method. Where appropriate, statistical analyses of data were performed by using SAS GLM (SAS Institute, Cary, NC). Data are expressed as the geometric means $\pm$ standard error. Data from multiple groups were analyzed by analysis of variance to establish whether significant differences ($p < .05$) were present, in which case p values for pairwise differences between groups were calculated by least significant difference (LSD) tests of models appropriate for the experimental designs.

Experimental Protocols

Effect of This SPA on L-692,585-Stimulated GH Response

On day 1 and day 4 in a crossover design four male castrates received either saline or this SPA at 0.1 or 0.4 mg/kg at −10 minutes and L-692,585 at 0.1 mg/kg body weight at 0 minute. The SPA:L-692,585 molar ratio was either 0.4:1 or 1.6:1. To achieve a SPA:L-692,585 molar ratio of 4:1, animals received either saline or this SPA at 0.1 mg/kg at −10 minutes and L-692,585 at 0.01 mg/kg body weight at 0 minute.

Effect of CP-99,994 on L-692,585-Stimulated GH Response

Four animals were treated with CP-99,994 (0.2 mg/kg) at −10 minutes and L-692,585 (0.01 mg/kg) at 0 minute to determine if this potent substance P antagonist could inhibit the L-692,585 stimulated GH response. Another four animals were treated with saline at −10 minutes and L-692,585 (0.01 mg/kg) at 0 minute to serve as full response control.

Results

Effect of This SPA on L-692,585-Stimulated GH Response

The effect of this SPA on L-692,585-stimulated GH responses is summarized in Table 8.1. At a 0.4:1 (SPA:L-692,585) molar ratio, SPA reduced L-

TABLE 8.1. GH AUC and peak in response to sequential intravenous administrations of SPA ([D-Arg^1,D-Phe^5,D-Trp^{7,9},Leu^{11}]-substance P), CP-99,994, and L-692,585 at various molar ratios in male castrate swine.

Treatment[a] 1. (−10 min) 2. (0 min)	GH AUC[b]		GH peak	
	ng-min/ml	% change	ng/ml	% change
1. SPA: L-692,585 at a 0.4:1 molar ratio				
Saline + L-692,585 (0.1 mg/kg)	5,167 ± 999	—	131 ± 36	—
SPA (0.1 mg/kg) + L-692,585 (0.1 mg/kg)	3,197 ± 619	−38 (NS)	108 ± 30	−18 (NS)
2. SPA: L-692,585 at a 1.6:1 molar ratio				
Saline + L-692,585 (0.1 mg/kg)	4,915 ± 823	—	134 ± 32	—
SPA (0.4 mg/kg) + L-692,585 (0.1 mg/kg)	2,069 ± 331	−58 ($p < .01$)	59 ± 16	−56 ($p < .05$)
3. SPA: L-692,585 at a 4.0:1 molar ratio				
Saline + L-692,585 (0.01 mg/kg)	1,666 ± 331	—	58 ± 16	—
SPA (0.1 mg/kg) + L-692,585 (0.01 mg/kg)	570 ± 113	−66 ($p < .01$)	14 ± 4	−76 ($p < .01$)
4. CP-99,994:L-692,585 at a 33:1 molar ratio				
Saline + L-692,585 (0.01 mg/kg)	2,757 ± 379	—	69 ± 15	—
CP-99,994 (0.2 mg/kg) + L-692,585 (0.01 mg/kg)	2,331 ± 321	−15 (NS)	80 ± 18	+16 (NS)

[a]Sequential treatments administered at −10 (treatment #1) and 0 min (treatment #2), respectively.
[b]GH AUC for SPA:L-692,585 molar ratios of 0.4:1 and 1.6:1 based on 0 to +180 min (longer peak shoulder), while GH AUC for SPA:L-692,585 molar ratio of 4:1 based on 0 to +90 min (shorter peak shoulder).
NS, not statistically significant ($p > .05$).

692,585-stimulated GH peak and GH area under the curve (AUC) by 18% (NS) and 38% (NS), respectively (Fig. 8.2A). At a 1.6:1 (SPA:L-692,585) molar ratio, SPA reduced L-692,585-stimulated GH peak and GH AUC by 56 ($p < .05$) and 58% ($p < .01$), respectively (Fig. 8.2B). At a 4:1 (SPA:L-692,585) molar ratio, SPA reduced L-692,585-stimulated GH and GH AUC by 76% and 66% (both at $p < .01$), respectively (Fig. 8.2C).

$\longrightarrow$

FIGURE 8.2. Effect of this SPA on L-692,585-stimulated GH response in male castrate swine. Mean plasma GH concentrations in response to this SPA at 0.1 or 0.4 mg/kg dose at −10 minutes and L-692,585 at 0.01 or 0.1 mg/kg administered intravenously at 0 minute are shown in the figure. The SPA:L-692,585 molar ratios evaluated were 0.4:1 (A), 1.6:1 (B), and 4:1 (C). The values for each dose level were means from three to four animals for each time point.

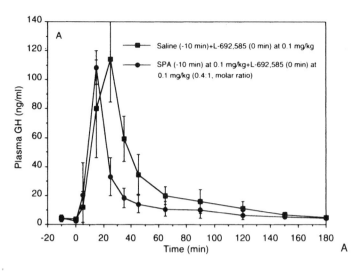

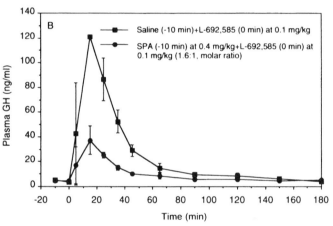

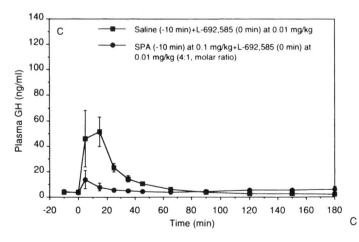

Effect of CP-99,994 on L-692,585-Stimulated GH Response

Relative to GH response following L-692,585 treatment, administration of CP-96,345 10 minutes prior to L-692,585 treatment at a molar ratio of 33:1 (CP-99,994:L-692,585) did not significantly affect the GH response (Fig. 8.3; Table 8.1).

Discussion

This SPA has been reported to be a weak substance P antagonist (IC_{50} = 20 μM) (1). Previous studies with rats indicated that this SPA significantly reduced both GHRP-6- and L-692,429-stimulated GH response, but had no effect in inhibiting GHRF-stimulated GH response. As demonstrated in this study over the SPA:L-692,585 molar ratio range of 0.4:1 to 4:1, this SPA reduced L-692,585-stimulated GH release in a dose-related manner. In contrast, at a 33:1 molar ratio (CP-99,994:L-692,585) CP-99,994, a potent substance P antagonist (IC_{50} = 0.2 nM) (12), did not inhibit L-692,585-

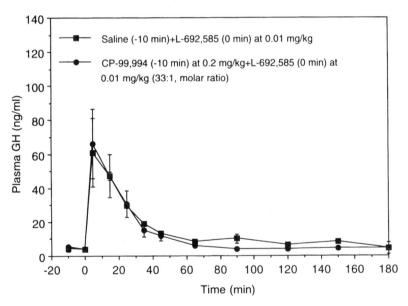

FIGURE 8.3. Effect of CP-99,994 on L-692,585-stimulated GH response in male castrate swine. Plasma GH concentrations in response to intravenous administrations of saline at −10 minutes and L-692,585 (at 0.01 mg/kg) at 0 minute, or CP-99,994 (at 0.2 mg/kg) at −10 minutes and L-692,585 treatment (at 0.01 mg/kg) at 0 min. The CP-99,994:L-692,585 molar ratio evaluated was 33:1. The values for each treatment group were means from four animals for each time point.

stimulated GH response. Results of the present study indicate that this SPA is a potent inhibitor of L-692,585-induced GH secretion. The results further suggest that the inhibiting activity of this SPA on L-692,585-induced GH response is not mediated through the substance P receptor.

Acknowledgments. We thank Drs. Matt Wyvratt, Jr. and Gary Hom for providing L-692,585 and CP-99,994, respectively, for the studies; Ms. Eva Szekely, Mr. Al Rogers, and other animal specialist staff for their assistance in the animal phase of the studies; and Mr. David Krupa and Ms. Donghui Zhang for assistance in statistical analyses.

References

1. Cheng K, Wei L, Chaung L-Y, Chan WW-S, Butler B, Pong SS, Smith RG. Substance P antagonist inhibits L-692,429- and GHRP-6-stimulated rat growth hormone release both *in vitro* and *in vivo*. Proceedings of the 76th Annual Meeting of The Endocrine Society, 1994;662(abstr).
2. Bowers CY, Momany FA, Reynolds GA, Hong A. On the in vitro and in vivo activity of a new synthetic hexapeptide that acts on the pituitary to specifically release growth hormone. Endocrinology 1984;114:1537–45.
3. Smith RG, Cheng K, Schoen WR, Pong S-S, Hickey G, Jacks T, et al. A nonpeptidyl growth hormone secretagogue. Science 1993;260:1640–3.
4. Chang CH, Rickes EL, McGuire L, Cosgrove S, Taylor JE, Schoen WR, et al. L-692,429, a novel non-peptidyl GH secretagogue, stimulates GH release in swine. J Anim Sci 1993;71(suppl 1):P134(abstr).
5. Cheng K, Chan WW-S, Butler B, Wei L, Schoen WR, Wyvratt M Jr, et al. Stimulation of growth hormone release from rat primary pituitary cells by L-692,429, a novel non-peptidyl GH secretagogue. Endocrinology 1993;132:2729–31.
6. Hickey GJ, Jacks T, Judith F, Taylor J, Schoen WR, Krupa D, et al. Efficacy and specificity of L-692,429, a novel nonpeptidyl growth hormone secretagogue, in beagles. Endocrinology 1994;134:695–701.
7. Gertz BJ, Barrett JS, Eisenhandler R, Krupa DA, Wittreich JM, Seibold JR, Schneider SH. Growth hormone response in man to L-692,429, a novel nonpeptide mimic of growth hormone releasing peptide (GHRP-6). J Clin Endocrinol Metab 1993;77:1393–7.
8. Jacks T, Hickey G, Judith F, Taylor J, Chen H, Krupa D, Feeney W, et al. Effects of acute and repeated intravenous administration of L-692,585, a novel nonpeptidyl growth hormone secretagogue, on plasma growth hormone, ACTH, cortisol, prolactin, thyroxin (T4), insulin and IGF-I levels in beagles. J Endocrinol 1994;143:399–406.
9. Hickey GJ, Baumhover J, Faidley T, Chang C, Anderson LL, Nicolich S, et al. Effect of hypothalamo-pituitary stalk transection in the pig on GH secretory activity of L-692,585. Proceedings of the 76th Annual Meeting of The Endocrine Society, 1994;366(abstr).

10. Schoen WR, Ok D, DeVita RJ, Pisano JM, Hodges P, Cheng K, et al. Structure-activity relationships in the amino acid sidechain of L-692,429. BioMed Chem Lett 1994;4:1117–22.
11. Sethumadhavan K, Veeraragavan K, Bowers CY. Demonstration and characterization of the specific binding of growth hormone-releasing peptide to rat anterior pituitary and hypothalamic membranes. Biochem Biophys Res Commun 1991;178:31–7.
12. Desai MC, Lefkowitz SL, Thadeio PF, Longo KP, Snider RM. Discovery of a potent substance P antagonist: recognition of the key molecular determinant. J Med Chem 1992;35:4911–3.
13. Desai MC, Lefkowitz SL. Conformationally restricted analogs of CP-99,994: synthesis of a spirocyclic amine. 1993;3(10):2083–6.

9

Growth Hormone Releasing Hormone Receptor in Human Breast Cancer Cell Line MCF-7

Grégoire Prévost, Nathalie Veber, Lucien Israël, and Philippe Planchon

Growth hormone secretion from the pituitary is regulated by somatostatin (SS) and growth hormone releasing hormone (GHRH) (1, 2). Interestingly, SS actions via specific receptor subtypes are widely described in peripheral normal and tumoral tissues (3, 4). In contrast, the synthesis of GHRH and its actions in peripheral tissues are poorly reported (5–7). In addition, the recently cloned GHRH receptors having seven membrane-spanning domains were only expressed in the pituitary, thus reinforcing an exclusive pituitary activity (8, 9). However, the handling of GHRH is technically challenging due to the very "sticky" GHRH and could lead to doubtful data, particularly when the receptor level is low.

The human peripheral cell line MCF-7 established from a human breast cancer possessed functional receptors for several hypothalamic peptides such as SS (10), gastrin releasing peptide (11), or luteinizing hormone releasing hormone (12). Here, to assess whether human GHRH (hGHRH) might also affect MCF7 cells, we studied its effects on adenosine $3',5'$-cyclic monophosphate (cAMP) production and then analyzed the presence of GHRH receptor by one modified cross-linking assay. Taken together, these data confirm that the role of GHRH might be extended to extrapituitary sites.

Materials and Methods

Radioimmunoassay of Intracellular cAMP

MCF-7 cells were cultured following the American Type Culture Collection instructions and plated at 2.10^5 cells/well in a 24-well plate (Falcon). After 48 hours, cells were treated for 30 min as indicated in Figure 9.1 and then

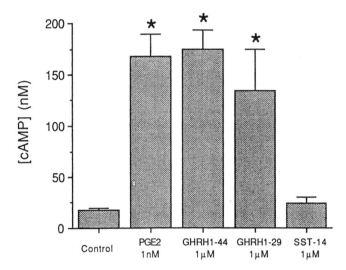

FIGURE 9.1. GHRH increases intracellular cAMP concentration in human breast cancer cell line MCF-7. Cells were treated with hGHRH(1-44)NH$_2$ (1 μM), hGHRH (1-29)NH$_2$ (1 μM), somatostatin-14 (1 μM), or PGE$_2$ (1 nM), and the cAMP concentration was measured by radioimmunoassay. Values are the mean of triplicate incubations of one representative experiment. Asterisk indicates a significant effect.

washed twice with 25 mM ice-cold N-[2-hydroxyethyl] piperazine-N'-[2-ethanesulfonic acid] buffer (HEPES, Gibco) followed by the addition of 0.1 N HCl (4°C), scraped, sonicated, and centrifuged (7,000 g for 15 min at 4°C). One hundred microliters of each supernatant was mixed with 4 μl of 3 M sodium acetate (Merck). The cAMP concentration was measured by radioimmunoassay (RIA-NEN cAMP assay kit, NEN-Dupont). Values are the mean of triplicate incubations. Statistical analyses were performed with Scheffe F-test. An effect was significant when the F-ratio presented a $p < .01$.

Chemical Cross-Linking Assay

Cells were twice rinsed with PBS, scraped, and pelleted. The pellet was resuspended at 1 g in 1 ml of buffer: 20% glycerol, 20 mM Tris-HCl pH 7.4, 1 mM phenylmethanesulfonyl fluoride (PMSF), and frozen in liquid nitrogen to break the cells. Membranes were isolated by two successive differential centrifugations (800 g, 10 min; 50,000 g, 30 min). Protein concentration was determined by a colorimetric method (Sigma kit) and then diluted at 1 μg/μl in binding buffer: 100 mM Tris-HCl pH 7.4, 5 mM MgCl$_2$, 1 mM PMSF. Fifty microliters of this solution and 100 μl of the labeled GHRH resuspended in the same buffer were incubated at 4°C for 30 min. Then, the cross-linking agent, ethylene glycol-bis (succinimidyl succinate) (Sigma,

France) (100 μM) was directly added for 15 min. This step was stopped by addition of 50 μl of loading buffer: SDS 10%, 125 mM Tris-HCl pH 7.4, 50% glycerol, 20% β-mercaptoethanol, 0.025% bromophenol blue. Proteins were heated at 95°C for 3 min and separated by electrophoresis on 12.5% SDS polyacrylamide gel. Gels were dried and autoradiographed overnight at −80°C.

Results

MCF-7 cells were treated in vitro by hGHRH(1-44)NH$_2$ and hGHRH(1-29)NH$_2$ (the minimal sequence required to stimulate pituitary GH secretion) (1) and cell extracts were assayed for intracellular cAMP content. Figure 9.1 shows that both hGHRHs (1 μM) enhanced the cAMP level in MCF-7 on the same order (10-fold over control level) ($p < .01$). A similar stimulation was obtained with the prostaglandin (PG)E$_2$ used as a positive inductor, whereas somatostatin-14 was without significant effect in this assay (p = .12) (13).

The ability of GHRH to enhance cAMP concentration prompted us to search for specific receptors in MCF-7 membranes. To limit the binding assay problems, (a) our binding assay was carried out in a unique tube without differential centrifugation, (b) following the binding period, the putative radiolabeled GHRH/receptor complexes were covalently linked by the homobifunctional agent ethylene glycol-bis (succinimidyl succinate) (EGS), and (c) all GHRH molecules (free and receptor-bound) were recovered from the tube by the final addition of detergent (SDS) before their separation by gel electrophoresis. Figure 9.2 shows that a low concentration (0.15 nM) of the labeled hGHRH(1-44)NH$_2$ allowed the detection of one major 23-kd complex and one minor 28-kd complex in MCF-7 membrane proteins (lane 1). Addition of unlabeled hGHRH(1-44)NH$_2$ or GHRH(1-29)NH$_2$ suppressed only the 23-kd complex (lanes 2 and 3, respectively), whereas unrelated peptides such as somatostatin-14 (lane 4), gastrin-releasing peptide, and growth hormone (GH) secretagogue hexapeptide did not (data not shown). These data demonstrated the specificity of the 23-kd complex. Thus, due to the presence of the 5.18-kd ligand, the apparent size of this receptor was estimated as 18-kd. In addition, the 23-kd complex formation was not sensitive to the presence of the nonhydrolyzable guanosine triphosphate (GTP) analogue: GTP gamma-S (lane 5), suggesting the absence of G-protein coupling. Conversely, the 23-kd complex was specifically suppressed in the presence of ethyleneglycoltetraacetic acid (EGTA) (lane 6) or ethylenediaminetetraacetic acid (EDTA) (data not shown), suggesting the need for divalent ions to form this binding. The efficiency of the cross-linking agent was preserved in the presence of the ion chelator as suggested by the formation of similar SS cross-linked complexes in the absence or the presence of ion chelators (data not shown).

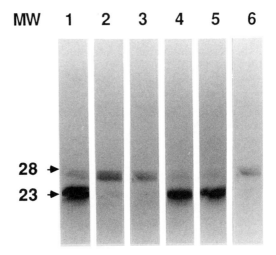

FIGURE 9.2. This autoradiogram shows one specific complex of 23 kd formed between the radioiodinated hGHRH(1-44)NH$_2$ and MCF-7 membrane proteins. The cross-linking assay was assessed with ^{125}I-tyr^{10}-GHRH(1-44) (10^5 dpm, 2000 Ci/mmol, Amersham, France) and unlabeled peptides (1 μM): hGHRH(1-44)NH$_2$ (lane 2), hGHRH(1-29)NH$_2$ (lane 3), somatostatin-14 (lane 4), the nonhydrolyzable nucleotide GTP gamma-S (30 μM) (lane 5), or the divalent ion chelator EGTA (1 mM) (lane 6). MW, molecular weight (kd).

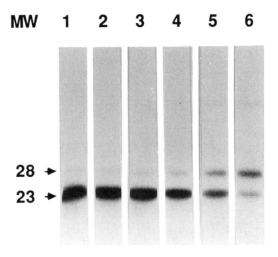

FIGURE 9.3. Displacement of the binding of ^{125}I-labeled GHRH to MCF-7 breast cancer cells by unlabeled GHRH. This cross-linking assay was carried out with a constant concentration of the labeled ligand (0.15 nM) mixed with 50 μg of MCF-7 proteins in the presence of increasing concentrations of cold human GHRH(1-44)NH$_2$ (lanes 1–6; 0, 10 nM, 50 nM, 100 nM, 500 nM, and 1 μM, respectively). MW, molecular weight (kd).

TABLE 9.1. Distribution of GHRH cross-linked complexes by comparative densitometry of various cells.[a]

Tissues	GHRH Cross-linked complex amount compared to the one of MCF-7
Cell lines	
MCF-7 (human breast tumoral epithelial cells)	100
MCF-7ras (human breast tumoral epithelial cells)	130
MG3 (human breast fibroblast)	69
T (human breast fibroblast)	64
NMSF (human breast fibroblast)	98
NCI-H69 (human small cell lung cancer)	82
IGR-N-835 (human neuroblastoma)	196
NIH-3T3 (mouse fibroblast)	65
Surgical biopsies	
Rat anterior pituitary	101
Human anterior pituitary	99
Human pituitary adenoma	100
Human digestive carcinoid (patient 1)	177
Human digestive carcinoid (patient 2)	158
Prokaryote cells	
Escherichia coli C600	0

[a] Surgical biopsy was first frozen in liquid nitrogen, pulverized with a hammer, and then treated like a cell pellet as described in Figure 9.3.

Figure 9.3 shows that increasing amounts of unlabeled hGHRH(1-44)NH_2 progressively reduced the formation of the 23-kd complex in MCF-7. The semiquantitative densitometry of this representative autoradiogram showed that the unlabeled GHRH concentration of 125 nM suppressed 50% of this specific binding in these conditions, demonstrating the high affinity of this receptor (data not shown).

The study of the 23-kd complex was extended to other cell types with this assay coupled to a semiquantitative densitometry (Table 9.1). To compare the 23-kd complex amounts present in the tested tissues, we arbitrarily fixed the value 100 for MCF-7 content. The values ranged from 0 for prokaryote cells (*Escherichia coli*) to 300 for the neuroblastoma cell line IGR-N-835.

Discussion

This study shows that (a) GHRH stimulates the intracellular cAMP production of the human breast cancer cell line MCF-7, (b) GHRH specifically binds one membrane protein of MCF-7, and (c) this GHRH binding protein can be detected in other human tissues.

The cAMP production was stimulated by hGHRH(1-44) in MCF-7 cells. This pathway of activation was largely described in the pituitary via specific receptors. In addition, GHRH(1-44) was also able to increase the cAMP

production in the human peripheral kidney 293 cells, which were transfected by the cloned pituitary GHRH receptor (100-fold compared with the
basal level with 1 μM) (8). This latter experiment suggested that one
GHRH receptor can be efficiently coupled to one G protein activating one
adenylyl cyclase in peripheral cells. Using a photoaffinity cross-linking assay with GHRH(1-32)NH$_2$ in these cells and in the ovine pituitary, Gaylinn
et al. (14) reported one complex with a size of 55 kd. In our study, the
receptor was characterized by a size of 18 kd. However, in our assay, even
a 100-fold increase of labeled GHRH(1-44)NH$_2$ concentration has not led
to the detection of other specific complexes in peripheral tissues as well as
in pituitary samples (data not shown). The different methodologies including ligands, cross-linking agents, and tissues might be responsible of this size
discrepancy. However, this size of 18 kd appeared too low to be compatible
with other G protein–coupled GHRH receptors (8, 9, 14). In addition,
the 23-kd complex formation was not sensitive to the presence of the
nonhydrolyzable GTP analogue, GTP gamma-S (lane 5), suggesting the
absence of G-protein coupling. The use of the divalent ion chelator was
required for binding experiments of the pituitary GHRH receptor (8, 9, 14).
Conversely, in our assay, the 23-kd complex was specifically suppressed in
its presence, suggesting the need for divalent ions to form this binding. The
molecular and pharmacologic characterizations of these complexes must be
continued in further research. Thus, our binding assay can be used to
develop new selective peptides for GHRH receptor characterization. The
more efficient of these molecules might directly act in peripheral tissues in
several clinical peptide-based therapies (15, 16).

Despite the differences of size and amino acid sequences reported for rat
and human GHRH (16), similar complexes were labeled by hGHRH in
tissues from these species, indicating that this GHRH target was well conserved. Furthermore, receptors for SS-14 can be detected in these tested
tissues, except in the prokaryote cells (3, 17 and data not shown), which
demonstrate the simultaneous presence of both main GH secretion modu-
lating receptors. In addition, the GHRH receptor pattern was not modified
in regard to the physiopathologic state of the tissues when normal and
tumoral tissues were compared. The role of GHRH, well described in
pituitary, remains to be fully characterized in peripheral tissues. Since
GHRH, as well as SS, are present in extrapituitary sites such as pancreas
(16), breast (10), placenta (19), or testis (6), the locally synthesized peptides
could regulate various functions in physiopathologic conditions.

Acknowledgments. Our work was supported by the French Association de
Recherche sur le Cancer and Cis-Bio-International laboratories (Les Ulis,
France). We thank Prof. Jean-Luc Moretti for helpful discussion and Dr.
Peter Eden in preparing the manuscript.

References

1. Guillemin R, Brazeau P, Bohlen P, Esch F, Ling N, Wehrenberg W. Growth hormone releasing factor from a human pancreatic tumor that caused acromegaly. Science 1982;218:585–7.
2. Brazeau P, Vale W, Burgas R, Ling N, Butscher M, Rivier J, Guillemin R. Hypothalamic polypeptides that inhibit the secretion of immunoreactive pituitary growth hormone. Science 1973;179:77–9.
3. Prévost G, Lanson M, Thomas F, Veber N, Gonzalez W, Beaupain R, Starzec A, Bogden AE. Molecular heterogeneity of somatostatin analog BIM23014C receptors in human breast carcinoma cells using the chemical cross-linking assay. Cancer Res 1992;52:843–50.
4. Lamberts SWJ, Krenning E, Reubi JC. The role of somatostatin and its analogs in the diagnosis and treatments of tumors. Endocr Rev 1991;12:450–81.
5. Weigent DA, Blalock JE. The production and function of growth hormone in the immune system. In: Growth hormone II: basic and clinical aspects, Serono Symposia USA, Norwell, MA, 1994 (abstr.): 1992:10.
6. Srivastava CH, Breyer PR, Rothorock JK, Peredo MJ, Pescovitz OH. A new target for growth hormone releasing hormone action in rat: the Sertoli cell. Endocrinology 1993;133:1478–81.
7. Moretti C, Bagnato A, Solan N, Frajesse G, Catt KJ. Receptor-mediated actions of growth hormone releasing factor on granulosa cell differentiation. Endocrinology 1990;127:2117–226.
8. Mayo KE. Molecular cloning and expression of a pituitary-specific receptor for growth hormone. Mol Endocrinol 1992;6:1734–44.
9. Gaylinn BD, Harisson JK, Zysk JR, Lyons CE, Lynch KR, Thorner MO. Molecular cloning and expression of a human anterior pituitary receptor for growth hormone releasing hormone. Mol Endocrinol 1993;7:77–84.
10. Prévost G, Provost P, Salle V, Lanson M, Thomas F. A cross-linking assay allows the detection of receptors for the somatostatin analogue Lanreotide in human breast tumours. Eur J Cancer 1993;29:1589–92.
11. Patel KV, Schrey MP. Activation of inositol phospholipid signaling and Ca efflux in human breast cancer cells by bombesin. Cancer Res 1990;50:235–9.
12. Miller WR, Scott WN, Morris R, Fraser HM, Sharpe RM. Growth of human breast breast cancer cells inhibited by a luteinizing hormone-releasing hormone agonist. Nature 1985;313:231–3.
13. Prévost G, Foehrlé E, Thomas F, Pihan I, Veber N, Starzec A, Israel L. Growth of human breast cancer cells inhibited by the somatostatin analog BIM23014C (Somatuline). Endocrinology 1991;128:323–9.
14. Gaylinn B, Lyons C, Zysk J, Clarke I. Thorner, M. Photoaffinity cross-linking to the pituitary receptor for growth hormone releasing factor. Endocrinology 1994;135:950–5.
15. Thorner MO, Spiess J, Vance ML, Rogol AD, Kaiser DL, Webster JD, Rivier J, Borges JL, Bloom SR, Cronin MJ, Evans WS, Macleod RM, Vale W. Human pancreatic growth hormone releasing factor selectively stimulates growth hormone secretion in man. Lancet 1983;1:24–8.

16. Bowers CC, Sartor AO, Reynolds GA, Bagder TM. On the actions of the growth hormone releasing hexapeptide GHRP. Endocrinology 1991;128: 2027–35.
17. Prévost G, Bourgeois Y, Mormont C, Lerrant Y, Véber N, Poupon MF, Thomas F. Characterization of somatostatin receptors and growth inhibition by the somatostatin analogue BIM23014 in small cell lung carcinoma. Life Sci 1994;55:155–62.
18. Ciocca DR, Puy LA, Fasoli LC, Tello O, Aznar JC, Gargo FE, Papa IS, Sonego R. Corticotropin releasing hormone, luteinizing hormone releasing hormone, growth hormone releasing hormone and somatostatin like immunoreactivities in biopsies from breast cancer patients. Breast Cancer Res Treat 1990;15: 175–84.
19. Gonzalez-Crespo S, Boronat A. Expression of the rat growth hormone releasing hormone gene in placenta is directed by an alternative promoter. Proc Natl Acad Sci USA 1991;88:8749–53.

Part III

Cellular and Molecular Properties of Growth Hormone Secretagogues

10

Cellular Physiology of Growth Hormone Releasing Hormone

Lawrence A. Frohman

This chapter reviews the mechanisms involved in the normal transduction of the signal generated by binding of growth hormone (GH) releasing hormone (GHRH) to its receptor on the somatotropes in relation to GH secretion, GH synthesis, and cellular proliferation. This chapter also discusses the consequences of altered signal transduction, citing models of both genetic and transgenic origin with both decreased and increased activity of the system.

Normal Transmission of the GHRH Signal

GHRH signal transduction is initiated by its binding to a specific receptor on the somatotroph cell membrane. The receptor is a single chain of 462 amino acids, and is a member of the G-protein superfamily of receptors containing seven transmembrane-spanning domains (1). The receptor is coupled to at least one heterotrimeric G protein that, in turn, is linked to effector proteins. Considerable evidence supports the activation and interplay of at least four intracellular pathways of signal transduction in mediating the responses to GHRH: (1) the adenylyl cyclase–adenosine 3′, 5′-cyclic monophosphate (cAMP)–protein kinase A system, (2) the Ca^{2+}-calmodulin system, (3) the inositol phosphate–diacylglycerol–protein kinase C system, and (4) the arachidonic acid–eicosanoid pathways (Fig. 10.1).

GH Secretion

The participation of cAMP as a second messenger in the stimulation of GH secretion preceded the isolation of GHRH by more than a decade and was based on studies with phosphodiesterase inhibitors and cAMP analogues (2). GHRH-induced GH release is coupled to a marked dose-dependent stimulation of adenylyl cyclase activity and cAMP production in

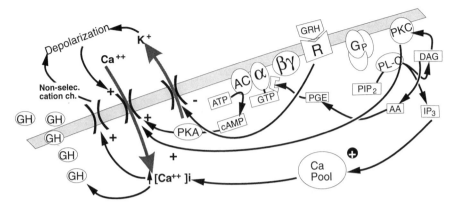

FIGURE 10.1. Model of the potential intracellular signal transduction pathways involved in the mechanism of action of growth hormone releasing hormone (GHRH). R, GHRH receptor; α and βγ, α, β, and γ subunits of G_s, the GHRH receptor-linked GTP binding protein that activates adenylate cyclase (AC); Gp, GHRH receptor-linked GTP binding protein that activates phospholipase C (PLC); cAMP, adenosine 3′,5′-cyclic monophosphate; PKA, protein kinase A; DAG, 1,2-diacylglycerol; PKC, protein kinase C; IP_3, 1,4,5-inositol phosphate; PIP_2, phosphatidylinositol diphosphate; Ca^{2+}, calcium; AA, arachidonic acid; PGE, prostaglandin E_1 and E_2.

somatotrophs, with levels of cAMP production detectable within 1 min after exposure to GHRH. Peak concentrations are reached within 15 min, after which levels decline. As occurs in other tissues, binding of GHRH to its receptor activates a stimulatory G protein (G_s). The membrane bound G_s protein, in its inactive state, consists of three subunits closely bound to one another and to guanosine diphosphate (GDP). Activation catalyzes the binding of guanosine triphosphate (GTP) to the α subunit, which then dissociates from the βγ complex and activates the catalytic subunit of adenylyl cyclase. In the presence of adenosine triphosphate (ATP), adenylyl cyclase catalyzes the formation of cAMP. The G_s-α subunit also contains intrinsic guanosine triphosphatase (GTPase) activity and the bound GTP is rapidly converted to GDP, followed by the reassociation of the heterotrimeric G_s complex and a return to the inactive state. cAMP activates protein kinase, which then phosphorylates a large variety of proteins, likely including ion channels. A comparison of the dose-response curves of GHRH on GH release and intracellular cAMP accumulation in primary pituitary cell cultures reveals that a half maximal GH response occurs with barely detectable increases in cAMP, and that the maximal GH response occurs prior to half maximal stimulation of cAMP levels.

Ion channels also exert a major role in GH secretion. The three families of identified ion channels—Na^+, Ca^{2+}, and K^+—have all been identified in somatotrophs and determine the electrical activity of the cells (3). Two forms of Na^+ channels are recognized, although their action is the least well understood and are probably the least important in the regulation of GH secretion. A tetradotoxin-sensitive and voltage-gated channel appears to participate in regulating the basal level of GH secretion, but is not altered by GHRH. In contrast, the tetradotoxin-insensitive channel is essential for the increase in $[Ca^{2+}]_i$ and the GH response to GHRH. It is believed to be stimulated by cAMP, presumably by phosphorylation mediated by protein kinase A. The function of the Na^+ channels is believed to be a triggering of spontaneous action potentials, leading to an influx of Ca^{2+}.

Two of the three recognized types of Ca^{2+} channels have been identified in somatotrophs and, together with Na^+ channels, are involved in the depolarizing phase of action potentials. The influx of Ca^{2+} during these action potentials leads to an increase in $[Ca^{2+}]_i$ and GH release. GHRH increases the amplitude of Ca^{2+} currents and also depolarizes the cell membrane potential in the presence of extracellular Ca^{2+}. This action is mediated by cAMP and protein kinase A and is associated with phosphorylation of the channel protein. The action of $[Ca^{2+}]_i$ is modulated by Ca^{2+} binding proteins, such as calmodulin. Binding of Ca^{2+} to calmodulin leads to an activated complex that, in turn, regulates the activity of calmodulin-dependent protein kinases. These kinases are believed to modulate secretory vesicle membranes and/or the function of peptides on the cytoplasmic surface of cell membranes.

In a resting state, most of the current across the somatotroph membrane is carried by K^+ ions. GHRH transiently decreases membrane K^+ conductance, which leads to depolarization.

In summary, GHRH increases Na^+ and Ca^{2+} currents while decreasing K^+ currents in the somatotroph. The resultant depolarization further increases Ca^{2+} currents (influx of Ca^{2+} from the extracellular environment).

Activation of the inositol phosphate-diacylglycerol-protein kinase C system is also important in the mediation of GHRH responses. Binding of GHRH to its receptor initiates the hydrolysis of phosphatidyl-biphosphate by phospholipase C through an intermediary GTP binding protein (G_p). The products of hydrolysis, diacylglycerol, and inositol triphosphate (IP_3) serve as endogenous substrates for activation of protein kinase C (PKC) and for mobilization of Ca^{2+} stores from the endoplasmic reticulum, respectively. Diacylglycerol induces translocation of PKC from the cytosol to the cell membrane and increases its affinity for Ca^{2+}. The activated PKC leads to an influx of extracellular Ca^{2+}, possibly mediated by an increase in cAMP by an increase in G_s-α activity or an inhibition of G_i activity.

Arachidonic acid metabolites of the cyclooxygenase pathway (prostaglandins E_1 and E_2 and prostacyclins) bind to specific receptors that are coupled to G_s and lead to an increase in cAMP production and GH secretion. GHRH increases PGE_2 release from pituitary cells, an effect that is blocked by the cyclooxygenase inhibitors, indomethacin, and aspirin.

GH Synthesis

In addition to its effects on GH secretion, GHRH also stimulates the synthesis of GH. This effect has been demonstrated at the level of total GH content in freshly dispersed pituitary cells and in GH-secreting cell lines (Fig. 10.2). In both cell lines and freshly dispersed cells, an increase in GH messenger RNA (mRNA) has been demonstrated (4, 5) and increased nuclear transcription rates have also been observed, indicating a true stimulation of GH mRNA synthesis (6). Using selective inhibitors of secretion and synthesis, the two processes have been shown to be independent of one another (7). The second messengers involved in the effect on synthesis are not fully understood, although cAMP plays a pivotal role, since forskolin, a direct activator of adenylyl cyclase, exerts effects on synthesis that are comparable to those of GHRH. cAMP response elements have been identified in the GH promoter and also on the Pit-I promoter, which is required for normal GH gene transcription.

Somatotroph Proliferation

GHRH also exhibits proliferative effects on the somatotroph. Thymidine incorporation into somatotrophs is increased by GHRH as is cell number in

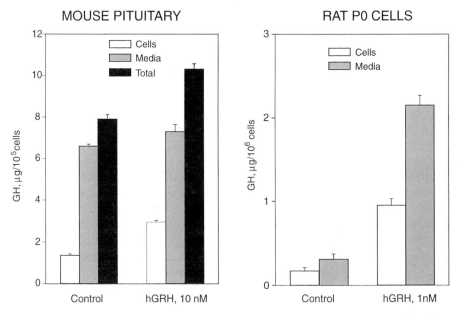

FIGURE 10.2. Stimulation of GH synthesis by 24-hour exposure to GHRH in primary cultures of mouse pituitary cells (left) and in a rat somatomammotropic cell line (right).

primary cultures of somatotrophs (8). As with secretion and synthesis, cAMP serves as a critical mediator of this effect. GHRH also stimulates *fos* gene expression, as reflected by *fos* mRNA and *fos* protein (9). Similar effects on *fos* have also been observed using forskolin as the stimulus.

Alterations in Signal Transduction

Impaired Signal Transduction

Two genetic models offer insight into deficient GHRH signaling: the *lit* mouse and the *dw* rat. The *lit* mouse, which is markedly GH deficient and decreased in body size, fails to exhibit any increase in either GHRH or cAMP in response to simulation with GHRH (10). However, responses to cholera toxin, an activator of G_S-α, forskolin, a direct stimulator of adenylyl cyclase, and to dibutyryl cAMP are completely intact. These findings led us to propose the presence of a GHRH receptor defect. This has recently been identified (11, 12) as a point mutation leading to an amino acid change in the extracellular portion of the receptor that is highly conserved in the superfamily of receptors to which the GHRH receptor belongs. We origi- nally observed that *lit* somatotrophs were also completely resistant to stimulation by GHRP-6 (13). While this was initially believed to be incon- sistent with the concept that GHRP-6 acts through a GHRH receptor- independent mechanism, the accumulating experience with other experimental and clinical models suggests that some as yet undefined effect of GHRH within the somatotrophs is essential for GHRP-6 action. The absence of such action in *lit* somatotrophs could thus explain their failure to respond to GHRP-6.

The *dw* rat exhibits a different defect in GHRH signal transduction. This animal also has reduced pituitary GH content and secretion, but does respond minimally to GHRH in vivo (14). When studied in vitro, *dw* somatotrophs exhibit a slightly impaired response to GHRH (when cor- rected for their markedly diminished GH content) but a nearly complete loss of cAMP stimulation (15). Using other probes of the signal transduc- tion system, we found intact responses to dibutyryl cAMP and forskolin, but moderately impaired responses to cholera toxin and phorbol myristic acid (PMA), an agent that mimics the action of diacylglycerol, an activator of PKC, and a nearly complete loss of response to prostaglandin E_1. These findings excluded the GHRH receptor as a site of the defect and pointed to an abnormality in G_S-α or in its interaction with adenylyl cyclase. A detailed investigation into G_S-α in *dw* rats pituitaries revealed normal levels of both mRNA and protein, normal function, as determined by cholera toxin- induced adenosine diphosphate (ADP) ribosylation, and a normal cDNA sequence (16). This leaves the possibility of an altered G_S-$\beta\gamma$ complex or a defect in the G_S-α binding site of adenylyl cyclase to be explored. Total pituitary cell number is also decreased by about 40% in *dw* rats and is entirely attributable to a marked decrease in somatotrophs (from 30% to 6% of the pituitary cell population). This defect is seen as early as fetal day 18 and persists through development (17). The overall results are consistent with a failure of normal somatotroph commitment early in gestation due to the consequence of the specific defect, which is likely to be mediated by the reduced levels of cAMP. Once somatotrophs are committed, however, they appear to have the capacity to increase in number at a normal rate. Somatotrophs from *dw* pituitaries also exhibit an impaired response to GHRP-6, to the same extent as that observed to GHRH (18). The parallelism in the impaired response to these two agents further underscores the importance of some component of GHRH action within the somatotroph (possibly cAMP generation) for normal GHRP-6 action.

Excessive Signal Transduction

hGHRH Transgenic Mice

Increased GHRH signal transduction is most dramatically demonstrated by changes in cellular proliferation. Mice expressing the mouse metallothionein (mMT)-hGHRH transgene exhibit marked increases in pituitary size (Fig. 10.3) (19). Cell number is increased and composed almost entirely of somatotrophs and somatomammotrophs. When examined at 4 months of age, the histologic pattern is that of pure hyperplasia. However, by 10 months of age, numerous foci of adenomatous change can be observed. It is still unknown whether this is a long-term consequence of GHRH stimulation alone or whether the hyperplastic cells are more susceptible to acquiring other mutation(s) that are responsible for the neoplastic transformation.

Cholera Toxin Transgenic Mice

Targeting of cholera toxin to somatotrophs using a GH promoter causes constitutive activation of G_S-α, leading to increased cAMP production. Cholera toxin-transgenic mice exhibit markedly enlarged pituitaries and, at 6 months of age, profound somatotroph hyperplasia (20). In contrast to hGHRH transgenic mice, those expressing the cholera toxin transgene do not have an increase in somatomammotrophs and do not demonstrate evidence of tumor formation. Because of the lack of information over longer time periods, it is not known whether such changes would have occurred with a longer observation period or whether they are the consequence of cAMP-independent effects of GHRH.

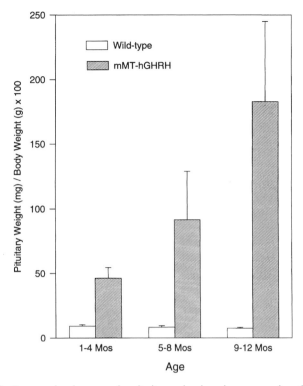

FIGURE 10.3. Progressive increase in pituitary size in mice expressing the hGHRH transgene. Shown are mean ± S.E.M from three to six mice per group. Data from ref. 19.

Human Models of Enhanced GHRH Signal Transduction

Patients with excessive secretion of GHRH, whether from ectopic sources, or from within the central nervous system, exhibit markedly increased GH secretion and somatotroph hyperplasia (21). In those patients in whom the excess GHRH secretion can be reversed (e.g., by removal of an extrapituitary tumor), pituitary size reduction has been observed (22). However, with long-standing GHRH hypersecretion, evidence for adenomatous transformation has been reported (23), as occurs in hGHRH transgenic mice.

Approximately 40% of patients with GH-secreting tumors have one of two somatic point mutations in the G_s-α gene from their tumors (24, 25). The resulting single amino acid changes occur either at the cholera toxin binding site or at a site required for the molecule's GTPase activity. Both result in a constitutively activated G_s-α, increased cAMP production, and adenoma formation. A mosaic germline mutation at the identical sites in the G_s-α gene has been described in the McCune-Albright syndrome (26). Children with this defect also develop somatotroph hyperplasia and eventually somatotroph tumor formation.

Summary

The rapid pace of research on GHRH signal transduction has led to major advances in our understanding of this process and to the identification of several disorders of impaired or enhanced activity in man and experimental animals. While there are several interrelated pathways that participate in the transduction of the GHRH signal, the role of cAMP is clearly crucial and responsible for the regulation of GH secretion and synthesis and for somatotroph proliferation. Given the many intracellular modulators of cAMP action, it is not unreasonable to postulate that genetic alterations leading to modified affinity for any one of the second messengers could result in mild to major changes in the secretion of GH. In this context, the heterogeneity of GH responses to GHRH and/or GHRP could be manifestations of subtle changes in the GHRH receptor, G proteins, protein kinase(s), ion channels, etc. As these systems are better defined, it can be anticipated that specific molecular mechanisms will be uncovered and the basis for the variations in responses clarified. It is also conceivable that future gene transfer technology may offer novel therapeutic approaches to the correction of some of these disorders.

Acknowledgments. The studies from the author's laboratory were supported by USPHS Grant DK 30667 and an award from the Bane Foundation.

References

1. Mayo KE. Molecular cloning and expression of a pituitary-specific receptor for growth hormone-releasing hormone. Mol Endocrinol 1992;6:1734–44.
2. Frohman LA, Downs TR, Chomczynski P. Regulation of growth hormone secretion. Front Neuroendocrinol 1992;13:344–405.
3. Chen C, Vincent JD, Clarke IJ. Ion channels and the signal transduction pathways in the regulation of growth hormeone secretion. TEM 1994;5: 227–33.
4. Chomczynski P, Brar A, Frohman LA. Growth hormone synthesis and secretion by a somatomammotroph cell line derived from normal adult pituitary of the rat. Endocrinology 1988;123:2275–83.
5. Gick GG, Zeytin FN, Ling NC, Esch FS, Bancroft C, Brazeau P. Growth hormone-releasing factor regulates growth hormone mRNA in primary cultures of rat pituitary cells. Proc Natl Acad Sci USA 1984;81:1553–5.

6. Barinaga M, Yamamoto G, Rivier C, Vale W, Evans R, Rosenfeld MG. Transcriptional regulation of growth hormone gene expression by growth hormone-releasing factor. Nature 1983;306:84–5.
7. Barinaga M, Bilezikjian LM, Vale WW, Rosenfeld MG, Evans RM. Independent effects of growth hormone releasing factor on growth hormone release and gene transcription. Nature 1985;314:279–81.
8. Billestrup N, Swanson LW, Vale W. Growth hormone-releasing factor stimulates proliferation of somatotroph *in vitro*. Proc Natl Acad Sci USA 1986;83:6854–7.
9. Billestrup N, Mitchell RL, Vale W, Verma IM. Growth hormone-releasing factor induces c-*fos* expression in cultured primary pituitary cells. Mol Endocrinol 1987;1:300–5.
10. Jansson J, Downs TR, Beamer WG, Frohman LA. Receptor-associated resistance to growth hormone-releasing factor in dwarf "little" mice. Science 1986;232:511–2.
11. Lin SC, Lin CR, Gukovsky I, Lusis AJ, Sawchenko PE, Rosenfeld MG. Molecular basis of the little mouse phenotype and implications for cell type-specific growth. Nature 1993;364:208–13.
12. Godfrey P, Rahal JO, Beamer WG, Copeland NG, Jenkins NA, Mayo KE. GHRH receptor of little mice contains a missense mutation in the extracellular domain that disrupts receptor function. Nat Genet 1993;4:227–32.
13. Jansson J, Downs TR, Beamer WG, Frohman LA. The dwarf "little" (lit/lit) mouse is resistant to growth hormone (GH)-releasing peptide (GH-RP-6) as well as to GH-releasing hormone (GRH). Program 68th Annual Meeting of the Endocrine Society, 1986;#397(abstr).
14. Charlton HM, Clark RG, Robinson IC, Porter-Goff AEP, Cox BS, Bugnon C, et al. Growth hormone-deficient dwarfism in the rat: a new mutation. J Endocrinol 1988;119:51–8.
15. Downs TR, Frohman LA. Evidence for a defect in growth hormone-releasing factor signal transduction in the dwarf (dw/dw) rat pituitary. Endocrinology 1991;129:58–67.
16. Zeitler PA, Downs TR, Frohman LA. (Gs-alpha MS). Impaired growth hormone releasing-hormone signal transduction in the dwarf (dw) rate is independent of a generalized defect in the stimulatory G-protein, Gs-alpha. Endocrinology 1993;133:2782–6.
17. Zeitler PA, Downs TR, Frohman LA. Development of pituitary cell types in the spontaneous dwarf (*dw*) rat: evidence for an isolated defect in somatotroph differentiation. Endocrine 1994;2:729–33.
18. Pinhas-Hamiel O, Zeitler P. Impaired response to GHRP-6 in somatotrophs from dwarf (*dw*) rats. Program 76th Annual Meeting of the Endocrinc Society 1994;#663(abstr).
19. Lloyd RV, Jin L, Chang A, Kulig E, Camper SA, Ross BD, et al. Morphological effects of hGRH gene expression on the pituitary, liver, and pancreas of MT-hGRH transgenic mice. An in situ hybridization analysis. Am J Pathol 1992;141:895–906.
20. Burton FH, Hasel KW, Bloom FE, Sutcliffe JG. Pituitary hyperplasia and gigantism in mice caused by a cholera toxin transgene. Nature 1991;350:74–7.

21. Frohman LA, Thominet JL, Szabo M. Ectopic growth hormone releasing factor syndromes. In: Raiti S, Tolman R, eds. Human growth hormone. New York: Plenum, 1986:347–60.
22. Caplan R, Koob L, Abellera RM, Pagliara AS, Kovacs K, Randall R. Cure of acromegaly by operative removal of an islet cell tumor of the pancreas. Am J Med 1978;64:874–82.
23. Ezzat S, Asa SL, Stefaneanu L, Whittom R, Smyth HS, Horvath E, et al. Somatotroph hyperplasia without pituitary adenoma associated with a longstanding GHRH-producing bronchial carcinoid tumor. J Clin Endocrinol Metab 1994;78:555–60.
24. Landis CA, Harsh G, Lyons J, Davis RL, McCormick F, Bourne HR. Clinical characteristics of acromegalic patients whose pituitary tumors contain mutant Gs protein. J Clin Endocrinol Metab 1990;71:1416–20.
25. Spada A, Arosio M, Bochicchio D, Bazzoni N, Vallar L, Bassetti M, et al. Clinical, biochemical, and morphological correlates in patients bearing growth hormone-secreting pituitary tumors with or without constitutively active adenylyl cyclase. J Clin Endocrinol Metab 1990;71:1421–6.
26. Weinstein LS, Shenker A, Gejman PV, Merino MJ, Friedman E, Spiegel AM. Activating mutations of the stimulatory G protein in the McCune-Albright syndrome. N Engl J Med 1991;325:1688–95.

11

Mechanism of Action of GHRP-6 and Nonpeptidyl Growth Hormone Secretagogues

Roy G. Smith, Kang Cheng, Sheng-Shung Pong, Reid Leonard,
Charles J. Cohen, Joseph P. Arena, Gerard J. Hickey, Ching H.
Chang, Tom Jacks, Jennifer Drisko, Iain C.A.F. Robinson,
Suzanne L. Dickson, and Gareth Leng

Growth hormone (GH) secretion from the pituitary gland is regulated by the hypothalamic peptide hormones growth hormone releasing hormone (GHRH) and somatotropin release inhibiting factor (SRIF) (Scheme 11.1). The factor controlling the episodic nature of GH release is unknown but its effects are probably mediated by feedback loops involving the positive effector GHRH and the negative regulator SRIF (1). In 1984, Bowers and Momany and coworkers (2, 3) described the synthesis and properties of a series of small peptide GH secretagogues that were based on the structure of Leu and Met enkephalins. Growth hormone releasing peptide (GHRP)-6 was the most potent of these peptides and was subsequently shown to be active in man (4, 5). Because of the limited oral bioavailability of peptides we sought a class of GH secretagogues more amenable to chemical modification so that oral bioavailability and pharmacokinetic properties could be optimized. Implicit in establishing assays to identify new small molecules was an understanding of the mechanisms regulating GH release from the anterior pituitary gland. Based on its size, GHRP-6 was considered a potential template for a small molecule peptide mimetic. Our approach was based on screening selected structures in functional and mechanism based assays. Following identification of a benzolactam lead structure, L-692,429 was synthesized and used as a prototype to investigate specificity and efficacy in clinically relevant target populations (6–11).

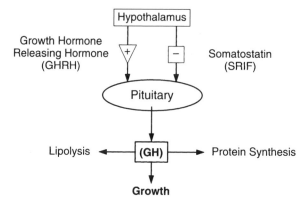

SCHEME 11.1. Regulation of GH release by SRIF and GHRH.

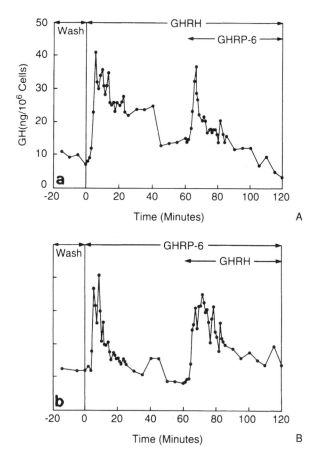

FIGURE 11.1. Response of rat pituitary cells to sequential treatment with GHRH and GHRP-6 using perifusion technique (5).

Mechanism In Vitro

Evidence That GHRP-6 and GHRH Act on Different Receptors

A perifusion system consisting of primary cultures of rat pituitary cells attached to sepharose beads was used to investigate the dynamics of GHRP-6 and GHRH on GH secretion (12). During constant perifusion of GHRH, GH release was maximal within the first minute and was sustained for about 40 min. By contrast, the effects of GHRP-6 were maintained for only 15 min. When GHRH was perifused until GH secretion fell to basal levels, and GHRP-6 was then added to the perifusate, a second increase in GH secretion was observed (Fig. 11.1A). Similar observations were made when the order of exposure to GHRH and GHRP-6 was reversed (Fig. 11.1B). These results illustrated that the decrease in GH secretion observed with prolonged infusion of either secretagogue was not explained by depletion of GH pools in somatotrophs but was associated with desensitization of distinct receptors or their signal transduction pathways.

To investigate resensitization of pituitary cells, GHRP-6 was introduced into the perifusate for 1 min at 15-, 30-, or 60-min intervals. When the first 1-min pulse of GHRP-6 was followed by a 15-min infusion of culture medium and a second 1-min pulse of GHRP-6, the amount of GH released in response to the second pulse was reduced by approximately 50% (Fig. 11.2). After four such pulses at 15-min intervals GH secretion in response to GHRP-6 was no longer evident (Fig. 11.2). Perifusion with medium for 30-min periods between GHRP-6 pulses was inadequate to maintain a full response to repeated GHRP-6 administration (data not shown). Complete recovery required 60 min (Fig. 11.2). These studies demonstrated that continuous exposure to GHRP-6 resulted in desensitization of the GHRP-6 receptor and that resensitization required removal of GHRP-6 for 60 min.

GHRP-6, L-692,429, and L-692,585 Are Potent GH Secretagogues That Apparently Act Through the Same Receptor

The structures of the nonpeptide secretagogues are shown in Figure 11.3. In a dose-dependent manner GHRP-6, L-692,429, and L-692,585 cause an increase in GH secretion from primary cultures of rat pituitary cells with median effective concentrations (EC_{50}s) of 10, 60, and 3 nM, respectively (13). As reported previously, the antagonists His-DTrp-DLys-Trp-DPhe-LysNH2 and L-692,400 inhibit the activity of GHRP-6 and L-692,429 but are ineffective in preventing GHRH-induced GH release (4). Moreover,

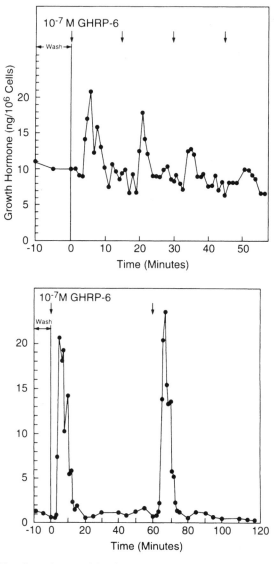

FIGURE 11.2. Kinetics of resensitization of rat pituitary cells to 1-min pulses of 100 nM GHRP-6. Methods according to Blake and Smith (12).

when rat pituitary cells are treated with maximal stimulatory concentrations of GHRP-6, L-692,429, and L-692,585 in combination, no
additional GH release occurs. However, in combination with GHRH,
GHRP-6, L-692,429, and L-692,585 potentiate the effects of GHRH on GH
secretion.

His-D-Trp-Ala-Trp-D-Phe-Lys-NH$_2$

GHRP-6

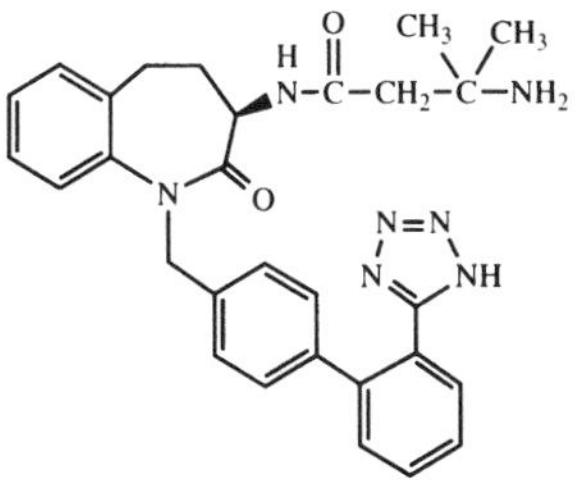

L-692,429

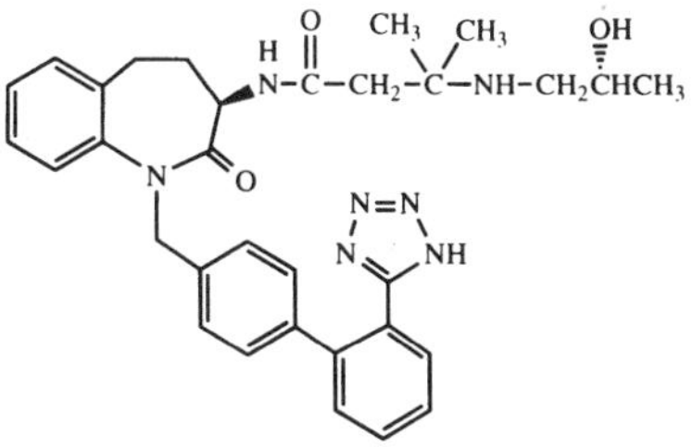

L-692,585 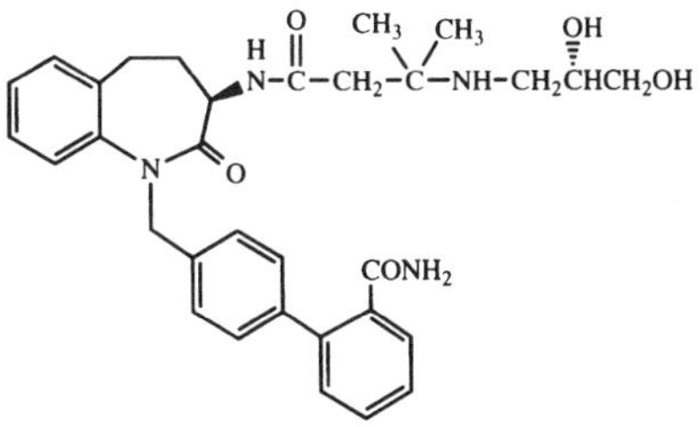   **L-700,653**

FIGURE 11.3. Structures of nonpeptide GH secretagogues.

Signal Transduction

A comparison of the effects of GHRH and GHRP-6 on intracellular aden-osine 3′,5′-cyclic monophosphate (cAMP) levels in rat pituitary cells is illustrated in Figure 11.4. Clearly, in contrast to GHRH, GHRP-6 does not increase cAMP. However, when the secretagogues are used in combination, the GHRH-induced increase in cAMP is potentiated (6, 14). Phorbol myristate acetate (PMA) mimics these effects of GHRP-6 and L-692,429 (6, 15).

To investigate the effects of desensitization of the kinase C pathway on secretagogue-induced GH release, rat pituitary cells were exposed to PMA overnight and then treated with fresh culture medium containing PMA, GHRP-6, and GHRH. Under these conditions PMA was inactive and the effects of GHRP-6 were attenuated, but the cells were still fully responsive to GHRH (Fig. 11.5), demonstrating that the effects of GHRP-6, but not GHRH, were partially dependent on protein kinase C (15). Moreover, phospholipase C mimicked the effects of GHRP-6, suggesting that GHRP-6 and the nonpeptides activate a G protein–coupled receptor that activates protein kinase C through phospholipase C (15).

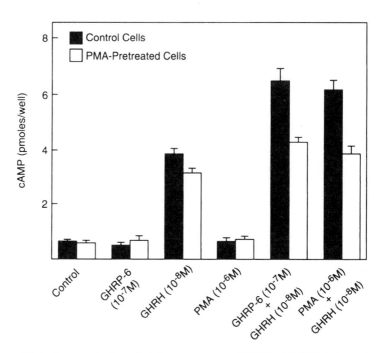

FIGURE 11.4. Demonstration that in contrast to GHRH, GHRP-6 alone has no effect on cAMP levels but in combination with GHRH potentiates the GHRH effect. Methods according to Cheng et al. (14).

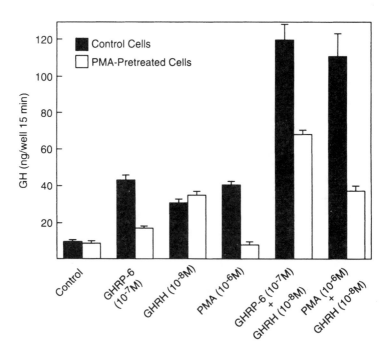

FIGURE 11.5. The phorbol ester PMA mimics the effects of GHRP-6 on stimulating GH release (14).

GHRP-6, L-692,429, and L-692,585 Cause Depolarization of Rat Pituitary Cells

By using the membrane potential sensitive fluorescent dye bis-oxonol, we showed that addition of the secretagogues GHRP-6, L-692,429, or L-692,585 causes depolarization of the plasma membrane of pituitary cells (16). Somatostatin reverses the depolarization. To allow investigation of the effects of GHRP-6 and L-692,429 on somatotrophs directly, somatotrophs were identified using a reverse hemolytic plaque assay (17). Effects on electrical activity were monitored with the perforated patch clamp technique, which maintains soluble second messengers in the cell (18). The secretagogues caused excitation and depolarization of the plasma membrane. These effects were not observed with L-692,428, the inactive enantiomer of L-692,429.

It seemed likely that depolarization caused by GHRP-6 or L-692,429 might explain potentiation of the GHRH response. To test this hypothesis rat pituitary cells were depolarized by addition of the sodium channel agonist veratridine. Veratridine alone caused small increases in GH secretion but markedly potentiated the effects of GHRH (Fig. 11.6); both effects were blocked by the sodium channel antagonist tetrodotoxin. Veratridine

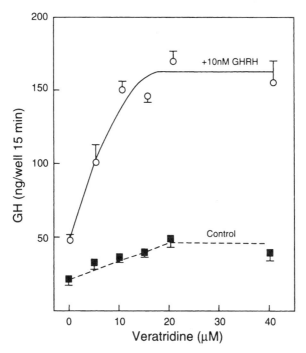

FIGURE 11.6. The sodium channel agonist veratridine potentiates GHRH-induced GH release. Experimental design as described previously (14).

had no additional effect on GHRP-6 or L-692,429-induced GH release. Depolarization caused by elevation of extracellular potassium, or by addition of the potassium channel blockers tetraethylammonium (TEA) and 4-aminopyridine, elicited GH secretion. The magnitude of GH release caused by GHRP-6 was always greater than that observed with depolarization alone, yet addition of GHRP-6 in the presence of depolarizing agents did not increase GH release. Interestingly, GH release induced by PMA could be augmented by depolarizing agents.

Ion Channels Involved in GH Secretion

Electrophysiologic studies in vitro were used to determine whether depolarization of somatotrophs is mediated through sodium or potassium channels. TTX-sensitive sodium channels were observed by whole cell patch clamp (18), but substitution of sodium chloride with choline chloride in the pituitary cell culture medium did not prevent secretagogue-induced GH release. This suggests that sodium channels are not involved in GHRP-6 induced GH release (Table 11.1). Voltage-gated potassium channels are inhibited by concentrations of L-692,429 and L-692,585 active in inducing GH secretion (18–21). Modulation of K channels was observed when soluble second messengers were maintained by use of the perforated-patch voltage clamp technique or with on-cell single channel current recordings; modulation was not observed when the cells were dialyzed by use of the whole cell configuration of the patch electrode voltage clamp technique. Thus, a soluble second messenger apparently mediates the effect of receptor activation. Together with the pharmacology, these results are consistent with involvement of K channels in the regulation of GH secretion by GHRP-6, L-692,429, and L-692,585.

TABLE 11.1. Na^+ ions do not regulate GHRH, GHRP-6, forskolin, and PMA stimulated GH secretion.

	GH secretion (ng/well/15 min)	
	Normal medium	Na^+-free medium[a]
Basal	20.8 ± 1.7	19.7 ± 0.8
GHRH (10 nM)	77.8 ± 2.5	40.3 ± 1.6^b
GHRP-6 (100 nM)	98.4 ± 9.1	103.6 ± 7.2 ND
PMA (500 nM)	49.3 ± 3.7	53.8 ± 4.2 ND
Forskolin (10 μM)	101.2 ± 6.5	90.3 ± 8.2 ND

Value are mean $\pm$ SEM, $n = 4$.
[a] 120 mM of choline chloride was used to substitute NaCL in medium.
[b] Significant difference vs. normal medium.
ND, no significant difference vs. normal medium within each group.

Block of K channels by GHRP-6 elicits electrical spiking activity, thereby enhancing Ca entry through voltage-gated channels. Somatotrophs have two types of voltage-gated Ca channels, commonly referred to as T-type, or low-voltage activated, and L-type. GH release from rat pituitary cells induced by GHRP-6 or L-692,429 is inhibited by blockers of L-type calcium channels, such as nifedipine, nitrendipine, or diltiazem. The peptide toxin ω-agatoxin IIIA, which blocks L-type but not T-type Ca channels, also blocks the secretagogue activity of GHRP-6 and L-692,429 (6, 19). Collectively, these data show that activation of L-type calcium channels is necessary for stimulation of GH release.

Mechanism In Vivo

Pulsatile GH Release

When the activities of a series of structurally different nonpeptide GH secretagogues were evaluated in vivo it became obvious that analogues with long half-lives gave multiple peaks of GH release (data not shown). In pursuing these observations we monitored the GH secretory profile during constant infusion of GH secretagogues. Infusion into pigs of a related secretagogue, L-700,653 (Fig. 11.3) (22), and infusion of L-692,585 into guinea pigs (23) confirmed that these compounds induced pulsatile GH release similar to that observed with GHRP-6 (23, 24). Intriguingly, while the effects of GH secretagogues on pituitary cells were not sustained in vitro, in animals the observation that amplification of episodic GH release continued during infusion suggests that in vivo, desensitization and resensitization of the signal transduction pathway occurs in the continued presence of the ligand.

GHRP-6, L-692,429, and L-692,585 Act on the Arcuate Nucleus

A series of electrophysiology studies showed that GHRP-6 administered intravenously to rats caused excitation of secretory neurons in the arcuate nucleus (25). These effects were reproduced with L-692,585 (26). Similarly, GHRP-6, L-692,429, and L-692,585, but not GHRH, activated c-fos expression selectively in the arcuate nucleus (26); the activation was associated with neurosecretory neurons (27). C-fos expression was concentration dependent and well correlated with the relative potency of L-692,429, and L-692,585 in releasing GH. The biologically inert enantiomer L-692,428 was ineffective, illustrating the stereoselectivity of the activation. Thus GHRP-6, L-692,429, and L-692,585 act directly or indirectly on the arcuate nucleus perhaps targeting GHRH containing neurons.

Effects of L-692,585 Is Attenuated in Hypothalamic-Pituitary Stalk Sectioned Pigs

The effects of L-692,585 were examined in hypothalamic-pituitary stalk sectioned pigs to determine whether the higher levels of GH release observed in vivo compared with in vitro was explained by the presence of hypothalamic factors such as GHRH (28). Prior to surgery, animals were treated with L-692,585, GHRH, and L-692,585 + GHRH, and blood was collected for measurement of GH. The magnitude of GH release was higher in the L-692,585 than in GHRH-treated animals. When L-692,585 and GHRH were injected together the GH response was more than additive. The pigs were next separated into two groups, one group was subjected to hypothalamic-pituitary stalk transection and the other to mock transection. Each group was again treated with the secretagogues and their GH response measured. The magnitude of GH secretion was markedly diminished in the stalk transected animals treated with L-692,585, but when GHRH was replaced exogenously L-692,585 provided GH release of magnitude indistinguishable from controls (28), suggesting that GHRH is released by the hypothalamus in response to L-692,585. When the same stalk transected pigs were treated with gonadotrophin releasing hormone and corticotrophin releasing hormone, a normal luteinizing hormone (LH) and adrenocorticotropic hormone (ACTH) response was measured demonstrating that surgical intervention had not compromised the functional integrity of the pituitary gland (28, 29).

Effects of L-692,585 on GHRH in Hypophysial Portal Vessels

Sheep were cannulated to allow introduction of L-692,585 into the jugular vein and to allow collection of blood from peripheral and portal vessels. L-692,585 was infused for 15 min and blood samples were collected at 10-min intervals for 2 hours. GH levels were measured in peripheral blood, and GHRH and SRIF were measured in portal blood. The results shown in Figure 11.7 are representative of results obtained in two sheep but are considered preliminary because for reasons not yet understood this effect was not reproduced in all sheep. However, these results are indeed consistent with those reported by others (30). Figure 11.7 shows that coincident with injection of L-692,585, GHRH levels begin to increase in the portal vessels, which slightly precedes an increase of GH in the peripheral circulation.

Discussion

The studies described above indicate that GHRP-6, L-692,429, and L-692,585 act through a common second messenger pathway distinct from

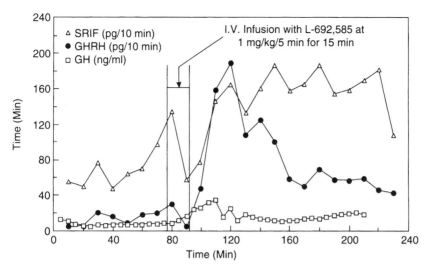

FIGURE 11.7. Effects of intravenous infusion of L-692,585 on peripheral GH levels and concentrations of SRIF and GHRH in the portal vessels of conscious sheep. Surgery and cannulation was performed with the assistance of Fred Karsch's group, and radioimmunoassays (RIAs) for determination of SRIF and GHRH were performed by Larry Frohman's group.

that regulated by GHRH. However, in common with all agents capable of stimulating GH release from somatotrophs, activation of L-type Ca-channels is an essential convergent step in the pathway (10, 19). In contrast to GHRH, GHRP-6 and the nonpeptides act through the protein kinase C pathway and cause depolarization of the somatotroph plasma membrane (10, 18). Scheme 11.2 summarizes the biochemical pathways regulated by GHRP-6, GHRH, and SRIF.

The pituitary receptor for GHRP-6 and the nonpeptides appears to be linked to phospholipase C. Adams et al. (31) recently reported that in human pituitary tumor cells L-692,429 and GHRP-6 markedly increase the rate of phosphatidylinositol turnover. This pathway produces diacylglycerol and inositol triphosphate (IP_3) as second messengers. The involvement of diacylglycerol and its activation of protein kinase C is consistent with our earlier studies (6) and those recently described by Mau et al. (32) also implicating kinase C and IP_3 in the signal transduction pathway. A role for IP_3 is supported by electrophysiology and Ca^{2+} imaging studies showing that an early event in the action of GHRP-6 is an increase in free intracellular Ca^{2+} caused by redistribution from intracellular stores (33, 34). The redistribution of intracellular Ca^{2+} is not inhibited by SRIF and is therefore probably not sufficient to cause GH release; however, we speculate that the release of intracellular Ca^{2+} prior to activation of L-type Ca^{2+} channels synchronizes fusion of GH containing secretory granules to the plasma membrane facilitating GH secretion (35, 36).

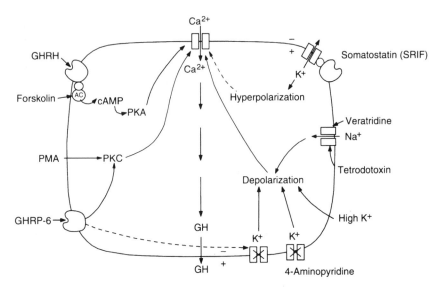

SCHEME 11.2. Pathways involved in the control of GH release from somatotrophs.

GHRP-6 and L-692,429 depolarize the membranes of somatotrophs by inhibiting K$^+$ channels (18, 20). Similar coupling between membrane receptors and K$^+$ channels has been described for the m1 muscarinic receptor, where G$_\alpha$ activates phospholipase C to cause tyrosine phosphorylation of a K$^+$ channel resulting in inhibition of the channel (37). Depolarization alone does not account for the magnitude of GH release induced by GHRP-6 and the nonpeptides, suggesting that other events such as initiation of electrical spiking activity or the phosphorylation of L-type Ca-channels are important for secretagogue activity (38, 39). Augmentation of GHRH-induced increases in intracellular cAMP levels is perhaps mediated by crosstalk between the G$_{\beta\gamma}$ subunits associated with the GHRP-6 receptor and G$_{\alpha s}$ of the GHRH receptor complex to potentiate activation of adenylate cyclase (40). Therefore, activation of the signal transduction pathway associated with phospholipase C explains the observed effects of GHRP-6, L-692,429, and L-692,585 on GH release and their potentiation of the GHRH pathway.

Although GHRP-6, L-692,429, and L-692,585 cause GH release by acting directly on somatotrophs in the pituitary gland, these secretagogues also activate neurosecretory cells of the arcuate nucleus of the hypothalamus. Treatment of hypothalamic/pituitary stalk sectioned pigs with L-692,585 showed that maximal response to L-692,585 was dependent upon a hypothalamic factor(s), and since coadministration of GHRH provided a response identical to that in intact animals it appears that GHRH is an essential hypothalamic factor. Moreover, the increases in GHRH measured in the portal vessels of sheep following treatment with hexarelin (30) and L-

692,585 support the notion that L-692,585 causes GHRH release from the hypothalamus.

Most intriguing is the demonstration that constant infusion of GHRP-6 in man causes increases in the amplitude of pulsatile GH release (41). Similarly, infusion of GHRP-6 and L-692,585 into conscious rats or guinea pigs initiates and amplifies episodic GH release (23, 24). These observations are consistent with the role of GHRP-6, L-692,429, and L-692,585 as GHRH releasing hormones and/or amplifiers of GHRH action, since constant infusion of GHRH also results in pulsatile GH release (24, 42). Based on their observed biologic properties we define GHRP-6, L-692,429, and L-692,585 as GHRH amplifier hormone(s) (GHRH/AH). Scheme 11.3, based on in vitro and in vivo findings, explains the action of GHRH/AH on GH release from the pituitary gland. At three levels—antagonism of SRIF on the arcuate nucleus and somatotroph, release of GHRH, and potentiation of GHRH action on the somatotroph—GHRH/AH synchronizes the hypothalamic/pituitary axis to amplify the signals involved in GH release.

In Scheme 11.4 we propose that pulsatile GH release is regulated by feedback loops involving GHRH/AH, GHRH, and SRIF. At the hypothalamic level, it has been suggested that through short loop feedback pathways, SRIF inhibits GHRH release from the arcuate nucleus and GHRH inhibits SRIF secretion from the periventricular nucleus (43, 44); GH also stimulates SRIF release from the hypothalamus and SRIF prevents GH release from the pituitary gland (45). Since GHRH/AH behaves as a func-

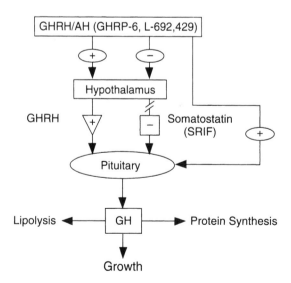

SCHEME 11.3. The role of GHRH/AH (GHRP-6, L-692,429, and L-692,585) on the hypothalamic/pituitary axis.

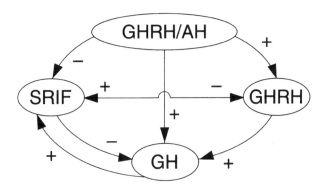

SCHEME 11.4. Proposed feedback pathways that modulate episodic GH release.

tional antagonist of SRIF (4, 10), we suggest that acute treatment with GHRH/AH immediately interrupts SRIF action. The interruption in SRIF tone in the hypothalamus and pituitary gland simultaneously allows GHRH release from the arcuate nucleus and sensitization of the somatotroph to both GH secretagogues (GHRH and GHRH/AH). The biologic activity of the receptors for each of these regulatory hormones will be governed by the relative concentrations of the hormones as a function of time, the kinetics of binding, and the ratio of active to inactive receptor. Crucial to the model is the fact that SRIF antagonizes the concentration-dependent desensitization of pituitary cells to GHRH/AH in the continued presence of ligand (Cheng et al., in preparation) and regulates the activity of GHRH/AH in the hypothalamus (46). Therefore, while initiation of pulsatility is dependent upon activation of GHRH/AH receptors, the maintenance of frequency and amplitude of GH secretion is dependent upon SRIF and its receptors. Ongoing studies are directed at measuring the kinetic properties of the GHRH/AH, GHRH, and SRIF receptors and the relative levels of GHRH and SRIF in hypophysial portal vessels as a function of time during GHRH/AH infusion. This information will be used to mathematically test the validity of our proposed model.

Acknowledgments. We greatly acknowledge the assistance of Fred Karsch, University of Michigan, Ann Arbor, MI and Larry Frohman, University of Illinois, Chicago, IL and their research groups for their invaluable assistance with the measurement of GHRH and SRIF in the portal vessels of conscious sheep.

References

1. Plotsky PM, Vale W. Patterns of growth hormone-releasing factor and somatostatin into the hypophysial portal circulation of the rat. Science 1985;230: 461–3.

2. Bowers CY, Momany FA, Reynolds GA, Hong A. On the in vitro and in vivo activity of a new synthetic hexapeptide that acts on the pituitary to specifically release growth hormone. Endocrinology 1984;114:1537–45.

3. Momany FA, Bowers CY, Reynolds GA, Hong A, Newlander K. Conformational energy studies and in vitro and in vivo activity data on growth hormone-releasing peptides. Endocrinology 1984;114:1531–6.

4. Bowers CY, Reynolds GA, Durham D, Barrera CM, Pezzoli SS, Thorner MO. Growth hormone (GH)-releasing peptide stimulates GH release in normal man and acts synergistically with GH-releasing hormone. J Clin Endocrinol Metab 1990;70:975–82.

5. Hartman ML, Farello G, Pezzoli SS, Thorner MO. Oral administration of growth hormone (GH)-releasing peptide stimulates GH secretion in normal men. J Clin Endocrinol Metab 1992;74:1378–84.

6. Smith RG, Cheng K, Pong S-S, Hickey G, Jacks T, Butler B, et al. A nonpeptidyl growth hormone secretagogue. Science 1993;260:1640–3.

7. Schoen WR, Wyvratt MJ, Smith RG. Growth hormone secretagogues. Ann Rep Med Chem 1993;28:177–86.

8. Hickey GJ, Jacks T, Judith F, Taylor J, Clark JN, Smith RG. In vivo efficacy and specificity of L-692,429, a novel nonpeptidyl growth hormone secretagogue in beagles. Endocrinology 1994;134:695–701.

9. Gertz BJ, Barrett JS, Eisenhandler R, Krupa DA, Wittreich JM, Seibold JR, et al. Growth hormone response in man to L-692,429, a novel nonpeptide mimic of growth hormone-releasing peptide-6. J Clin Endocrinol Metab 1993;77:1393–7.

10. Aloi JA, Gertz BJ, Hartman ML, Huhn WC, Pezzoli SS, Wittreich JM, et al. Neuroendocrine responses to a novel growth hormone secretagogue, L-692,429, in healthy older subject. J Clin Endocrinol Metab 1994;79:943–9.

11. Bowers CY. Editorial: On a peptidomimetic growth hormone-releasing peptide. J Clin Endocrinol Metab 1994;79:940–2.

12. Blake AD, Smith RG. Desensitization studies using perifused rat pituitary cells show that growth hormone releasing hormone and His-D-Trp-Ala-Trp-D-Phe-Lys-NH2 stimulate growth hormone release through distinct receptor sites. J Endocrinol 1991;129:11–19.

13. Schoen WR, Ok D, DeVita RJ, Pisano JM, Hodges P, Cheng K, et al. Structure-activity relationships in the amino acid sidechain of L-692,429. Biorg Med Chem Lett 1994;4:1117–22.

14. Cheng K, Chan W-S, Barreto A, Convey EM, Smith RG. The synergistic effects of His-DTrp-Ala-Trp-DPhe-Lys-NH$_2$ on GRF stimulated growth hormone release and intracellular cAMP accumulation in rat primary pituitary cell cultures. Endocrinology 1989;124:2791–7.

15. Cheng K, Chan W-S, Butler B, Barreto A, Smith RG. Evidence for a role of protein kinase-C in His-DTrp-Ala-Trp-DPhe-Lys-NH$_2$-induced growth hormone release from rat primary pituitary cells. Endocrinology 1991;129:3337–42.

16. Pong S-S, Chaung L-YP, Smith RG. GHRP-6 (His-D-Trp-Ala-Trp-D-Phe-Lys-NH$_2$) stimulates growth hormone secretion by depolarization in rat pituitary cell cultures. Proceedings of the 73rd Annual Meeting of the Endocrine Society, Program and Abstracts. Bethesda, MD: Endocrine Society, 1991:88.

17. Neil JD, Frawley LS. Detection of hormone release from individual cells in mixed populations using a reverse hemolytic plaque assay. Endocrinology 1983;112:1135–7.

18. Leonard RJ, Chaung L-YP, Pong S-S. Ionic conductances of identified rat somatotroph cells studied by perforated patch recording are modulated by growth hormone secretagogues. Biophys J 1991;59:a254.

19. Pong S-S, Chaung L-Y, Smith RG, Ertel E, Smith MM, Cohen CJ. Role of calcium channels in growth hormone secretion induced by GHRP-6 (His-D-Trp-Ala-Trp-D-Phe-Lys-NH$_2$) and other secretagogues in rat somatotrophs. Proceedings of the 74th Annual Meeting of the Endocrine Society, Program and Abstracts. Bethesda, MD: Endocrine Society, 1992:255.

20. McGurk JF, Pong SS, Chaung LY, Gall M, Butler B, Arena JP. Growth hormone secretagogues modulate potassium currents in rat somatotrophs. Soc Neurosci Abstr 1993;19:1559.

21. Pong S-S, Chaung L-YP, Leonard RJ. The involvement of ions in the activity of a novel growth hormone secretagogue L-692,429 in rat pituitary cell culture. Proceedings of the 75th Annual Meeting of the Endocrine Society, Program and Abstracts. Bethesda, MD: Endocrine Society, 1993:172.

22. Chang CH, Rickes EL, Marsilio F, McGuire L, Cosgrove S, Taylor J, et al. Activity of a novel non-peptidyl growth hormone secretagogue L-700,653 in swine. Endocrinology 1995;136:1065–71.

23. Fairhall KM, Mynett A, Smith RG, Robinson ICAF. Consistent responses to respeated injections of GHRP-6 and a new, non-peptide GH secretagogue L-692,585. J Endocrinol 1995; 145:417–26.

24. Clark RG, Carisson LMS, Trojnar J, Robinson ICAF. The effects of a growth hormone-releasing peptide and growth hormone-releasing factor in conscious and anaesthetized rats. J Neuroendocrinol 1989;1:249–55.

25. Dickson SL, Leng G, Robinson ICAF. Systemic administration of growth hormone-releasing peptide (GHRP-6) activates hypothalamic arcuate neurones. Neuroscience 1993;53:303–6.

26. Dickson SL, Leng G, Dyball REJ, Smith RG. Central actions of peptide and non-peptide growth hormone secretagogues in the rat. Neuroendocrinology 1995;61:36–43.

27. Doutrelant-Viltart O, Dickson SL, Dyball REJ, Leng G. Expression of Fos protein following growth hormone-releasing peptide (GHRP-6) injection in rat arcuate neurones retrogradely labelled by systemic fluorogold administration. J Physiol 1995;483:49–50.

28. Hickey GJ, Baumhover J, Faidley T, Chang C, Anderson LL, Nicolich S, et al. Effect of hypothalamo-pituitary stalk transection in the pig on GH secretory activity of L-692,585. Proceedings of the 76th Annual Meeting of the Endocrine Society, Program and Abstracts. Bethesda, MD: Endocrine Society, 1994:366.

29. Chang CH, Drisko J, Anderson LL, Rickes EL, McGuire LA, Faidley T, et al. Evidence that L-692,585 stimulates GH and ACTH responses primarily through a central nervous system action in swine. Proceedings of the 77th Annual Meeting of the Endocrine Society, Program and Abstracts. Bethesda, MD: Endocrine Society, 1995:484.

30. Guillaume V, Magnan E, Cataldi M, Dutour A, Sauze N, Renard M, Razafindraibe H, Conte-Devolx B, Deghenghi R, Lenaerts V, Oliver C. Growth hormone (GH)-releasing hormone secretion is stimulated by a new GH-releasing hexapeptide in sheep. Endocrinology 1994;135:1073–5.

31. Adams EF, Petersen B, Lei T, Buchfelder M, Fahlbusch R. The growth hormone secretagogue, L-692,429, induces phosphatidylinositol hydrolysis and hor-

mone secretion by human pituitary tumors. Biochem Biophys Res Commun 1995;208:555–61.

32. Mau SE, Witt MR, Bjerrum OJ, Saermark T, Vilhardt H. Growth hormone releasing hexapeptide (GHRP-6) activates the inositol (1,4,5)-trisphosphate/diacylglycerol pathway in rat anterior pituitary cells. J Recept Signal Transduct Res 1995;15:311–23.

33. Herrington J, Hille B. Growth hormone-releasing hexapeptide elevates intracellular calcium in rat somatotrophs by two mechanisms. Endocrinology 1994;135:1100–8.

34. Bresson-Bépoldin L, Dufy-Barbe L. GHRP-6 induces a biphasic calcium response in rat pituitary somatotrophs. Cell Calcium 1994;15:247–58.

35. Blondel O, Bell GI, Seino S. Inositol 1,4,5-trisphosphate receptors, secretory granules and secretion in endocrine and neuroendocrine cells. Trends Neurosci 1995;18:157–61.

36. Littleton JT, Bellen HJ. Synaptotagmin controls and modulates synaptic-vesicle fusion in a Ca^{2+}-dependent manner. Trends Neurosci 1995;18:177–83.

37. Huang X-Y, Morielli AD, Peralta EG. Tyrosine kinase-dependent suppression of a potassium channel by the G protein-coupled m1 muscarinic acetylcholine receptor. Cell 1993;75:1145–56.

38. Albert PR, Wolfson G, Tashjian AH. Diacylglycerol increases cytosolic free Ca^{2+} concentration in rat pituitary cells. J Biol Chem 1987; 262:6577–81.

39. Mantegazza M, Fasolato C, Hescheler J, Pietrobon D. Stimulation of single L-type calcium channels in rat pituitary GH3 cells by thryrotropin-releasing hormone. EMBO J 1995;14:1075–83.

40. Tang W-J, Gilman AG. Type specific regulation of adenylyl cyclase by G protein βγ subunits. Science 1991;254:1500–3.

41. Huhn WC, Hartman ML, Pezzoli SS, Thorner MD. 24-hour growth hormone (GH)-releasing peptide (GHRP) infusion enhances pulsatile GH secretion and specifically attenuates the response to a subsequent GHRP-6 bolus. J Clin Endocrinol Metab 1993;76:1202–8.

42. Vance ML, Kaiser DL, Evans WS, Furlanetto R, Vale W, Rivier J, Thorner MD. Pulsatile growth hormone secretion in normal man during a continuous 24-hour infusion of human growth hormone releasing factor (1-49): evidence for intermittent somatostatin secretion. J Clin Invest 1985;75:1584–90.

43. Katakami H, Arimura A, Frohman LA. Growth hormone (GH)-releasing factor stimulates hypothalamic somatostatin release: an inhibitory feedback effect on GH secretion. Endocrinology 1986;118:1872–7.

44. Aguila MC. Growth hormone-releasing factor increases somatostatin release and mRNA levels in the rat periventricular nucleus via nitric oxide by activation of guanylate cyclase. Proc Natl Acad Sci USA 1994;91:782–6.

45. Bertherat J, Bluet-Pajot MT, Epelbaum J. Neuroendocrine regulation of growth hormone. Eur J Endocrinol 1995;132:12–24.

46. Fairhall KM, Mynett A, Robinson ICAF. Central effects of growth hormone-releasing hexapeptide (GHRP-6) on growth hormone release are inhibited by central somatostatin action. J Endocrinol 1995;144:555–60.

12

Metabolic Regulation of Growth Hormone Secretagogue Gene Expression

MICHAEL BERELOWITZ AND JOHN F. BRUNO

Growth hormone (GH), secreted from somatotrophs in the anterior pituitary, is a highly regulated hormone that stimulates longitudinal growth and has multiple effects on cellular metabolism. Plasma GH levels vary widely at different times of the day and night, with profound influences of diet, exercise, sleep, and stress in a given individual. GH levels also differ with sex, age, and body habits, providing a challenge to those defining "normal" GH physiology (1–3).

In young healthy adults, GH is secreted episodically with pulses occurring at greater frequency at night in relation to early slow wave sleep. The pattern of episodic GH secretion is different in males than in females. The male pattern is characterized by high GH spikes with low (essentially absent) GH levels during trough periods. In females, GH episodic secretion has a lower pulse amplitude with higher trough levels (2).

Studies in experimental animals have helped define the mechanisms of GH regulation with indirect indications that the same processes apply to the human situation.

GH releasing factor (GHRF) is the hypothalamic peptide that stimulates GH synthesis and secretion/release (4, 5). GHRF is synthesized in neurons within the arcuate nucleus of the hypothalamus from where it is transported to nerve endings in the median eminence and released into the hypophysial-portal vessels (6). Anterior pituitary somatotrophs have plasma membrane receptors for GHRF with the characteristic seven transmembrane spanning domain structure of other G protein–linked receptors (7). Binding of GHRF to its receptor activates adenylyl cyclase, stimulates adenosine $3', 5'$-cyclic monophosphate (cAMP), initiates a transmembrane signal transduction cascade that stimulates de novo GH synthesis with increased GH gene transcription and translation, and stimulates GH secretion (4, 5). GHRF is trophic for somatotrophs, and in the presence of constitutively activated G_s

or prolonged GHRF exposure (transgenic models or ectopically produced) somatotroph hyperplasia adenoma formation is seen (8).

Somatostatin [somatotropin release inhibiting factor (SRIF)] is the hypothalamic peptide inhibitory to GH secretion. SRIF messenger RNA (mRNA) and peptide expression is ubiquitous within the CNS with highest concentrations in the hypothalamus (9, 10). The hypothalamic SRIF pathway involved in GH regulation has its nuclear bodies in the rostral periventricular nucleus with a neural pathway passing through the retrochiasmatic area to terminate on nerve endings in the median eminence (6). The pituitary possesses an array of SRIF receptor subtypes sst1–5 (11, 12). Binding of SRIF to its receptor(s) results in activation of G_i with inhibition of adenylyl cyclase activity and cAMP levels. Receptor binding also leads to ion channel activation that results in membrane hyperpolarization (13, 14). These actions effectively shut off GH secretion due to any stimulus tested to date with accumulation of unreleased GH in secretory granules within the somatotroph. No good evidence exists that SRIF directly influences GH gene expression.

Pituitary GH secretion at any particular point in time seems to depend on the relative ambient tone of GHRF and SRIF (1–3). In rats episodic GH secretion results from episodic hypothalamic GHRF and SRIF influences that are 180° out of phase such that SRIF tone is low when GHRF is released, and SRIF tone is high with a GHRF trough (15, 16). Studies in humans given GHRF at various times during the day suggest that states of higher and lower SRIF tone/GHRF receptivity occur in humans as well (4, 5). Studies in sheep reveal a more complex arrangement that has not fully supported this model (17). Episodic GH secretion appears to be a fundamental property occurring even in the presence of continuous GHRF infusion or in the presence of a somatotroph adenoma secreting high levels of GH. Episodic GH secretion may be required for optimal GH-induced growth and metabolic effects (4, 5).

The orchestration of GHRF and SRIF secretion from the hypothalamus in a manner that permits episodic GH is obviously complex and not fully understood. There does, however, appear to be cross talk of SRIF on GHRF-secreting neurons and of GHRF on SRIF release (ultrashort loop feedback) (4).

GH exerts its peripheral actions through a plasma membrane receptor of the helix-bundle-helix configuration that is tightly linked to a tyrosine kinase activity (3). The actions of GH on growth and metabolism are beyond the scope of this review except insofar as they apply to GH regulation. GH exerts negative feedback (short-loop feedback) on the hypothalamus (although the means by which it effects these actions are unclear) (2, 3). GH does, however, stimulate steady-state levels of SRIF mRNA and inhibits GHRF mRNA levels in the periventricular nucleus and arcuate nucleus, respectively, and, in parallel, effects SRIF and GHRF peptide content in their hypophysiotropic pathways (4, 10).

The peripheral actions of GH are mediated through the actions (hormonal or local) of insulin-like growth factor-I (IGF-I) (previously known as somatomedin C) (18). Hormonal IGF-I circulates at levels dependent on ambient GH concentration and likely originates from the liver where its synthesis is under GH control. IGF-I is synthesized in many other tissues, however, and in these areas (including the CNS and pituitary) is also under GH regulation (19). GH can, thus, influence its peripheral actions through circulating or local IGF-I effects. IGF-I in vitro or in vivo inhibits GH secretion (long-loop feedback) through a direct effect on pituitary and indirectly through the hypothalamus. Pituitary somatotrophs possess receptors for IGF-I that act to inhibit GH gene expression, synthesis, and secretion. Hypothalamic SRIF secretion is stimulated in vitro by IGF-I, although the mechanism for this action remains unclear (20, 21).

From this outline it can be seen that the hypothalamic GHRF/SRIF-pituitary GH-peripheral IGF-I axis is complex with multiple potential regulatory sites. The final metabolic outcome can be influenced by modulation of any of the signals, their receptors or signal transduction systems, or their binding proteins (which have not been discussed). This chapter provides evidence that in states of metabolic perturbation any (or perhaps all) of these steps in the GH regulatory axis can be influenced.

Metabolic Influences on GH Secretion

Under normal circumstances GH secretion is regulated by metabolic substrates in a predictable and well-described manner (1, 2, 22–24). In humans GH release occurs rapidly in response to insulin-induced hypoglycemia—a response that is the basis for a test of pituitary GH reserve. GH stimulation depends on hypothalamic-pituitary integrity and is due to the glycopenia-induced, not the insulin, stimulus, since maintaining euglycemia during insulin administration prevents the response. Glucose administration results in the converse effect: suppression of GH release. While the effect of glucose on basal GH has been, in the past, beyond sensitivity of the assay, suppression of episodic or stimulated GH is described. An inability of glucose to inhibit GH in subjects with suspected acromegaly provides clinical utility to this effect (2, 23). In the rat the effect of glucose is opposite to that in humans—hypoglycemia inhibits GH in the rat. The mechanism whereby this effect occurs, however, may provide a useful model with which to study the human response. In the rat in vitro hypothalamic SRIF release is profoundly stimulated in medium-containing low glucose concentrations and inhibited by medium glucose concentrations above those seen in fasting rat plasma (25). The effect of glucopenia on SRIF release can be duplicated by inhibitors of hypothalamic glucose uptake or metabolism, indicating that hypothalamic glucose recognition is dependent on its neural metabolism and (likely) on energy production (25). The end result of hypothalamic glucopenia in humans vs. rats may be distinguished by the resulting GHRF/SRIF ratio.

Amino acids, particularly by iv administration stimulate GH secretion, an effect that has also been utilized as a test of pituitary function in growth disorders (22, 23). L-arginine, L-leucine, L-tryptophan, and ornithine have all been shown to stimulate GH secretion, the response to iv L-arginine being the best described. Oral amino acids have also been shown to influence basal and sleep-related GH secretion. It is not clear what the mechanism is for this effect, however, in the case of arginine, nitric oxide production or metabolism to agmatine (with binding to α_2-adrenergic or imidazoline receptors) could be responsible. Free fatty acids induced in humans by infusion of heparin plus lipid solution inhibit GH secretion.

Altered metabolic states are associated with abnormal GH regulation. Thus, GH secretion is elevated in poorly controlled type I diabetes mellitus, anorexia nervosa, starvation, and protein-calorie malnutrition, and suppressed in morbid obesity (2, 4, 23, 24, 26).

In poorly controlled type I diabetes mellitus GH secretion is elevated with increased basal levels, GH episodes, and elevated nocturnal sleep-related GH secretion (2, 23). This elevation in nocturnal GH, probably through its antagonistic effect on insulin action, appears to be responsible for the so-called dawn phenomenon of fasting hyperglycemia (27). Elevated GH levels in this state of abnormal glucose homeostasis thus serve to worsen the glucose homeostasis. Injection of GH into patients with well-controlled type I diabetes mellitus at levels sufficient to mimic those nocturnal GH elevations seen in poorly controlled patients induce the dawn phenomenon, and somatostatin analogue therapy at night in patients with poor control decreases fasting plasma glucose (27). Elevated GH levels in poorly controlled patients with type I diabetes mellitus are normalized by optimal metabolic regulation (2, 23).

States of starvation or malnutrition are also associated with elevated GH levels (1–3, 24). In these patients a state of GH resistance exists where, despite high levels of GH, IGF-I production remains low. The level of dietary protein appears to be critical in mediating GH elevation, with caloric intake playing an important role.

Patients with morbid obesity have suppressed GH levels with reduced episodes of GH secretion (3, 4, 23, 24). GH responses to insulin-induced hypoglycemia and to GHRF are reduced. However, with weight reduction, GH secretory responsiveness returns such that a correlation exists between percent desirable body weight and GH response to provocative stimuli. Low GH with elevated insulin levels in obese individuals provides a hormonally lipogenic milieu in patients who have difficulty in losing weight. Understanding the mechanism of abnormal GH in these patients and therapeutically improving GH secretion may help subjects with weight reduction, although exogenous GH administration has been of little long-term benefit in these patients.

Metabolic Regulation of GH Secretagogues

Metabolic influences on GH secretion must depend on an interplay between hypothalamic neuropeptide influences on GH synthesis and secretion, direct effects on pituitary GH regulation, and "long-loop" effects mediated by the IGF-I system. Studies in human subjects can provide some insights into the mechanisms of GH dysregulation—peripheral levels of regulators can be measured and aspects of these systems can be inferred using pharmacologic influences. Our laboratory, however, has taken the route of studying animal models of metabolic influence. Hypotheses generated from animal data can then be tested in human subjects.

Diabetes Mellitus

Insulinopenic diabetes mellitus in the rat, whether arising spontaneously or after streptozotocin administration, results in suppression of GH secretion with loss of high-amplitude secretory pulses (28, 29) together with decreased pituitary GH synthesis and content and accompanying growth failure (30). Decreased GH synthesis and secretion in the diabetic rat could result from decreased stimulatory influences on the pituitary (decreased GHRF) or increased inhibitory tone (excessive SRIF or IGF-I).

To study the mechanisms involved in abnormalities of GH regulation, we used the model of streptozotocin (STZ) 100 mg/kg BW -induced diabetes mellitus in Sprague-Dawley rats. Animals were studied 17 to 20 days after injection of either STZ (with metabolic confirmation of diabetes) or sodium citrate diluent (control). STZ-diabetic rats were studied after insulin therapy (insulin-treated) or saline injections (diabetic)(31).

Diabetic rats demonstrated marked hyperglycemia together with a failure of weight gain compared with control rats that was partly restored by insulin therapy. Serum GH levels were reduced in diabetic rats together with decreased GH content in whole pituitary and dispersed anterior pituitary cells. Insulin treatment partly reversed these changes.

Monolayer cultures of anterior pituitary cells from STZ-diabetic rats released less GH than control but similar amounts of prolactin. This suggested that diabetes did not cause a uniform pituitary secretory defect. GHRF-responsiveness of pituitary cells from STZ-diabetic rats was similar to that in controls (normalized for lower basal release). In vivo insulin treatment of STZ-diabetic rats restored basal GH secretion from cultured pituitary cells, supporting the concept that insulin deficiency or its metabolic consequences were the cause of the GH disturbances. SRIF inhibited GHRF-stimulated GH release from pituitary cells of control and diabetic rats; however, the dose-inhibition curve for cells from diabetic rats was shifted to the right by approximately one log order. This suggested that there was resistance to SRIF action on the pituitary in experimental diabetes (31).

Somatostatin and Somatostatin Receptors in Diabetic Rats

Prepro SRIF mRNA expression and SRIF content in extracts of medial-basal hypothalamus of diabetic rats was similar to that in controls (31). In preliminary experiments using needle-punch microdissection, however, SRIF content was reduced in the somatostatinergic regions of the hypothalamus most involved in GH regulation (periventricular nucleus, retrochiasmatic area, and median eminence) but unaffected in other nuclei.

Somatostatin receptor binding was markedly reduced in pituitary membrane preparations from STZ-diabetic rats compared with control, with partial restoration in insulin-treated diabetic rats. Scatchard analysis of binding data suggested that the reduction in SRIF receptor binding was due to a decrease in binding capacity rather than any change in affinity (31). Expression of pituitary SRIF receptor subtype mRNA showed an isoform-specific alteration in diabetic rats with reduction of mRNA for sst1, sst2, sst3, and sst5 but no change in sst4. Insulin treatment restored sst5 completely and sst1 partially, but was without influence on sst2, 3, and 4. Expression of mRNA for SRIF receptor subtypes in hypothalamus was unaffected in diabetic rats (32).

GHRF Expression in Diabetic Rats

GHRF content in extracts of medial basal hypothalamus of diabetic rats was no different than in controls or the insulin-treated. Prepro GHRF mRNA expression, however, was markedly reduced in hypothalami of STZ-diabetic rats compared with controls and was largely normalized in insulin-treated STZ-diabetic rat hypothalami (31).

IGF-I in Diabetic Rats

Serum IGF-I levels were markedly reduced in STZ-diabetic rats with restoration to control levels in insulin-treated diabetic animals. Tissue prepro IGF-I mRNA expression and IGF-I content, particularly in liver, was also reduced and similarly restored by insulin treatment. In contrast, pituitary prepro IGF-I mRNA expression was unaffected by STZ-diabetes, and IGF-I content was *increased* compared with control rats with normalization in pituitaries of insulin-treated rats (33).

Summary

In the STZ-diabetic rat model, a state of insulinopenia in which pituitary GH synthesis and secretion are reduced and circulating GH levels suppressed, the GH regulatory axis shows many perturbations.

1. Hypothalamic GHRF synthesis (and presumably release) are decreased, explaining low GHRF mRNA expression and normal GHRF content.
2. Hypothalamic SRIF tone is increased, based on the evidence for increased hypophysiotropic SRIF content and downregulated pituitary SRIF receptor mRNA and binding.
3. Increased intrapituitary IGF-I content with no decline in IGF-I mRNA in a generally IGF-I–deficient diabetic state supports a local (paracrine/autocrine) role of IGF-I in inhibition of GH synthesis and/or secretion.

Food Deprivation

Food deprivation in the rat results in an inhibition of GH secretory episodes with a reduction in high amplitude bursts that occurs as early as 24 hours after removal of food, with a progressive reduction in amplitude and duration of GH secretory episodes after 48 and 72 hours (26). Decreased GH secretion in food-deprived rats could result, as in diabetes, from decreased GHRF or increased inhibitory SRIF or IGF-I influences. Circumstantial evidence has implicated SRIF, at least in part, based on in vivo immunoneutralization studies in which low GH secretion in food-deprived rats was partly restored after administration of specific SRIF antiserum (34, 35). These data do not provide evidence for a direct role of SRIF, however, because of the intricate functional relationships between regulatory peptides.

Our studies were performed in Sprague-Dawley rats deprived of food, but allowed free access to water, for time periods up to 72 hours (food-deprived, FD) or in control, animals allowed free access to food and water over the same time period (fed). FD or fed rats were refed various diets for 72 hours in experiments evaluating the role of various components of the normal rat diet in mediating perturbations of GH secretion. Test diets for refeeding experiments consisted of either normal rat chow, rat chow deficient in one or another of the major food groups, and chow containing various protein concentrations or various amino acid replacements. Food-deprived rats lost weight over the course of the experiment; thus, in some experiments weight-matched controls were employed. On certain of the

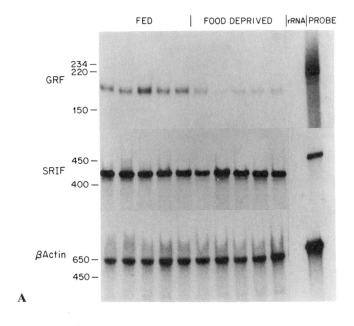

A

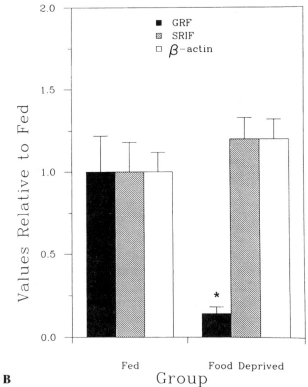

B

diets FD rats ate less than fed controls. In such studies calorically matched diets was provided to control rats (32, 36–39).

Somatostatin and Somatostatin Receptors in FD Rats

Prepro SRIF mRNA expression was similar in hypothalamus of FD and fed rats (Fig. 12.1) as was SRIF peptide content in whole hypothalamus and in microdissected regions of the hypophysiotropic somatostatinergic pathway (36).

SRIF receptor subtype mRNA expression was also similar in hypothalamus of FD rats compared with fed rats. In contrast, in pituitary of FD rats, expression of SRIF receptor subtypes sst1, sst2, and sst3 was markedly reduced compared with their expression in fed rats, while sst4 and sst5 were unchanged. This reduction in pituitary SRIF receptor subtype mRNA expression was accompanied by a decrease in pituitary plasma membrane SRIF receptor binding. Scatchard analysis suggested that decreased binding was due to a reduction in receptor number with no change in affinity (32).

IGF-I and IGF-I Receptors in FD Rats

Serum IGF-I levels were markedly reduced in FD rats compared with fed, and were restored by refeeding. This decrease in circulating IGF-I was associated with a decrease in IGF-I content and prepro IGF-I mRNA levels in liver and kidney. In pituitary IGF-I levels were also reduced, with no change, however, in prepro IGF-I mRNA (37).

In FD rats, plasma membrane receptor binding of IGF-I was increased in pituitary, as was the case in other tissues in this low IGF-I state. No change in IGF-I receptor mRNA was detected in pituitary, however, unlike other tissues in which receptor binding was upregulated (37).

GHRF in FD Rats

GHRF content in extracts of medial basal hypothalamus of FD rats was unchanged from that in fed controls. Prepro GHRF mRNA expression was,

◄ ───

FIGURE 12.1. Effect of 72-hour food deprivation on hypothalamic expression of prepro-GHRF, prepro-SRIF, and β-actin mRNA. (A) Autoradiogram of nuclease protection analysis for GHRF, SRIF, and β-actin mRNA in hypothalamic extracts from fed or 72-hour FD rats. rRNA lane represents ribosomal RNA control. Probe lane indicates RNase undigested cRNA probe. (B) Densitometric quantification of mRNA levels in fed ($n = 15$ animals) and 72-hour FD ($n = 15$) rats. Shown are mean relative densitometric values ($\pm$ SEM) obtained from autoradiograms of nuclease protection analyses of three independent experiments. *$p < .001$ vs. fed. Reprinted with permission from Bruno et al. (36). © 1990 The Endocrine Society.

however, markedly reduced (Fig. 12.1). This reduction in hypothalamic GHRF mRNA was progressive over the duration of food deprivation from a 30% decrease at 24 hours to 80% decrease at 72 hours (Fig. 12.2). Refeeding normal rat chow to FD rats restored hypothalamic GHRF mRNA expression over the next 72 hours such that GHRF mRNA expres-

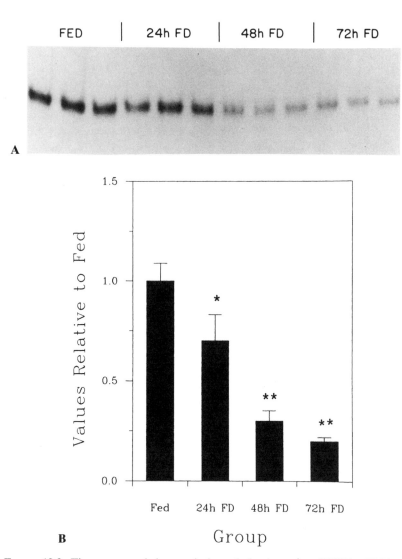

FIGURE 12.2. Time course of changes in hypothalamic prepro-GHRF mRNA after food deprivation. (*A*) Nuclease protection analysis for prepro-GHRF mRNA. (*B*) Histogram of relative mean densitometric values (± SEM) from nuclease protection analyses (*n* = 6 rats/group). *p* < .05, **p* < .001 vs. fed. Reprinted with permission from Bruno et al. (36). © 1990 The Endocrine Society.

FED | 72h FD | 24h RF | 48h RF | 72h RF

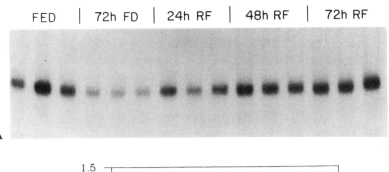

A

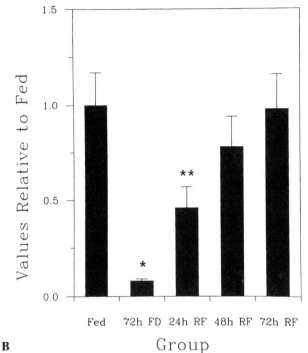

B   Group

FIGURE 12.3. Effect of refeeding after 72-hour FD on hypothalamic prepro-GHRF mRNA. (*A*) Nuclease protection analysis for prepro-GHRF mRNA. (*B*) Histogram of relative mean densitometric values ($\pm$ SEM) from nuclease protection analysis (*n* = 6 rats/group). *$p < .001$, **$p < .05$ vs. fed. Reprinted with permission from Bruno et al. (36). © 1990 The Endocrine Society.

sion was identical to that of fed controls after 72 hours of refeeding (Fig. 12.3) (36).

Utilizing exclusion diets we set about attempting to determine which dietary component(s) might be particularly relevant to regulation of hypothalamic GHRF mRNA expression. Refeeding diets that were complete, fat free, or carbohydrate free given to 72 hour FD rats restored to fed levels the decreased GHRF mRNA seen in FD rats. On the other hand, protein-free refeeding was unable to do so and was without effect in restoring low hypothalamic GHRF mRNA expression in FD rats (Fig. 12.4). Various control experiments suggested this was a specific protein effect. The effect of dietary protein was dose related, with a progressive increase in GHRF mRNA expression seen as dietary protein refeeding was increased from 4% to 12% (normal content in rat chow) (Fig. 12.5). Feeding normal rats a diet specifically deficient in protein caused a marked reduction in hypothalamic GHRF mRNA to levels seen in FD rats over the same 72 hour period (38).

The inability of protein-free diets to restore hypothalamic GHRF mRNA expression in FD rats provided a model system in which the role of different amino acids could be determined as dietary regulators of GHRF mRNA expression. When tyrosine, tryptophan, or glutamic acid were added to

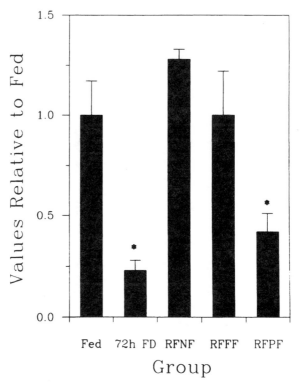

FIGURE 12.4. Effect of selective nutrient refeeding on hypothalamic prepro-GHRF mRNA. Relative mean densitometric values ($\pm$ SEM) from nuclease protection analysis from fed, 72 hour FD rats, and 72 hour FD rats refed normal (RFNF), fat-free (RFFF), or protein-free (RFPF) diets ($n = 6$/group). *$p < .05$ vs. fed. Reprinted with permission from Bruno et al. (38). © 1991 The Endocrine Society.

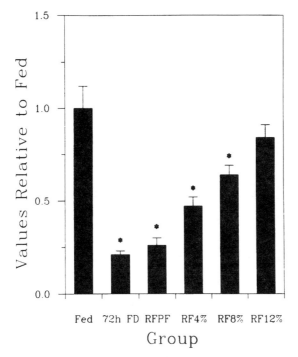

FIGURE 12.5. Effect of graded protein refeeding after 72-hour FD on hypothalamic prepro-GHRF mRNA. Histogram of relative mean densitometric values (± SEM) from nuclease protection analysis from 72 hour FD rats refed protein free (RFPF) diet or diet containing 4%, 8%, or 12% protein (n = 6 rats/group). *p < .05 vs. fed. Reprinted with permission from Bruno et al. (38). © 1991 The Endocrine Society.

protein-free chow, they had no effect on GHRF mRNA expression. Addition of histidine, in contrast, restored GHRF mRNA expression in FD rats to levels not significantly different than those in fed rats. Since histidine is a precursor of histamine, its role as neurotransmitter precursor was evaluated by treating normal fed rats with the histidine decarboxylase inhibitor αFMH, thereby producing a state of neural histamine depletion. While αFMH was without effect on hypothalamic levels of SRIF or NPY mRNA expression, GHRF mRNA was reduced by 40% (39).

Summary

Once again a metabolic state in the rat resulting in GH suppression, in this series of studies food deprivation, was associated with multiple perturbations of the GH regulatory axis.

1. Evidence for increased SRIF tone on the pituitary is provided by our findings (decreased SRIF receptor binding and SRIF receptor mRNA

expression), but we were unable to demonstrate that this SRIF had a hypothalamic source.

2. IGF-I and IGF-I receptor alterations in pituitary of FD rats were similar to those seen in other tissues and appear to reflect GH deficiency rather than provide an explanation for it.

3. Reduction of hypothalamic GHRF mRNA expression was the most startling finding in FD rats. This seemed to result from protein deficiency and could largely be explained according to our data by the deficiency of specific amino acid neurotransmitter precursors, specifically histidine.

Conclusion

GH is a hormone with profound effects on metabolic homeostasis—direct and via IGF-I mediation. For reasons as yet unclear, episodic GH secretion is important to its function, and multiple layers of regulation exist that provide for this to occur. Metabolic influences normally regulate GH secretion and, in turn, abnormal metabolic states result in aberrant GH secretion. Based on our studies in rat models of metabolic states that occur commonly in humans, it seems clear that no single regulatory system accounts for alterations of GH synthesis and secretion.

In STZ-diabetic rats an increase in hypothalamic SRIF tone and an increase in local IGF-I together provide powerful negative influences on pituitary GH synthesis and secretion and pituitary SRIF receptor expression/binding. These defects are accompanied by decreased GHRF mRNA expression, which would certainly augment the GH deficient state. Which defect is primary (if any) is unclear, but certainly all respond well to insulin therapy.

In food-deprivation, GHRF deficiency appears to be the primary defect resulting in suppression of GH synthesis/secretion. The perturbations of SRIF receptor binding and mRNA expression seen in FD rats could be accounted for by increased SRIF originating in the peripheral circulation (for example, the GI tract). Similarly, pituitary IGF alterations appear to result from GH deficiency rather than account for it.

Further mechanistic insights will require in vitro systems and studies based on these in vivo findings. Models that more closely mimic human GH responses to metabolic disturbances must also be identified in order to extend these findings and establish their relevance to the perturbations of GH secretagogues seen in human diabetes mellitus, malnutrition, and other metabolic disturbances.

References

1. Quabbe HJ. Hypothalamic control of GH secretion: pathophysiology and clinical implications. Acta Neurochir 1985;75:60–71.

2. Martin JB, Reichlin S. Regulation of growth hormone secretion and its disorders. In: Martin JB, Reichlin S, editors. Clinical neuroendocrinol, 2nd ed. Philadelphia: FA Davis, 1987:233–94.
3. Strobl JS, Thomas MJ. Human growth hormone. Pharmacol Rev 1994;46:1–34.
4. Frohman LA, Jansson J-O. Growth hormone-releasing hormone. Endocr Rev 1986;7:223–53.
5. Gelato MC, Merriam GR. Growth hormone releasing hormone. Annu Rev Physiol 1986;48:569–91.
6. Gomez-Pan A, Rodriguez-Arnao MD. Somatostatin and growth hormone releasing factor: synthesis, location, metabolism and function. Clin Endocrinol Metab 1983;12:469–507.
7. Mayo KE. Molecular cloning and expression of a pituitary-specific receptor for growth hormone-releasing hormone. Mol Endocrinol 1992;6:1734–44.
8. Mayo KE, Hammer RE, Swanson LW, Brinster RL, Rosenfeld MG, Evans RM. Dramatic pituitary hyperplasia in transgenic mice expressing a human growth hormone-releasing factor gene. Mol Endocrinol 1988;2:606–12.
9. Reichlin, S. Somatostatin. N Engl J Med 1983;309:1495–1501, 1556–63.
10. Patel YC, Srikant CB. Somatostatin mediation of adenohypophysial secretion. Annu Rev Physiol 1986;48:551–67.
11. Bruno JF, Berelowitz M. Somatostatin receptor: orphan that found family and function. Mol Cell Neurosci 1993;4:307–9.
12. Bruno JF, Xu Y, Song J, Berelowitz M. Tissue distribution of somatostatin receptor subtype messenger ribonucleic acid in the rat. Endocrinology 1993;133:2561–7.
13. Patel YC, Murthy KK, Escher EE, Banville D, Spiess J, Srikant C. Mechanism of action of somatostatin: an overview of receptor function and studies of the molecular characterization and purification of somatostatin receptor proteins. Metabolism 1990;39:63–9.
14. Rens-Domiano S, Reisine T. Biochemical and functional properties of somatostatin receptors. J Neurochem 1992;58:1987–96.
15. Tannenbaum GS, Ling N. The interrelationship of growth hormone (GH)-releasing factor and somatostatin in generation of the ultradian rhythm of GH secretion. Endocrinology 1984;115:1952–7.
16. Plotsky PM, Vale WW. Patterns of growth hormone-releasing factor and somatostatin secretion into the hypophyseal-portal circulation of the rat. Science 1986;230:461–3.
17. Frohman LA, Downs TR, Clarke IJ, Thomas GB. Measurement of growth hormone-releasing hormone and somatostatin in hypothalamic-portal plasma of unanesthetized sheep. Spontaneous secretion and response to insulin-induced hypoglycemia. J Clin Invest 1990;86:17–24.
18. Van Wyk JJ. The somatomedins: biological actions and physiological control mechanisms. In: Li CH, ed. Hormonal proteins and peptides. New York: Academic Press, 1984;12:82–125.
19. Froesch ER, Schmid C, Schwander J, Zapf J. Actions of insulin-like growth factors. Annu Rev Physiol 1985;47:443–67.
20. Berelowitz M, Szabo M, Frohman LA, Firestone S, Chu L, Hintz RL. Somatomedin-C mediates growth hormone negative feedback by effects on both the hypothalamus and the pituitary. Science 1981;212:1279–81.

21. Morita S, Yamashita S, Melmed S. Insulin-like growth factor 1 action on rat anterior pituitary cell: effects of intracellular messengers on growth hormone secretion and messenger ribonucleic acid levels. Endocrinology 1987;122: 2204–10.

22. Delitala G, Tomasi P, Virdis R. Neuroendocrine regulation of human growth hormone secretion. Diagnostic and clinical applications. J Endocrinol Invest 1988;11:441–62.

23. Dieguez C, Page MD, Scanlon MF. Growth hormone neuroregulation and its alterations in disease states. Clin Endocrinol 1988;28:109–43.

24. Press M. Growth hormone and metabolism. Diabetes Metab Rev 1988;4: 391–414.

25. Berelowitz M, Dudlak D, Frohman L. Release of somatostatin-like immunoreactivity from incubated rat hypothalamus and cerebral cortex. J Clin Invest 1982;69:1293–6.

26. Tannenbaum GS, Rorstad O, Brazeau P. Effects of prolonged food deprivation on the ultradian growth hormone rhythm and immunoreactive somatostatin tissue levels in the rat. Endocrinology 1979;104:1733–8.

27. Campbell PJ, Bolli CB, Cryer PE, Gerich JE. Pathogenesis of the dawn phenomenon in patients with insulin-dependent diabetes mellitus. Accelerated glucose production and impaired glucose utilization due to nocturnal surges in growth hormone secretion. N Engl J Med 1985;312:1473–8.

28. Tannenbaum GS. Growth hormone secretory dynamics in streptozotocin diabetes: evidence of a role for endogenous circulating somatostatin. Endocrinology 1981;108:76–81.

29. Tannenbaum GS, Colle E, Gurd W, Wanamaker L. Dynamic time-course studies of the spontaneously diabetic BB Wistar rat. I. Longitudinal profiles of plasma growth hormone, insulin, and glucose. Endocrinology 1981;109:1872–9.

30. Gonzalez C, Jolin T. Effect of streptozotocin diabetes and insulin replacement on growth hormone in rats. J Endocrinol Invest 1985;8:7–11.

31. Olchovsky D, Bruno J, Wood T, Gelato M, Leidy J, Gilbert J, Berelowitz M. Altered pituitary growth hormone (GH) regulation in streptozotocin-diabetic rats: a combined defect of hypothalamic somatostatin and GH-releasing factor. Endocrinology 1990;126:53–61.

32. Bruno J, Xu Y, Song J, Berelowitz M. Pituitary and hypothalamic somatostatin receptor subtype messenger ribonucleic acid expression in the food-deprived and diabetic rat. Endocrinology 1994;135:1787–92.

33. Olchovsky D, Bruno J, Gelato M, Song J, Berelowitz M. Pituitary insulin-like growth factor-I content and gene expression in the streptozotocin-diabetic rat: evidence for tissue-specific regulation. Endocrinology 1991;128:923–8.

34. Tannenbaum GS, Epelbaum J, Colle E, Brazeau P, Martin JB. Antiserum to somatostatin reverses starvation-induced inhibition of growth hormone but not insulin secretion. Endocrinology 1978;102:1909–14.

35. Hugues JN, Enjalbert A, Moyse E, Shu C, Voirol MJ, Sebaoun J, Epelbaum J. Differential effects of passive immunization with somatostatin antiserum on adenohypophysial hormone secretions in starved rats. J Endocrinol 1986; 109:169–74.

36. Bruno J, Olchovsky D, White J, Leidy J, Song J, Berelowitz M. Influence of food deprivation in the rat on hypothalamic expression of growth hormone-releasing factor and somatostatin. Endocrinology 1990;127:2111–6.

37. Olchovsky D, Song J, Gelato M, Sherwood J, Spatola E, Bruno J, Berelowitz M. Pituitary and hypothalamic insulin-like growth factor-I (IGF-I) and IGF-I receptor expression in food-deprived rats. Mol Cell Endocrinol 1993;93: 193–8.
38. Bruno J, Song J, Berelowitz M. Regulation of rat hypothalamic prepro growth hormone-releasing factor messenger ribonucleic acid by dietary protein. Endocrinology 1991;129:1226–32.
39. Bruno J, Song J, Xu Y, Berelowitz M. Regulation of hypothalamic prepro growth hormone-releasing factor messenger ribonucleic acid expression in food-deprived rats: a role for histaminergic neurotransmission. Endocrinology 1993;133:1377–81.

Part IV

Physiology of Growth Hormone Secretagogues

13

The Role of Growth Hormone Releasing Hormone (GHRH) and Growth Hormone (GH) in the Onset of Puberty and During Glucocorticoid-Altered Growth

Lisa K. Conley, Michel L. Aubert, Andrea Giustina, and William B. Wehrenberg

Pulsatile growth hormone (GH) release is under the control of two cyclically released hypothalamic peptides—growth hormone releasing hormone (GHRH) and somatostatin. Areas of the hypothalamus responsible for the control of GH release are the somatostatin neuron-containing periventricular nucleus (PeV) (1–3) and the GHRH neuron-containing arcuate nucleus (4–6).

GHRH has been found to stimulate the synthesis and nuclear export of GH messenger RNA (mRNA) (7) from pituitary somatotrophs, as well as somatotroph synthesis and release of GH (7–9). The role of somatostatin in the control of GH secretion, on the other hand, appears to involve primarily inhibition of GH mRNA nuclear export (10) and GH release (11–13) perhaps via inhibition of GHRH-stimulated GH release from the pituitary (9). Plotsky and Vale (14) in 1985 demonstrated that maximal hypophysial portal GHRH concentrations regularly occurred during periods of expected GH peaks. This suggested that peak portal GHRH concentrations may be responsible for the periods of peak GH concentrations in peripheral plasma. In contrast, it has been shown that high hypophysial portal somatostatin concentrations are most likely responsible for the maintenance of low plasma GH concentrations characteristic of GH trough periods, as passive immunization with somatostatin antiserum increases basal GH concentrations in plasma (15). An interaction between GHRH and somatostatin in the generation of characteristic pulsatile GH release was supported by the finding that peak hypophysial portal GHRH concentrations are accompanied by moderate reductions in portal somatostatin levels (14) and that peripheral administration of GHRH antiserum completely inhibits the pulsatile secretion of GH (16). It is of interest that somatostatin administration always results in a decrease in circulating GH (11), while GHRH is not always effective in inducing pituitary GH release (17). This suggests that somatostatin is the predominant modulator of pituitary GH release and precipitates the current model for the control of GH. That is, the two neuropeptides, GHRH and somatostatin, are both cyclically released at intervals 180° out of phase from each other. The presence of peak portal GHRH concentrations in the face of low portal somatostatin tone determines the characteristics of the peripheral GH peaks, while peak portal somatostatin concentrations are responsible for plasma GH trough periods.

One of the major tools used by us and others to elucidate the mechanisms involved in the neuroendocrine regulation of GH secretion has been the passive immunization against endogenous GHRH and somatostatin (16–18). In light of the fact that passive immunization is noninvasive, highly specific, and reversible, we have expanded on our initial studies to investigate the importance of GHRH and the GH axis in the onset of puberty and to investigate the mechanism(s) behind alterations in somatic growth associated with glucocorticoid treatment. This chapter provides an overview of our recent advances in these two areas.

Involvement of the Growth Axis in the Onset of Puberty

Puberty refers to the gradual series of complex biologic and maturational events that lead to reproductive maturity. Although the somatic and physiologic changes that are associated with puberty in both males and females have been well documented (19–22), the exact neuroendocrine mechanisms that act to initiate the onset of puberty have yet to be elucidated. Although it is generally accepted that sexual maturation is primarily under the control of the pituitary gonadotropins, an interaction of the reproductive axis with other neuroendocrine components as a means of "triggering" puberty onset has not been ruled out. Among these possibilities, a potential interaction between the somatotropic and gonadotropic axes is compelling.

The existence of an interaction between these two axes is reinforced by observations that changes in the body's androgen milieu can affect the growth axis. This is most strongly suggested by the fact that GH secretion in the adult rat is sexually dimorphic (23). For example, adult male rats exhibit a predictable GH secretory pattern characterized by GH surges that occur regularly every 3 to 4 hours, with high peaks (200–300 ng/ml) separated by low troughs (<5 ng/ml) (24). Adult female rats, on the other hand, exhibit a less-predictable GH secretory pattern with shorter times between bursts, higher baseline plasma GH concentrations (10 ng/ml), and lower mean peak amplitudes (<100 ng/ml) (24, 25). By 90 days of age, these peaks have been found to occur without any regular timing (24, 26).

The fact that these sex differences become apparent only after the onset of puberty provides strong support for the notion that sex steroids are important in the development and maintenance of these sexually dimorphic adult GH secretory patterns (24, 27). Jansson and Frohman (28) demonstrated that neonatal orchidectomy of male rats shifted the adult male secretory pattern toward that of an adult female. In addition, they have also shown that neonatal gonadectomy of female rats increases the GH plasma pulse heights during puberty and decreases GH plasma baselines postpuberty (29). In fact, the development of male-pattern GH release in the female rat is inhibited by the presence of intact ovaries (30). Sexually dimorphic pituitary responsiveness to sex steroids has also been well documented (31). For example, estrogen treatment of and castration of male rats lowers pituitary GH content to levels similar to those found in untreated females (feminizes), and exposure of females to exogenous testosterone elevates pituitary GH content (masculinizes) (27).

An interaction of the two axes is also supported by observations that age-related changes in androgen secretion affect somatic growth (32, 33). Ojeda and Jameson (34), in their mapping of developmental changes in plasma and pituitary GH in the female rat, were among the first to suggest that the growth axis might play a role in the progression of sexual maturation. They suggested that fluctuations noted in circulating GH concentrations, which precede the gonadotropin surge associated with the onset of puberty, may play a physiologic role in the control of pubertal processes. In addition, GH deficiency in humans, which results in small body sizes for age, is often accompanied by a delay in the timing of puberty (35). This relationship again suggests that factors controlling growth or growth rate may have an effect upon the gonadotropin axis.

A possible role for GH and growth in the timing of puberty is also suggested by a large, but rather controversial, body of work that suggests that the onset of puberty is triggered by a signal that alerts brain centers controlling reproductive function that the body is large enough or is properly composed to support reproductive function (36, 37). These have been termed the "critical body size" or "critical body composition" theories and have been precipitated by a number of observations. For example, the timing of puberty across species occurs when the body is reaching adult proportions and composition (20, 38–40). In addition, the need for "adult sized" energy stores or fat-to-protein ratios appear important for the onset of puberty and crucial for the maintenance of reproductive function (20, 41).

Several approaches have been used to decrease GH release in order to study its involvement in the onset of puberty. These included median eminence GH implantation (42), pharmacologic GH treatment (43), and infection with the growth factor–secreting plerocercoid larvae of the tape worm *Spirometra mansonoides* (44). In most cases, a delay in the onset of puberty (as assigned by a delay in vaginal opening) was observed in the female model studied, suggesting that normal circulating levels of GH were requisite for the timely onset of puberty. It was not always clear, however, what confounding effects occurred in these studies. For example, the growth factor from *S. mansonoides* has been found to have insulin-like actions and to be structurally related to GH, not to insulin-like growth factors (IGFs). Therefore, because the factor is capable of cross-reacting with GH receptors (45), exaggerated rather than reduced GH effects may have been imposed. It is also not known what nonspecific effects were triggered by this infection or by the production of this growth factor. Problems with nonphysiologic conditions, nonspecificity of treatment, and CNS damage are also apparent with the other methods.

Our group has found passive immunization against rat GHRH to be a useful tool for investigating the role of GH secretion in somatic growth under various conditions (16, 46, 47). Figure 13.1 demonstrates the effec-

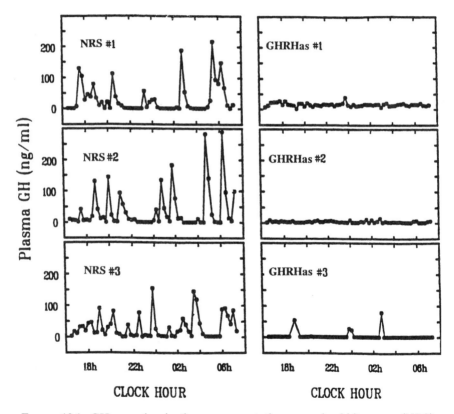

FIGURE 13.1. GH secretion in three representative normal rabbit serum (NRS)-treated control female rats (left panels) and in three growth hormone releasing hormone antiserum (GHRHas)-treated female rats (right panels). Blood was collected every 15 min from an indwelling jugular catheter using an automated blood-collecting machine. Reprinted with permission from Gruaz et al. (82).

tiveness of such treatment in female rats. The left column depicts the characteristic female secretory pattern noted in three intact rats. The right column, on the other hand, represents the secretory pattern of GH secretion in three rats that received antiserum against GHRH (GHRHas). It can be concluded from Figure 13.1 that GHRHas treatment is efficient in abolishing pulsatile GH release. We hypothesized that GHRHas treatment would allow for the study of a possible direct role of GHRH and GH in the course of sexual maturation by means of a specific and direct inhibition of GHRH action at the pituitary level. This inhibition then results in a temporary reduction in circulating GH. Moreover, passive immunization against GHRH allows one to precisely control the period and duration of GH deficiency. We have subsequently investigated the effect of GH deficiency so induced on the sexual development of both male and female rats and will review these data here beginning with our studies conducted in males.

The GH Axis and the Onset of Puberty in Male Rats

We investigated the effect of GH deficiency on sexual maturation in the male rat by administering GHRHas ($250\,\mu l$ sc, every other day) from 15 to 39 days of age. Sexual maturation in the male rat is a continuous process that starts at about 25 days and is nearly complete at about 55 to 60 days of age (48). Forty days of age can be considered an important midpoint of this process, so it was felt that the time period between 15 and 40 days would be appropriate for inducing transient GH deprivation in order to study the effects of such conditions on the early stages of sexual maturation. In addition, treatment regimens similar to the one chosen had been found to be effective in reducing growth rates in rats in prior studies (46). The progress of sexual maturation can be assessed by observing weight gains of the sexual organs or tissues, pituitary synthesis of gonadotropins, plasma levels of gonadotropins or androgens, as well as the progress of spermatogenesis. Because no clear-cut marker for dating sexual maturation, such as vaginal opening in the female rat, exists for the male rat, a number of the aforementioned parameters were followed.

As was suggested by earlier studies (46, 47), the growth rates of male rats that received GHRHas (Fig. 13.2) were significantly lower than those rates noted in their vehicle-treated counterparts. What was especially striking was the observation that the effect of GHRHas treatment was evident 3 days after the initiation of treatment (at 18 days of age). The difference in body weight between the two groups continued to increase until about 45 days, and by 50 days of age, the mean body weight of the GHRHas-treated rats was 64% that of the vehicle-treated animals. As we have seen in the past (47), no catch-up growth was observed in the GHRHas-treated rats following cessation of treatment. Following cessation of antisera treatment, the growth rates of all groups were similar, however.

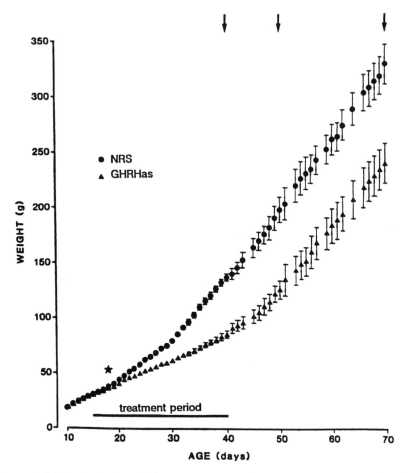

FIGURE 13.2. Daily body weight measurements in normal rabbit serum (NRS)-treated (●) and GHRH antiserum (GHRHas)-treated (▲) male rats. GHRHas (250 μl sc) was administered every second day between days 15 and 39 as indicated by the solid line. The star (★) indicates the first day with a significant difference in weight between NRS- and GHRHas-treated rats. Data are presented as mean ± SE. Groups of rats were sacrificed at 40, 50, and 70 days as indicated (arrows). Reprinted with permission from Arsenijevic et al. (83).

Besides the noted decrease in growth rate in rats that received GHRHas, a partial or complete blockade of GH synthesis can be inferred from the very low pituitary GH content noted at 40 and 50 days (Fig. 13.3). This suggests the existence of a defect in pituitary hormone synthesis and possibly a delayed differentiation of some of the pituitary cell types due to passive immunization against GHRH. Plasma insulin-like growth factor (IGF)-I concentrations were also depressed in GHRHas-treated rats at 40 and 50 days of age, although concentrations still represented about 30% of

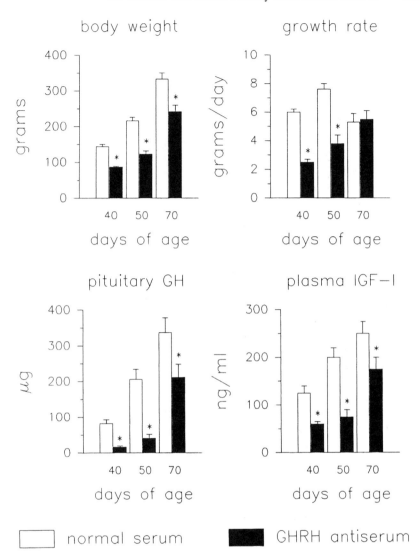

FIGURE 13.3. Comparison of the effects of normal rabbit serum (normal serum) or GHRH antiserum (GHRHas) on growth and maturation of GH axis parameters seen at 40, 50, and 70 days of age in male rats. GHRHas treatment resulted in a significant reduction in body weight, pituitary GH content, and plasma IGF-I concentrations for the duration of the study. GHRHas reduced the growth rate at 40 and 50 days of age, but growth rates between GHRHas- and NRS-treated rats were not different by 70 days of age. (*) indicates significant differences ($p < .05$). Selected data used with permission from Arsenijevic et al. (83).

those noted in the control animals (Fig. 13.3). It has been suggested that IGF-I production can serve as a marker of both GH action and normal ongoing growth (49, 50). It was of interest in this study, then, that IGF-I production was reduced by two-thirds in the face of a slow growth process and absent pulsatile GH secretion. The residual production of IGF-I might represent basal or GH-independent release. Recovery of normal plasma IGF-I concentrations was slow, as shown by the reduced IGF-I levels noted at 70 days of age (Fig. 13.3). An important question, then, is why the animals that received GHRHas continued to grow, albeit at a slower rate than their vehicle-treated counterparts. It could be suggested that neutralization of GHRH allows a small amount of basal GHRH-independent GH secretion. This basal release, although too small to be identified by current assay techniques, may have contributed to the residual growth noted in these animals. Alternatively, GH-independent growth mechanisms may have been involved.

As mentioned above, sexual maturation was assessed by several factors. At 40 days of age, animals treated with GHRHas demonstrated clear signs of delayed sexual maturation. This could be inferred from incomplete spermatogenesis (Fig. 13.4), delayed growth of the testes and seminal vesicles, and reduced pituitary gonadotropin content (Fig. 13.5) as compared with those markers in control animals. There is room for argument that difficulties abound in the correct establishment of a delay in growth of tissues such as gonads and seminal vesicles in animals suffering from a growth defect. For example, it could be argued that organ size should be corrected for total body size, as small rats would be expected to have small testicles. However, we found this type of correction misleading as the induced body growth deficit is so important that GHRHas-treated animals appeared to have relatively large testicles. The same was true for the pituitary content of the gonadotropins. Thus, reporting pituitary gonadotropin content in relation to pituitary weights resulted in the observation that luteinizing hormone (LH) pituitary content of animals treated with GHRHas appeared normal in the face of the obvious conclusion that pituitary somatotrophs, and possibly other pituitary cells, were atrophic due to the lack of GHRH stimulation or due to the reduction of other trophic factors such as IGF-I. It was also of interest that the induced delay in sexual maturation was only transient with recovery of normal spermatogenesis at 50 days of age in GHRHas-treated animals. This recovery occurred despite the fact that body weights of the rats receiving GHRHas were still drastically reduced (Fig. 13.2). It was also striking that follicle stimulating hormone (FSH) synthesis and secretion were the most affected reproductive parameters in GHRHas-treated animals (Fig. 13.5). This could be one factor contributing to the delayed development of testicular functions seen here as the trophic function of FSH in testicular maturation and in the progress of spermatogenesis are well known (51).

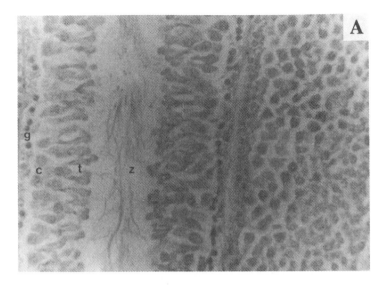

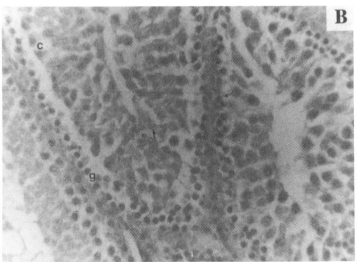

FIGURE 13.4. Spermatogenesis at 40 days of age in NRS- and GHRHas-treated male rats. NRS-treated rats (*A*) demonstrated normal and complete spermatogenesis: spermatogonia and spermatocytes are distributed in regular layers with numerous spermatids. Mature spermatozoa are present in the lumen. With GHRHas treatment (*B*), however, spermatogonia and spermatocytes are present but only scattered spermatids and spermatozoa are observed. g, spermatogonia; c, spermatocytes; t, spermatids; z, spermatozoa. Reprinted with permission from Arsenijevic et al. (83).

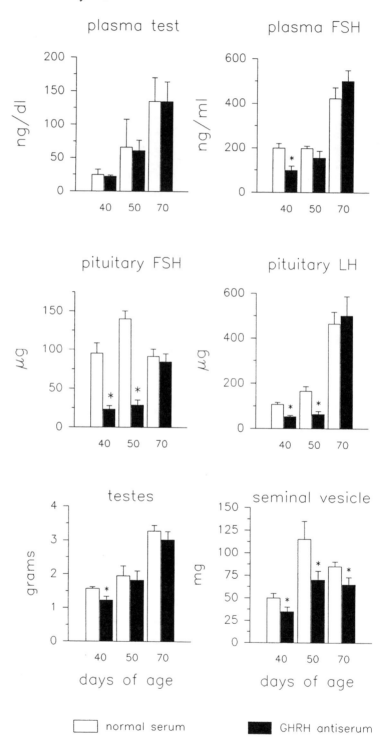

In summary, it has been postulated that GH plays at least a permissive role in the process of sexual maturation in the male rat (48, 52). Our study provides some support for the importance of the growth axis in the course of sexual maturation in the male rat. As for the possibility of a direct action of GH in this process, the delay of sexual maturation in male rats treated with GHRHas may have been the result of a lack of paracrine action of GH with the pituitary gonadotropins, or a lack of permissive action of GH on the maturation of Leydig and/or Sertoli cells. For example, it has been shown that GH has an effect on LH receptor content in hypophysectomized rats (53). In addition, since central effects of GH have been reported (54, 55), the lack of GH secretion during development may have had an effect upon gonadotropin-releasing hormone (GnRH) secretion directly. And finally, the observed delay of sexual maturation could also be attributable to the low levels of IGF-I noted during the very slow growth rate observed during GHRHas treatment. This possibility is supported by our observation that, between 40 and 50 days, during the time of increased IGF-I release, growth accelerated and sexual maturation occurred.

The GH Axis and the Onset of Puberty in Female Rats

After demonstrating a transient delay in puberty onset in male rats treated with GHRHas, we were naturally curious as to whether or not similar treatment of female rats would have the same effect. Thus, the effects of GH deprivation and low IGF-I plasma concentrations on sexual maturation were studied in female rats again using passive immunization against GHRH starting at 15 days of life.

As was noted with male rats, subcutaneous treatment with GHRHas resulted in a prompt reduction in growth rate with body weight already significantly decreased 2 days after the initiation of treatment (Fig. 13.6). At 35 days of age, the mean body weight of the GHRHas-treated rats was 36.2% lower than that noted in control rats. In spite of this significant decrease, the occurrence of vaginal opening was identical in the two groups (Fig. 13.6). As seen in the study using male rats, no period of compensatory growth was noted following cessation of antiserum treatment. Besides a significant effect on body weight, passive immunization against GHRH also

◀——————————————————————————————

FIGURE 13.5. Comparison of the effects of NRS or GHRHas treatments on growth and maturation of various reproductive axis parameters as seen at 40, 50, and 70 days of age in male rats. GHRHas treatment resulted in a significant reduction in plasma FSH concentrations at 40 days and pituitary FSH and LH content at 40 and 50 days of age. Seminal vesicle size was significantly reduced by the GHRHas treatment at all time points analyzed. Testes weight was only affected at 40 days. Plasma testosterone levels were unaffected. (*) indicates significant differences ($p < .05$). Selected data used with permission from Arsenijevic (83).

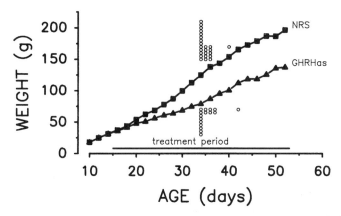

FIGURE 13.6. Body weights of NRS-treated (■) and GHRHas-treated (▲) female rats. GHRHas (250 μl sc) was administered every second day from days 15 through day 53 of age as indicated by the solid line. Data are the mean ± SE. Open circles represent the occurrence of vaginal opening for individual rats. Redrawn with permission from Gruaz et al. (82).

resulted in a significant decrease in plasma IGF-I concentrations, as well as pituitary GH content (Fig. 13.7).

In contrast to the definite effects of passive immunization on the GH axis, GH deprivation via GHRHas treatment did not produce many significant changes in the course of sexual maturation in female rats (Fig. 13.8). Most striking was the fact that vaginal opening took place at the same age in the GHRHas- and the control-treated rats (Figs. 13.6 and 13.8). In addition, estrous cycle length as well as pituitary FSH and LH content were unaffected by antiserum treatment. Ovarian and uteri weights were diminished in rats treated with GHRHas (Fig. 13.8), although in a manner proportional to the induced reduction in body size.

In summary, the results of this study suggest that GH itself may not play a major role in determining the progression of sexual maturation in the female rat. Nevertheless, some evidence for a trophic action of GH on the ovary has been documented (56–58). This trophic action may be on the development of the ovaries or involved in the activation of functional processes. The role of GH at the ovary could be direct or via the generation of IGF-I, either hepatic or ovarian. The existence of an intraovarian IGF system involving IGF-I and -II, specific receptor subtypes, as well as several populations of IGF binding proteins has been documented (59, 60). At the ovary, however, the internal signaling system involving IGFs appears to specifically respond to gonadotropins and not to be dependent upon external influences such as GH or hepatic IGF-I. It is interesting, then, that vaginal opening in our study occurred in the face of low circulating IGF-I concentrations.

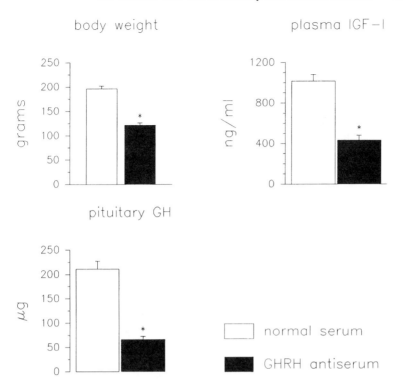

FIGURE 13.7. Effects of passive immunization against GHRH on several parameters of the somatotrophic axis. Animals were treated from day 15 to day 53 of age and sacrificed on the last day of treatment. GHRHas treatment resulted in significant reductions in body weight, plasma IGF-I concentrations, and pituitary GH content (*$p < .05$). Selected data used with permission from Gruaz et al. (82).

In conclusion, the involvement of GH in the progress of sexual matura-tion in the rat is most complex. The growth impairment induced during the juvenile period in the male rat produced a transient delay of sexual matura-tion. The induced transient decrease in circulating GH appears to have most affected the synthesis and secretion of FSH, which then resulted in a delay in testes growth and in a slowed germinal cell differentiation rate. Although it appears that IGF-I might be a major factor in determining the progression of sexual maturation in the male rat, neither pulsatile GH secretion nor normal plasma IGF-I concentrations appear necessary for the onset of puberty in the female rat. Of special note are our observations that growth rate or achieved body weight do not seem to be critical factors to initiate sexual maturation in either the male or female rat. It should be noted, however, that passive immunization against GHRH does not com-pletely halt somatic growth. Thus, the possibility remains that very low circulating plasma GH or IGF-I concentrations may be present in these

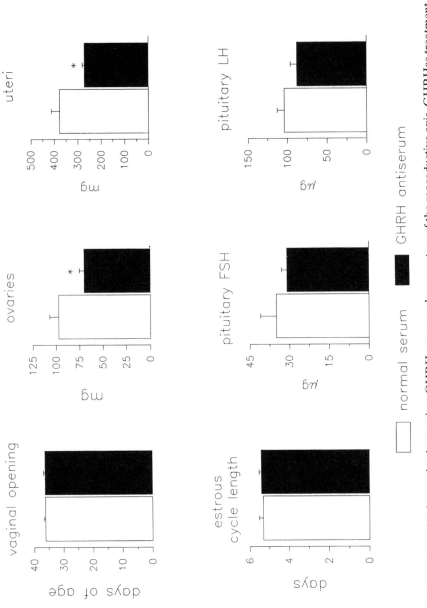

FIGURE 13.8. Effects of passive immunization against GHRH on several parameters of the reproductive axis. GHRHas treatment resulted in significant reductions in ovarian and uteri weights (*p < .05), although these were normal when adjusted for body weight. The other parameters measured were not affected by GHRHas treatment. Selected data used with permission from Gruaz et al. (82).

animals and that these circulating levels may be enough to interact with a
GH-hypersensitive ovary to allow puberty to proceed normally.

Altered Growth Induced by Glucocorticoid Treatment

It is well established that glucocorticoids are potent inhibitors of linear
growth in humans and animals (61, 62). Indeed, we have ourselves observed
such a phenomenon repeatedly in our work (Fig. 13.9). An immediate
conclusion is that the glucocorticoids somehow interact with the GH axis to
elicit such effects.

A relationship between these steroids and the GH axis was first established in vitro where glucocorticoid treatment was found to stimulate GH
secretion and synthesis (8, 63, 64). This enhanced GH secretion was found
following several days of co-incubation and was in fact amplified when cells
were first cultured for long periods without glucocorticoids. A relationship
between the two was also reinforced by work that demonstrated that, in
vitro, glucocorticoid treatment increases the GH responsiveness of cultured
somatotrophs to GHRH. For example, Vale et al. (8) in 1983 reported that
the GH response to GHRH is increased more than fivefold in rat pituitary
cells that were first pretreated for 24 hours with the long-acting glucocorti-

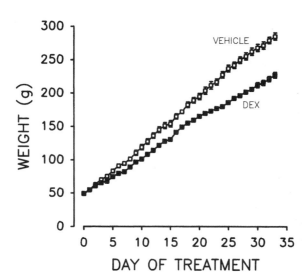

FIGURE 13.9. The effect of vehicle (□) or dexamethasone (DEX; ■) administration
on somatic growth in young male rats. DEX treatment significantly decreased
somatic growth as compared with vehicle-treated controls ($p < .05$). The error bars
(SE) are smaller than symbol sizes in most instances. Redrawn with permission from
Wehrenberg et al. (80).

coid dexamethasone. It has since been found that the mechanism of this action may involve a glucocorticoid-mediated increase in GHRH receptors on the somatotrophs (65). It is of interest that the responsiveness of the somatotrophs to GHRH under conditions of high glucocorticoid tone may be time dependent as it has been shown that GHRH- and forskolin-induced GH release are inhibited following less than 4 hours of exposure to dexamethasone, while following 18 hours of exposure, the somatotroph's responsiveness is enhanced (66). Similar time-dependent biphasic effects have been noted in vivo (67).

Collectively, these in vitro results suggest that in vivo, GH secretion would be suppressed during conditions of low glucocorticoid tone. Indeed, adrenalectomy without steroid replacement has been found to decrease the pituitary response to a submaximal dose of GHRH (68). Since this decreased response can be reversed following glucocorticoid replacement therapy and since a supramaximal GHRH dose fails to elicit a difference in GH response between normal and adrenalectomized rats, it would appear that adrenalectomy results in a decreased pituitary responsiveness to GHRH rather than a change in the releasable GH pool.

It also follows that under conditions of high glucocorticoid tone GH secretion would be increased. Our early work in anesthetized rats demonstrated that dexamethasone treatment of normal rats enhanced the GH response to GHRH (68). These findings from anesthetized animals were in conflict with a large body of data that demonstrate a suppression of GH secretion during conditions of high glucocorticoid tone and with the aforementioned decrease in somatic growth associated with glucocorticoid treatment. Indeed, these data show that in vivo, glucocorticoids decrease daily integrated (69) or nocturnal spontaneous GH secretion (70) in man, and inhibit the GH response to various stimuli, including GHRH, in both man and rats (71–75).

The work of Nakagawa et al. (76) in 1987 began to identify the source of these conflicts. Their work using a conscious animal model demonstrated that following chronic dexamethasone treatment, GH concentrations were suppressed, hypothalamic somatostatin content was increased, and hypothalamic GHRH content was decreased. We subsequently observed in conscious rats that trough or baseline GH values were lower for a longer time interval in glucocorticoid-treated rats than in normal animals (77). In addition, when GH pulses did occur, they were found to be significantly higher than those noted in normal animals (Fig. 13.10).

The conflicting results observed in anesthetized versus conscious rats, coupled with the fact that deep pentobarbital anesthesia has been reported to suppress the release of hypothalamic hormones (78), have precipitated the hypothesis that glucocorticoids are both inhibitors and stimulators of GH release. This theory suggests that the inhibitory effects may be mediated at the hypothalamic level by glucocorticoid enhancement of somatostatin tone, while the stimulatory effects may be mediated via the

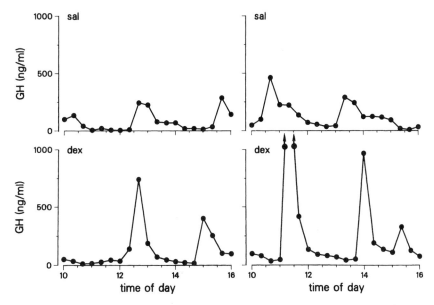

FIGURE 13.10. Four representative examples of plasma GH concentrations in conscious, freely moving male rats. Animals had received either saline (sal) or 40 μg dexamethasone (dex) for 7 days prior to blood sampling. Arrows indicate GH concentrations greater than 1000 ng/ml. Time points for upper and lower panels are identical. Reprinted with permission from Wehrenberg et al. (77).

pituitary where glucocorticoids act to enhance GH synthesis and the GH response to GHRH (79).

This hypothesis was supported by a series of experiments that we have performed in normal, conscious, and freely moving rats treated chronically with either saline or dexamethasone. Here, we found that the GH response to a physiologic dose of GHRH was significantly lower in glucocorticoid-treated rats than in saline-treated controls. When we passively immunized these animals with somatostatin antiserum, however, we saw the reverse. In other words, glucocorticoid-treated rats exhibited a significantly higher GHRH response to GHRH than did saline-treated animals (73). Thus, dexamethasone appeared to enhance hypothalamic release of somatostatin, which resulted in a decreased pituitary responsiveness to exogenous GHRH in normal rats.

To further test the above-mentioned hypothesis at the cellular and molecular levels, we have recently evaluated the effect of dexamethasone treatment on GHRH and somatostatin immunoreactive cells of the median eminence (Table 13.1). Here, we noted no dramatic effect of dexamethasone treatment on the intensity with which GHRH-immunoreactive cells stained or in the number of GHRH-immunoreactive cells located in the median eminence (ME). In marked contrast, however, dexamethasone

TABLE 13.1. Relative change in GHRH or SS immuno-cytochemical staining intensity and the number of cells stained in the median eminence in dexamethasone (Dex)-treated rats as compared with control-treated rats.

	Dex-treated staining intensity	Dex-treated # of cells stained
GHRH	↓	No change
SS	↑↑	↑↑↑

↑ indicates an increase relative to control. ↓ indicates a decrease relative to control. More than one arrow signifies a more pronounced difference from controls.

TABLE 13.2. Relative GH, GHRH, or somatostatin (SS) mRNA responses in dexamethasone (DEX)-treated rats as compared with control-treated rats.

	Dex-treated (mRNA)
GH	↑↑
GHRH	↓
SS	↑↑

↑ indicates an increase and ↓ a decrease in response relative to controls. More than one arrow signifies a more pronounced difference from controls.

treatment elicited a dramatic increase in immunoreactive somatostatin staining, especially when considering the number of somatostatin-immunoreactive cells in the ME. Considering these findings, we then used a ribonuclease (RNase) protection assay to look at what effect dexamethasone treatment has upon hypothalamic GHRH and somatostatin message, as well as on pituitary GH message content (Table 13.2). This evaluation revealed that dexamethasone treatment produces a large increase in pituitary GH and hypothalamic somatostatin mRNA content, while hypothalamic GHRH mRNA content is slightly decreased.

Coming full circle then, these results suggest that the inhibitory effects of glucocorticoids upon linear growth may be due, at least in part, to decreased GH secretion in response to enhanced somatostatin tone. This conclusion is further supported by a study in which we immunoneutralized endogenous somatostatin in rats chronically subjected to growth inhibition via dexamethasone treatment (80). Although the effect of passive immunization on growth was not complete, a partial reversal of the blunted somatic growth induced by dexamethasone treatment was observed (Fig. 13.11). The fact that the dexamethasone-induced inhibition of growth was only partially reversed suggests that the pathogenesis of glucocorticoid-mediated growth inhibition is multifactorial. For example, somatostatin is

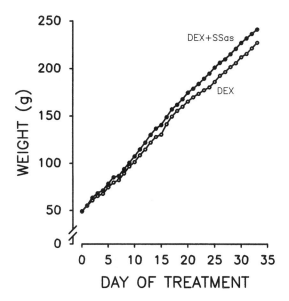

FIGURE 13.11. The effect of dexamethasone plus vehicle (DEX; ○) or DEX plus somatostatin antiserum (DEX + SSas; ●) administration on somatic growth in young male rats. DEX-induced somatic growth suppression was significantly reversed ($p < .01$) by concomitant SSas treatment. Data represent the means only as the error bars are smaller than the symbol sizes. Redrawn with permission from Wehrenberg et al. (80).

known to inhibit insulin secretion, and thus treatment with somatostatin antiserum may increase growth by increasing insulin release, which may in turn partially offset the catabolic effects of dexamethasone on peripheral tissues (81).

Collectively, the data discussed above provide support for the notion that the glucocorticoids exert effects both at the pituitary and at the hypothalamus. At the pituitary, they elicit an increase in GH message, as well as protein synthesis and release. Glucocorticoids also enhance pituitary responsiveness to GHRH. At the hypothalamic level, however, glucocorticoids cause a slight decrease in GHRH message synthesis, although it appears that GHRH protein content remains unaffected. The effect of the glucocorticoids on somatostatin message and protein synthesis are more dramatic, with conditions of high glucocorticoid tone eliciting marked increases in both hypothalamic somatostatin mRNA and protein content.

Conclusion

It appears that like pulsatile GH release itself, the involvement of the growth axis in the progress of sexual maturation in the rat exhibits sexually dimorphic trends. Growth impairment induced in prepubertal male rats produces a transient delay of sexual maturation. Here mechanisms regulating FSH synthesis, release, and perhaps action appear to be most sensitive to reductions in plasma GH, although sensitivity to reductions in circulating IGF-I may also be involved. In contrast, neither pulsatile GH secretion nor normal plasma IGF-I concentrations appear necessary for the onset of puberty in the female rat. Because a reduction in growth rate and maximal body size did not affect the onset of puberty, the theories that suggest that a critical body size or body composition are important for the onset of puberty in the female rat do not appear sound.

Current knowledge regarding interactions between the glucocorticoids and the growth axis supports a multifactorial mechanism of glucocorticoid action, especially when considering their effect upon somatic growth. Our model suggests that glucocorticoids act both as GH stimulatory and inhibitory agents. At the pituitary, they have a stimulatory effect on GH production and release. At the hypothalamus they precipitate the increased production and release of somatostatin, which in turn mediates an inhibition of pituitary GH synthesis and release. In the intact animal, it appears that the inhibitory hypothalamic pathway predominates as the stimulatory effects elicited via the pituitary become evident only after hypothalamic influence is removed.

References

1. Rogers KV, Vician L, Steiner RA, Clifton DK. The effect of hypophysectomy and growth hormone administration on pre-prosomatostatin messenger ribonucleic acid in the periventricular nucleus of the rat hypothalamus. Endocrinology 1988;122:586–91.
2. Epelbaum J, Tapia-Arancibia L, Alonso G, Astier H, Kordon C. The anterior periventricular hypothalamus is the site of somatostatin inhibition on its own release: an in vitro and immunocytochemical study. Neuroendocrinology 1986;44:255–9.
3. Crowley WR, Terry LC. Biochemical mapping of somatostatinergic systems in rat brain: effects of periventricular hypothalamic and medial basal amygdaloid lesions on somatostatin-like immunoreactivity in discrete brain nuclei. Brain Res 1980;200:283.
4. Millard WJ, Reppert SM, Sagar SM, Martin JB. Light-dark entrainment of the growth hormone ultradian rhythm in the rat is mediated by the arcuate nucleus. Endocrinology 1981;108:2394–6.
5. Okamura H, Murakami S, Chihara K, Nagatsu I, Ibata Y. Coexistence of growth hormone releasing factor-like and tyrosine hydroxylase-like immunoreactivities in neurons of the rat arcuate nucleus. Neuroendocrinology 1985;41:177–9.
6. McCarthy GF, Beaudet A, Tannenbaum GS. Colocalization of somatostatin receptors and growth hormone-releasing factor immunoreactivity in neurons of the rat arcuate nucleus. Neuroendocrinology 1992;56:18–24.

7. Miller HA, Rogers G, Frawley LS. Attenuation of growth hormone gene expression in desensitized somatotropes. Life Sci 1988;43:629–34.
8. Vale W, Vaughan J, Yamamoto G, Spiess J, Rivier J. Effects of synthetic human pancreatic (tumor) GH releasing factor and somatostatin, triiodothyronine and dexamethasone on GH secretion in vitro. Endocrinology 1983;112:1553–5.
9. Fukata J, Diamond DJ, Martin JB. Effects of rat growth hormone (rGH)-releasing factor and somatostatin on the release and synthesis of rGH in dispersed pituitary cells. Endocrinology 1985;117:457–67.
10. Morel G, Dihl F, Gossard F. Ultrastructural distribution of growth hormone (GH) mRNA and GH intron I sequences in rat pituitary gland: effects of GH releasing factor and somatostatin. Mol Cell Endocrinol 1989;65:81–90.
11. Brazeau P, Rivier J, Vale W, Guillemin R. Inhibition of growth hormone secretion in the rat by synthetic somatostatin. Endocrinology 1974;94:184–7.
12. Clark RG, Carlsson LMS, Robinson ICAF. Growth hormone (GH) secretion in the conscious rat: negative feedback of GH on its own release. J Endocrinol 1988;119:201–9.
13. Clark RG, Carlsson LMS, Rafferty B, Robinson ICAF. The rebound release of growth hormone (GH) following somatostatin infusion in rats involves hypothalamic GH-releasing factor release. J Endocrinol 1988;119:397–404.
14. Plotsky PM, Vale W. Patterns of growth hormone-releasing factor and somatostatin secretion into the hypophysial-portal circulation of the rat. Science 1985;230:461–3.
15. Jocovidou N, Patel YC. Antiserum to somatostatin-28 augments growth hormone secretion in the rat. Endocrinology 1987;121:782–5.
16. Wehrenberg WB, Brazeau P, Luben R, Bohlen P, Guillemin R. Inhibition of the pulsatile secretion of growth hormone by monoclonal antibodies to the hypothalamic growth hormone releasing factor (GRF). Endocrinology 1982;111:2147–8.
17. Wehrenberg WB, Ling N, Bohlen P, Esch F, Brazeau P, Guillemin R. Physiological roles of somatocrinin and somatostatin in the regulation of growth hormone secretion. Biochem Biophys Res Commun 1982;109:562–7.
18. Wehrenberg WB, Brazeau P, Luben R, Ling N, Guillemin R. A noninvasive functional lesion of the hypothalamo-pituitary axis for the study of growth hormone-releasing factor. Neuroendocrinology 1983;36:489–91.
19. Warren MP. Physical and biological aspects of puberty. In: Brooks-Gunn J, Peterson AC, eds. Girls at puberty: biological and psychological perspectives. New York: Plenum, 1983:3–28.
20. Frisch RE. Fatness, puberty, and fertility: the effects of nutrition and physical training on menarche and ovulation. In: Brooks-Gunn J, Petersen AC, eds. Girls at puberty: biological and psychosocial perspectives. New York: Plenum, 1983:29–49.
21. Sizonenko PC. Physiology of puberty. J Endocrinol Invest 1989;12(S3):59–63.
22. Vliet GV. Clinical aspects of normal pubertal development. Horm Res 1991;36:93–6.
23. Jansson JO, Eden S, Isaksson O. Sexual dimorphism in the control of growth hormone secretion. Endocr Rev 1985;6:128–50.
24. Eden S. Age- and sex-related differences in episodic growth hormome secretion in the rat. Endocrinology 1979;105:555–60.

25. Saunders A, Terry LC, Audet J, Brazeau P, Martin JB. Dynamic studies of growth hormone and prolactin secretion in the female rat. Neuroendocrinology 1976;21:193–203.
26. Clark RG, Carlsson LMS, Robinson ICAF. Growth hormone secretory profiles in conscious female rats. J Endocrinol 1987;114:399–407.
27. Birge CA, Peake GT, Mariz IK, Daughaday WH. Radioimmunoassayable growth hormone in the rat pituitary gland: effects of age, sex and hormonal state. Endocrinology 1967;81:195–204.
28. Jansson JO, Frohman LA. Differential effects of neonatal and adult androgen exposure on the growth hormone secretory pattern in male rats. Endocrinology 1987;120:1551–7.
29. Jansson JO, Ekberg S, Isaksson OGP, Eden S. Influence of gonadal steroids on age- and sex-related secretory patterns of growth hormone in the rat. Endocrinology 1984;114:1287–94.
30. Jansson JO, Frohman LA. Inhibitory effect of the ovaries on neonatal androgen imprinting of growth hormone secretion in female rats. Endocrinology 1987;121:1417–23.
31. Cronin MJ, Rogol AD. Sex differences in the cyclic adenosine $3':5'$-monophosphate and growth hormone response to growth hormone-releasing factor in vitro. Biol Reprod 1984;31:984–8.
32. Wehrenberg WB, Baird A, Ying SY, Ling N. The effects of testosterone and estrogen on the pituitary growth hormone response to growth hormone-releasing factor. Biol Reprod 1985;32:369–75.
33. Somana R, Visessuwan S, Samridtong A, Holland RC. Effect of neonatal androgen treatment and orchidectomy on pituitary levels of growth hormone in the rat. J Endocrinol 1978;79:399–400.
34. Ojeda SR, Jameson HE. Developmental patterns of plasma and pituitary growth hormone (GH) in the female rat. Endocrinology 1977;100:881–9.
35. Rechler MM, Nissley SP, Roth J. Hormonal regulation of human growth. N Engl J Med 1987;316:941–3.
36. Wilen R, Naftolin F. Pubertal food intake and body length, weight, and composition in the feed-restricted rat: comparison with well-fed animals. Pediatr Res 1977;12:263–7.
37. Glass AR, Serdloff RS. Nutritional influences on sexual maturation in the rat. Fed Proc 1980;39:2360–4.
38. Frisch RE, McArthur JW. Menstrual cycles: fatness as a determinant of minimum weight necessary for their maintenance or onset. Science 1974;185:949–51.
39. Bronson FH, Rissman EF. Biology of puberty. Biol Rev 1986;61:157–95.
40. Frisch RE, Hegsted DM, Yoshinaga K. Carcass components of first estrus of rats on high fat and low fat diets: body water, protein and fat. Proc Natl Acad Sci USA 1977;74:379–83.
41. Foster DL, Olster DH. Effect of restricted nutrition on puberty in the lamb: patterns of tonic luteinizing hormone (LH) secretion and competency of the LH surge system. Endocrinology 1985;116:375–81.
42. Advis JP, White SS, Ojeda SR. Activation of growth hormone short loop negative feedback delays puberty in the female rat. Endocrinology 1981; 108:1343–52.
43. Groesbeck MD, Parlow AF, Daughaday WH. Stimulation of supranormal growth in prepubertal, adult plateaued, and hypophysectomized female rats by

large doses of rat growth hormone: physiological effects and adverse consequences. Endocrinology 1987;120:1963–75.
44. Ramaley JA, Phares CK. Delay of puberty onset in females due to suppression of growth hormone. Endocrinology 1980;106(6):1989–93.
45. Salem MAD, Phares CK. In vitro insulin-like actions of the growth factor from the tapeworm, *Spirometra manosoides*. Proc Exp Biol Med 1989;190:203–10.
46. Wehrenberg WB. The role of growth hormone-releasing factor and somatostatin on somatic growth in rats. Endocrinology 1986;118:489–94.
47. Wehrenberg WB, Voltz DM, Cella SG, Muller EE, Gaillard RC. Long-term failure of compensatory growth in rats following acute neonatal passive immunization against growth hormone-releasing hormone. Neuroendocrinology 1992;56:509–15.
48. Ojeda SR, Andrews WW, Advis JP, White SS. Recent advances in the endocrinology of puberty. Endocr Rev 1980;1:228–57.
49. Froesch ER, Schmid C, Schwander J, Zapf J. Action of insulin-like growth factors. Annu Rev Physiol 1985;47:443–7.
50. D'Ercole AJ, Stiles AD, Underwood LE. Tissue concentrations of somatomedin C: further evidence for multiple sites of synthesis and paracrine or autocrine mechanisms of action. Proc Natl Acad Sci USA 1984;81:935-9.
51. Kula K. Induction of precocious maturation of spermatogenesis in infant rats by menopausal gonadotropin and inhibition by simultaneous administration of gonadotropins and testosterone. Endocrinology 1988;122:34–9.
52. Ramaley JA, Phares CK. Delay of puberty onset in males due to suppression of growth hormone. Neuroendocrinology 1983;36:321–9.
53. Zipf WB, Payne AH, Kelch RP. Prolactin, growth hormone, and luteinizing hormone in the maintenance of testicular luteinizing hormone receptors. Endocrinology 1978;103:595–600.
54. Tannenbaum GS. Evidence for autoregulation of growth hormone secretion via the central nervous system. Endocrinology 1980;107(6):2117–20.
55. Conway S, McCann SM, Krulich L. On the mechanism of growth hormone autofeedback regulation: possible role of somatostatin and growth hormone-releasing factor. Endocrinology 1985;117:2284–92.
56. Davoren JB, Hsueh JW. Growth hormone increases ovarian levels of immunoreactive somatomedian C/insulin-like growth factor I in vivo. Endocrinology 1986;118:888–90.
57. Jia X, Kalmijn J, Hsueh JW. Growth hormone enhances follicle-stimulating hormone-induced differentiation of cultured rat granulosa cells. Endocrinology 1986;118:1401–9. ·
58. Hernandez ER, Roberts CT Jr, LeRoith D, Adashi EY. Rat ovarian insulin-like growth factor I (IGF-I) gene expression is granulosa cell-selective: 5′-untranslated mRNA variant representation and hormonal regulation. Endocrinology 1989;125:572–4.
59. Adashi E, Resnick C, D'Ercole A, Svoboda M, Van Wyk J. Insulin-like growth factors as intraovarian regulators of granulosa cell growth and function. Endocr Rev 1985;6:400–20.
60. Adashi EY, Resnick CE, Hernandez ER, Hurwitz A, Rosenfeld RG. Follicle-stimulating hormone inhibits the constitutive release of insulin-like growth factor binding proteins by cultured rat ovarian granulosa cells. Endocrinology 1990;126:1305–7.

61. Ingle DJ, Higgins GM, Kendall EC. Atrophy of the adrenal cortex in the rat produced by administration of large amounts of cortin. Anat Rec 1938;71:363–72.
62. Blodgett FM, Burgin L, Iezzoni D, Gribetz D, Talbot NB. Effects of prolonged cortisone therapy on the statural growth, skeletal maturation and metabolic status of children. N Engl J Med 1956;254:636–41.
63. Kohler PO, Bridson WE, Rayford PL. Cortisol stimulation of growth hormone production by monkey adenohypophysis in tissue culture. Biochem Biophys Res Commun 1968;33:834–40.
64. Bridson WE, Kohler PO. Cortisol stimulation of growth hormone production by human pituitary tissue in culture. J Clin Endocrinol Metab 1970;30:538–40.
65. Seifert H, Perrin M, Rivier J, Vale W. Growth hormone-releasing factor binding sites in rat anterior pituitary membrane homogenates: modulation by glucocorticoids. Endocrinology 1985;117:424–6.
66. Ceda GP, Davis RG, Hoffman AR. Glucocorticoid modulation of growth hormone secretion in vitro. Evidence for a biphasic effect on GH-releasing hormone mediated release. Acta Endocrinol (Copenh) 1987;114:465–9.
67. Casanueva FF, Burguera B, Tome MA, Lima L, Tresguerres JAF, Devesa J, Dieguez C. Depending on the time of administration, dexamethasone potentiates or blocks growth hormone-releasing hormone-induced growth hormone release in man. Neuroendocrinology 1988;47:46–9.
68. Wehrenberg WB, Baird A, Ling N. Potent interaction between glucocorticoids and growth hormone-releasing factor in vivo. Science 1983;221:556–8.
69. Thompson RG, Rodriguez A, Kowarski A, Blizzard RM. Growth hormone: metabolic clearance rates, integrated concentrations, and production rates in normal adults and the effect of prednisone. J Clin Invest 1972;51:3193–9.
70. Krieger DT, Glick SM. Sleep EEG stages and plasma growth hormone concentration in states of endogenous hypercortisolemia or ACTH elevation. J Clin Endocrinol Metab 1974;39:986–1000.
71. Giustina A, Girelli A, Bossoni S, Legati F, Schettino M, Wehrenberg WB. Effect of galanin on growth hormone-releasing hormone-stimulated growth hormone secretion in adult patients with nonendocrine diseases on long-term daily glucocorticoid treatment. Metabolism 1992;41:548–51.
72. Giustina A, Bussi AR, Conti C, Doga M, Legati F, Macca C, Wehrenberg WB. Comparative effect of galanin and pyridostigmine on the growth hormone response to growth hormone releasing hormone in normal aged subjects. Horm Res 1992;37:165–70.
73. Wehrenberg WB, Janowski BA, Piering AW, Culler F, Jones KL. Glucocorticoids: potent inhibitors and stimulators of growth hormone secretion. Endocrinology 1990;126:3200–3.
74. Giustina A, Doga M, Bodini C, Girelli A, Legati F, Bossoni S, Romanelli G. Acute effects of cortisone acetate on growth hormone response to growth hormone-releasing hormone in normal adult subjects. Acta Endocrinol (Copenh) 1990;122:206–10.
75. Kaufmann S, Jones KL, Wehrenberg WB, Culler FL. Inhibition by prednisone of growth hormone (GH) response to GH-releasing hormone in normal men. J Clin Endocrinol Metab 1988;67:1258–61.
76. Nakagawa K, Ishizuka T, Obara T, Matsubara M, Akikawa K. Dichotomic action of glucocorticoids on growth hormone secretion. Acta Endocrinol (Copenh) 1987;116:165–71.

77. Wehrenberg WB, Baird A, Klepper R, Mormede P, Ling N. Interactions between growth hormone-releasing hormone and glucocorticoids in male rats. Regul Pept 1989;25:147–55.
78. Martin JB. Studies on the mechanism of pentobarbital-induced GH release in the rat. Neuroendocrinology 1973;13:339–50.
79. Giustina A, Wehrenberg WB. The role of glucocorticoids in the regulation of growth hormone secretion: mechanisms and clinical significance. Trends Endocrinol Metab 1992;3:306–11.
80. Wehrenberg WB, Bergman PJ, Stagg L, Ndon J, Giustina A. Glucocorticoid inhibition of growth in rats: partial reversal with somatostatin antibodies. Endocrinology 1992;127:2705–8.
81. Loeb JN. Corticosteroids and growth. N Engl J Med 1976;295:547–52.
82. Gruaz NM, Arsenijevic Y, Wehrenberg WB, Sizonenko P, Aubert ML. Growth hormone (GH) deprivation induced by passive immunization against rat GH-releasing factor (rGRF) does not disturb the course of sexual maturation and fertility in the female rat. Endocrinology 1994;135:509–19.
83. Arsenijevic Y, Wehrenberg WB, Conz A, Eshkol A, Sizonenko PC, Aubert ML. Growth hormone (GH) deprivation induced by passive immunization against rat GH-releasing factor delays sexual maturation in the male rat. Endocrinology 1989;124:3050–9.

14

Clinical Studies with Growth Hormone Releasing Hormone

Michael O. Thorner, Hal Landy, and Samir Shah

Growth hormone secretion is regulated by the interaction of two hypothalamic hormones, growth hormone releasing hormone (GHRH) and somatostatin (SRIH) (1, 2). The interaction of these two hypothalamic hormones regulates the pulsatile secretion of growth hormone, which mediates its biological actions. (Fig. 14.1)

The first studies with growth hormone releasing hormone in normal subjects demonstrated that a dose of 1 µg/kg intravenously stimulated growth hormone secretion, but the response was variable (3). This variability in the response is believed to be related to the tonic hypothalamic somatostatin secretion. Thus, if hypothalamic somatostatin secretion is high, then the growth hormone response to GHRH will be low; if hypothalamic somatostatin secretion is low, the response will be great.

Therapeutic Potential of GHRH

Early studies demonstrated that a large proportion of growth hormone deficient children responded to a single intravenous dose of GHRH with an increase in growth hormone secretion. Similarly, the immunostaining of pituitary cells from a growth hormone deficient patient of short stature demonstrated that immunoreactive growth hormone could be measured, thereby demonstrating that somatotropes were present in the pituitary. These data suggested that growth hormone deficiency is primarily the result of a hypothalamic disturbance and presumably the result of GHRH deficiency. The response to a single intravenous injection of 3.3 µg/kg of GHRH in 10 idiopathic growth hormone deficient children is shown in Figure 14.2 (4). Note that the mean GH response to arginine and L-dopa was much lower than the response to GHRH. In many children, there was a good response; in a few, there was a poor response or no response.

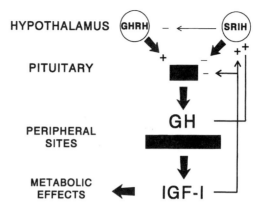

FIGURE 14.1. Schematic illustration of regulation of GH secretion by GHRH and somatostatin (SRIH). GH acts on multiple tissues to regulate metabolic functions and growth.

GHRH Therapy for Short Stature in Growth Hormone Deficient Children

We initially treated children with GHRH, which was administered subcutaneously every three hours by pump (5). The first two patients both demonstrated accelerated growth velocity, although in one the response was excellent (13.7 cm/yr) and in the other it was not as good (7.1 cm/yr). In a

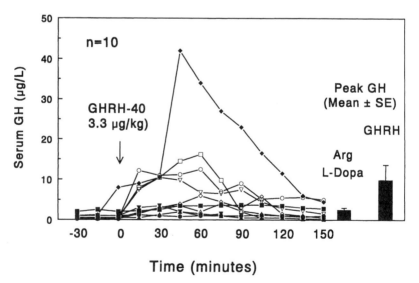

FIGURE 14.2. GH release in response to GHRH-40 in children with idiopathic GH deficiency (*n* = 10). Modified and reproduced with permission from Rogol et al. (4).

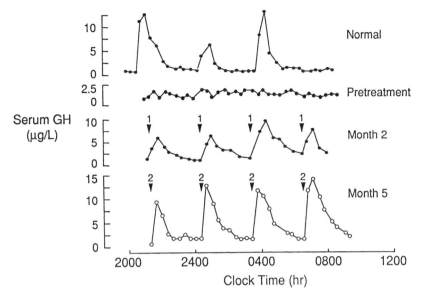

FIGURE 14.3. Serum GH profiles in a normal subject and in a GH deficient child. At pretreatment, two months (1 µg/kg/pulse) and five months (2 µg/kg/pulse) of GHRH treatment. Modified and reproduced with permission from Smith et al. (7).

subsequent multicenter study, 10 children received GHRH by pump every three hours with a dose of 1 to 3 µg/kg (6). Ten children received GHRH by pump every three hours only overnight (1 to 2 µg/kg per dose) and 4 children received twice daily injections of 4 to 8 µg/kg per dose. Figure 14.3 profiles the growth hormone secretion in a normal child and in one of the growth hormone deficient children treated by pump overnight (7). Note that each time the pump went off, there was an increase in growth hormone secretion. By the fifth month of treatment at a dose of 2 µg/kg per dose, there were peaks of growth hormone that reached at least 10 µg/L. In Figure 14.4, the growth velocities in the three groups of patients are shown. Note that in the children who received GHRH by pump every 3 hours throughout the 24-h period and in those who received twice daily injections, the growth velocity increased in all children. In those who received pump overnight, three did not have an increased growth velocity. Interestingly, the increase in growth velocity was proportional to the dose of GHRH that was administered.

Growth Hormone Releasing Hormone Administered as a Once Daily Injection

For GHRH to be a practical alternative to growth hormone, it would need to be administered no more frequently than once daily. Therefore, the

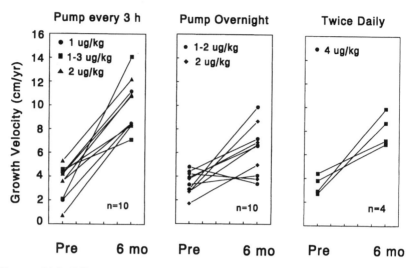

FIGURE 14.4. Effects of three different regimens of GHRH therapy on growth velocity: pump every three hours at doses of 1, 2, or 3μg/kg (n = 10, left panel); pump overnight at doses of 1 to 2μg/kg (n = 10, middle panel); twice daily injections of 4μg/kg (n = 4, right panel). Modified and reproduced with permission from Thorner et al. (6).

GEREF International Study Group performed a multicenter study to evaluate the safety and efficacy of once daily GHRH(1-29)NH$_2$ (GEREF) in the treatment of growth failure in idiopathic growth hormone deficient children. Children were studied in France, Italy, Spain, the United States, and Venezuela. One hundred and ten children were enrolled from 25 centers. Subjects were evaluated for six months prior to treatment. They then were treated with GHRH—30μg/kg administered subcutaneously once daily in the evening for 12 months. The inclusion criteria were prepubertal with a chronological age of less than 10 years in girls and less than 11 years in boys; no organic hypothalamic or pituitary lesion as demonstrated by CT or MRI; a height velocity of less than 10% for chronological age; a height of less than −2.5 standard deviations; a peak growth hormone response to two provocative tests of less than 10μg/L; and no previous growth hormone or GHRH treatment.

Of the evaluable patients, that is, all who met all eligibility criteria, there were 86 at baseline, 80 at 6 months, and 56 at 12 months. Analyses of all patients included eligible and major protocol violators. Because the results were similar for the two analyses, only the results in the patients who could be evaluated will be discussed. The pretreatment demographic data in these 86 patients were 63 boys and 23 girls with mean age 7.7 ± 2.2 yrs, height age 4.6 ± 1.7 yrs, bone age 5 ± 2 yrs, height velocity 4.1 ± 1 cm/yr, and height SDS −3.7 ± 0.9. At six months, the mean growth velocity was 8 ± 1.5 cm/ yr. Seventy-four percent of children had an increase in growth velocity of greater than 2 cm/yr. Fifty-six patients continued GHRH treatment for 12 months and the mean height velocity was 7.2 ± 1.3 cm/yr. Eighty-four percent of these 56 patients had an increased growth velocity of greater than 2 cm/yr. The ratio of the change in bone age to the change in height age after one year of GHRH treatment was 1.04 ± 0.58, which was no different from the expected ratio.

Pretreatment parameters were analyzed for predictors of good growth response to GHRH. The following appeared to predict a good response—children of a younger age, with a lower height velocity SDS, and a greater delay in bone age. The following criteria did not predict whether the response would be good or not: serum IGF-I; the growth hormone response to GHRH; and the growth hormone response to provocative tests.

Adverse Events

There was a total exposure to GHRH therapy of 117 patient years. Local injection reactions occurred in 14.5% of patients. One death occurred which was unrelated to treatment; there were two drop-outs due to local reactions; and seven patients were hospitalized for reasons unrelated to treatment. Five of 101 subjects were started on thyroxine because of a fall in serum thyroxine. Thirty-four of 102 patients developed antibodies at 6 months and 63 of 102 developed antibodies at any time. In eight patients, the antibodies reverted to negative in spite of continuing treatment. The antibodies were of low titer and had no impact on overall efficacy.

GH Treatment after GHRH Treatment

Children who failed to have an increase in growth velocity of greater than 2 cm/yr or at the end of GHRH treatment, had the option of switching to recombinant human growth hormone treatment (Saizen). The six children, who could be evaluated and switched to Saizen at six months, had a mean growth velocity before starting treatment of 4.6 ± 0.6 cm/yr. The mean growth velocity on GHRH at six months was 4.9 ± 0.7 and the mean growth velocity on growth hormone was 8.0 ± 3.6 cm/yr. Of these, three children did not respond to either GHRH or growth hormone, while three responded to growth hormone, but not to GHRH.

In the children who were treated with GHRH for a year or 18 months and then switched to growth hormone, the mean growth velocity on growth hormone treatment was similar to that on GHRH treatment. In the group that was switched after 12 months of GHRH treatment ($n = 8$), the pretreatment growth velocity was 3.9 ± 0.7 cm/yr, at 12 months was 6.1 ± 0.6 cm/yr, and at 6 months on growth hormone was 7.6 ± 3.3 cm/yr. In those switched to GH after 18 months of GHRH treatment ($n = 7$), the mean growth velocity before treatment was 3.3 ± 1.2. On GHRH it was 6.8 ± 0.5 at 18 months and after 6 months on growth hormone it was 6.4 ± 2.4 cm/yr.

Summary

Younger and more slowly growing children with a more delayed bone age are more likely to respond well to GHRH. Proportional changes in skeletal maturation and height imply preservation of final height potential. Antibody development was not clinically significant and GHRH(1-29)NH$_2$ was well tolerated.

Conclusions

Growth hormone deficiency is often due to GHRH deficiency. GHRH acutely stimulates growth hormone in normal and growth hormone deficient subjects. Once daily GHRH effectively accelerates growth velocity in growth hormone deficient patients. A sustained release GHRH preparation may restore pulsatile growth hormone secretion. Growth hormone releasing peptide (GHRP) or GHRP mimetics, in combination with GHRH, may optimize growth hormone secretion.

Acknowledgments. These studies were supported by Ares Serono, Geneva, Switzerland. Dr. Thorner is a consultant to Ares Serono and participated in the design and analysis of this trial. The authors thank the investigators who participated in this study, Patrick Engrand (Biometrics Department, Ares Serono, Geneva), who performed the statistical analysis, Anika Wüthrich (Ares Serono, Geneva) for the administration of the study, and the editorial assistance of Suzan Pezzoli and Sandra Saxe.

Geref International Study Group Participants

France:	Dr. Michel Colle (Bordeaux), Prof. Pierre Rochiccioli (Toulouse).
Italy:	Professor Lodovico Benso (Torino), Professor Luciano Cavallo (Bari), Professor Fabrizio De Matteis (L'Aquila), Professor Gaetano Lombardi (Napoli), Professor Giuseppe Saggese (Pisa), Professor Francesca Severi (Pavia)
Spain:	Dr. Manuel Pombo Arias (Santiago De Compostela), Dr. D. Garagorri Otero (Zaragoza), Dr. Carlos Pavia (Barcelona), Dr. Begona Sobradillo (Bilbao)
United Kingdom:	Mr. J. Leonard, Dr. M. Savage (London)

United States:	Dr. Barry B. Bercu (St. Petersburg), Dr. David Finegold (Pittsburgh), Dr. Joseph Gertner (New York), Dr. Jerome Grunt and Dr. Campbell Howard (Kansas City), Dr. G.J. Klingensmith (Denver), Dr. James F. Marks (Dallas), Dr. Ora Pescovitz (Indianapolis), Dr. Leslie P. Plotnick (Baltimore), Dr. Edward O. Reiter (Springfield), Dr. Paul Saenger (Bronx), Dr. Eric Smith (Cincinnati), Dr. Mark Wheeler (Davis)
Venezuela:	Dr. Eduardo Carrillo and Dr. Roberto Lanes (Caracas)

References

1. Guillemin R, Brazeau P, Bohlen P, Esch F, Ling N, Wehrenberg WB. Growth hormone-releasing factor from a human pancreatic tumour that caused acromegaly. Science 1982;218:585–7.
2. Rivier J, Spiess J, Thorner M, Vale W. Characterization of growth hormone-releasing factor from a human pancreatic islet tumour. Nature 1982;300:276–8.
3. Borges JLC, Blizzard RM, Gelato MC, Furlanetto R, Rogol AD, Evans WS et al. Effects of human pancreatic tumour growth hormone releasing factor on growth hormone and somatomedin-C levels in patients with idiopathic growth hormone deficiency. Lancet 1983;2:119–24.
4. Rogol AD, Blizzard RM, Johanson AJ, Furlanetto RW, Evans WS, Rivier J et al. Growth hormone release in response to human pancreatic tumor growth hormone releasing factor-40 in children with short stature. J Clin Endocrinol Metab 1984;59:580–6.
5. Thorner MO, Reschke J, Chitwood J, Rogol AD, Furlanetto R, Rivier J et al. Acceleration of growth in two children treated with human growth hormone-releasing factor. N Engl J Med 1985;312:4–9.
6. Thorner MO, Rogol AD, Blizzard RM, Klingensmith GJ, Najjar J, Misra R et al. Acceleration of growth rate in growth hormone-deficient children treated with human growth hormone-releasing hormone. Pediatr Res 1988;24:145–51.
7. Smith PJ, Brook CGD, Rivier J, Vale W, Thorner MO. Nocturnal pulsatile growth hormone releasing hormone treatment in growth hormone deficiency. Clin Endocrinol 1986;25:35–44.

15

Central and Peripheral Effects of Peptide and Nonpeptide GH Secretagogues on GH Release In Vivo

Keith M. Fairhall, Anita Mynett, Gregory B. Thomas, and Iain C.A.F. Robinson

Growth hormone (GH) secretion is controlled by the interplay between the secretion of hypothalamic GH releasing factor (GHRF) and somatostatin [somatotropin release inhibiting factor (SRIF)] into hypophysial portal blood (1). Quite separately from the elucidation of the naturally occurring GHRF, Bowers and his colleagues (2, 3) developed a different class of GH secretagogues [GH releasing peptides (GHRPs)] of which the most well studied is the hexapeptide GHRP-6. An important property of GHRP-like compounds is that they do not act as GHRF analogues, but show strong synergy with GHRF (4). The slight but significant oral activity of GHRP-6 (5, 6) adds to the potential of such compounds, and has no doubt encouraged the development of more potent peptide (7, 8) and nonpeptidyl GH secretagogues (9, 10). Since these secretagogues are active in man and synergize strongly with GHRF (11–13), they could prove clinically valuable to release GH in patients that respond poorly to GHRF (14).

If the secretagogues are to be given in the longer term, their GH releasing should be reasonably consistent, and GH release should be maintained with repeated exposure of secretagogues. However, the GH responses in many species elicited by single or repeated injections of the natural secretagogue GHRF are highly variable. In the rat, this is probably due to a regular variation in SRIF tone since it is much less pronounced in female rats, or in males given anti-SRIF antiserum (15, 16). A similar variability in GH response has been reported for GHRP-6 (2, 17, 18), which rapidly desensitizes in the face of prolonged infusion (18, 19). Several studies are currently testing the effects of daily injections of these secretagogues, but little is known about the reproducibility of the GH response and the factors that govern the desensitization to repeated or continuous exposure to GHRP-6.

Following the development of L-692,429, the first nonpeptide substituted benzolactam GH secretagogue by Smith and his colleagues (9, 20), a new analogue L-692,585 has been developed (21), which is 20-fold more potent than the earlier compound when tested in vitro on rat pituitary cells in culture. Much less is known about the in vivo properties of these new nonpeptide compounds including their relative potency with respect to GHRP-6, and their ability to stimulate release of GH upon frequent iv administration in the conscious animal. We have recently had the opportunity to compare the effects of repetitive or continuous administration of L-692,585 with those of GHRP-6 in vivo. The first part of this chapter reviews this series of studies performed in the conscious chronically cannulated guinea pig, using an automatic sampling and infusion system (22–24). GH release was stimulated by repeated iv challenge with GH secretagogues while monitoring the ongoing spontaneous secretory pulses of GH. The guinea pig was chosen as it is eminently suitable for long-term blood sampling studies, and we have previously demonstrated that the guinea pig responds sensitively and consistently to GHRP-6 (24, 25).

Because GHRP-6 can release moderate amounts of GH from pituitary cells in culture, most authors have considered that GHRP-6 acts primarily at the pituitary as a direct GH secretagogue (2, 8, 26, 27). On the other hand, we noticed that the effects of GHRP-6 are much more potent in the conscious animal in vivo, than in the anesthetized animal or from in vitro systems, and that much more subtle interactions were evident in the conscious animal with ongoing spontaneous GH secretion (18). This led us to propose that GHRP-6 also acted at a hypothalamic level, and that this was much more potent than the direct GH-releasing effects of GHRP-6 at the pituitary. Since that time, this idea has gained a degree of support from a variety of studies suggesting that GHRP-6 can also act at a hypothalamic site to release GH indirectly (28–30) and that in the absence of an intact hypothalamic connection to the pituitary, the effects of GHRP-6 are greatly reduced (31, 32).

We reasoned that if the central actions of GHRP-6 are important, and sensitive to small amounts of GHRP-6 following iv or even oral administration of this peptide, it should be possible to elicit GH release with much lower doses given by direct intracerebroventricular (icv) administration of the hexapeptide, which would not elicit a direct pituitary response. We have recently examined the GH response to icv injections of both GHRP-6 and the nonpeptide secretagogue L-692,585 in the anesthetized guinea pig (33), and GHRP-6 in the conscious sheep, and these studies are also discussed here.

Materials and Methods

For conscious sampling studies in guinea pigs (Hartley males, 300–500 g) animals were fitted with indwelling jugular venous cannulae and attached to the automatic sampling system described elsewhere (23, 24). This system delivers continuous iv infusions or pulsatile injections of GH secretagogues while withdrawing microsampling of blood (20 µl every 10 min) automatically. Where infusions and injections of different secretagogues were given, injections were performed manually between samples without interrupting either the infusion of the other secretagogue or the blood sampling procedure. In the conscious sheep studies, blood samples (5 ml) were withdrawn by hand every 10 min from indwelling jugular cannulae implanted previously in mature Scottish Blackface ewes (50–60 kg). For icv studies in guinea pigs, animals were anesthetized with urethane (1.5 mg/kg ip) and fitted with a jugular venous cannula and a cannula in a lateral ventricle, as previously described (34). The venous cannula was attached to the automatic system for withdrawing 20-µl blood samples automatically every 10 min. The secretagogues were then given iv between samples without interrupting the sampling procedure, as described above. For icv studies in sheep, we used conscious animals fitted with chronic lateral ventricular cannulae. The GHRF analogue used (hGHRF ^{27}Nle(1-29)NH$_2$) and GHRP-6 were synthesized by J. Trojnar (Ferring AB, Malmö). The substituted benzolactam GH secretagogue L-692,585 was kindly provided by R.G. Smith (Merck Research Laboratories), while the long-acting somatostatin analogue SMS 201-995 (SMS) was a gift from Sandoz (Basel). Guinea pig GH was measured by radioimmunoassay (RIA) directly on whole blood samples as previously described (24); ovine GH was measured on plasma samples. Unless otherwise stated, data are expressed as mean ± SEM. A more detailed description of these experiments will be found elsewhere (22, 33).

Results

Peripheral Administration of GH Secretagogues

When given by iv injection in the conscious guinea pig, both "585" and GHRP-6 released GH over a similar time course. When different doses of "585" and GHRP-6 were compared, "585" was found to be approximately 10-fold less effective on a weight basis than GHRP-6. Figure 15.1 compares the GH responses in two groups of seven guinea pigs to a single iv injection of 10 µg "585" or 1 µg GHRP-6. There is a marked synergism between GHRP-6 and GHRF (4, 14). If the nonpeptide secretagogue acts similarly, it should be possible to demonstrate this in vivo. Figure 15.2 shows that this is the case. The plasma peak GH response to a mixture of 10 µg "585" and 1 µg GHRF given iv was much greater than the responses to the two agents given separately.

Although single injections of "585" and GHRP-6 did release GH effectively in conscious guinea pigs, the amplitude of the responses was quite variable both within and between animals. This was similar to our earlier

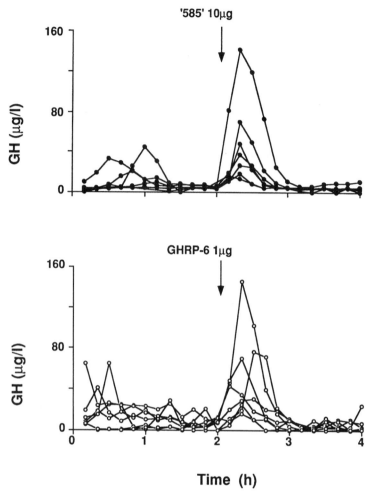

FIGURE 15.1. Blood samples were withdrawn from two groups of seven conscious guinea pigs and assayed for GH before and after the iv injection (arrow) of "585" (10 μg, top panel) or GHRP-6 (1 μg, bottom panel). Reproduced by permission of the Journal of Endocrinology Ltd. (22).

findings with GHRF in the rat, which were illuminated by giving serial injections at different frequencies (15). Accordingly, the GH responses to a series of injections of three different GH secretagogues were compared over a period of 30 hours of continuous sampling. Seven conscious animals were given serial injections of GHRF (2 μg) every hour for 19 hours, followed by a further six pulses of 2 μg GHRP-6 (Fig. 15.3). In the first 6 hours without injections, endogenous pulses of GH of 20 to 40 μg/L occurred in most of the animals. The first injection of GHRF elicited a large GH

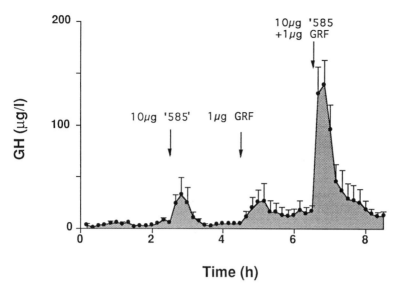

FIGURE 15.2. A group of five conscious guinea pigs were given iv injections (arrows) of "585" (10 μg), followed 2 hours later by GHRF (1 μg) and another 2 hours later by a mixture of both secretagogues. The combination of "585" and GHRF produced a much greater GH response than the sum of either secretagogue alone. Reproduced by permission of the Journal of Endocrinology Ltd. (22).

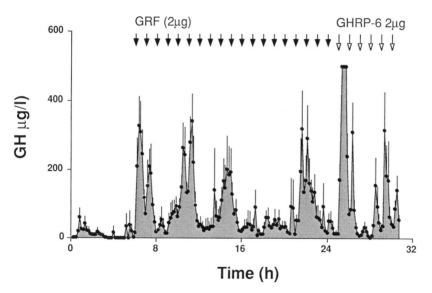

FIGURE 15.3. Growth hormone (GH) responses to serial injections of GHRF and GHRP-6. Blood samples were withdrawn from seven conscious guinea pigs for 31 hours. After 6 hours, iv injections of GHRF (^{27}Nle-hGHRF(1-29)NH$_2$, 2 μg) were given every hour for 19 hours (solid arrows), followed by 6 injections of GHRP-6 (2 μg, open arrows). Reproduced by permission of the Journal of Endocrinology Ltd. (22).

223

response. Subsequent GHRF injections continued to stimulate GH release, but a cyclic pattern of responsiveness and refractoriness emerged (Fig. 15.3), very similar to our results obtained with serial injections of GHRF in the conscious rat (15). For the last six injections, the secretagogue was switched from GHRF to GHRP-6 (Fig. 15.3). The first GHRP-6 injection produced a very large GH response in all the animals but the pattern of responsiveness clearly remained cyclic. A similar intermittent responsiveness was seen in response to the nonpeptide secretagogue "585" (10 μg) and GHRP-6 when given at 90-min intervals (22). Clearly, intermittent responsiveness is a property of the systems controlling GH release in the conscious animal, and is not peculiar to any one type of GH secretagogue.

In earlier studies with GHRF in the conscious rat, more reproducible GH responses were obtained when the pulse frequency was lengthened to match the endogenous rhythm of GH (15). This was also the case for the secretagogues, as shown in Figure 15.4. Consistent GH responses were obtained to both peptide and nonpeptide secretagogues at a 3-hour injection interval. For both "585" and GHRP-6, the GH release in each animal became entrained to the pulses, and spontaneous GH secretion, which was visible at the beginning of the sampling period, was markedly reduced or abolished between the peaks of GH induced by injections of either secretagogue.

Central Administration of GH Secretagogues

In preliminary experiments we compared the potency of GHRP-6 by icv vs. iv injection by crossover experiments in a group of six conscious ewes. Figure 15.5 shows that 5 μg injected icv resulted in a sevenfold increase in GH output, at least as much as that elicited when 500 μg was injected iv in the same animals on a different day. Central injections of saline (100 μl) did not affect GH release.

GHRP-6 was also effective in releasing GH after icv injection in anesthetized guinea pigs, although the magnitude of the plasma GH responses to any secretagogue are much smaller than in the conscious guinea pig. Figure 15.6 shows the GH responses in a group of five guinea pigs given 1 μg GHRP-6 icv. A sustained rise in plasma GH occurs, peaking 30 to 40 min after injection of GHRP-6. This response was not seen in a further five animals given saline injections icv nor in eight animals given icv injections of 2 μg GHRF (Fig. 15.6). GH release was obtained in response to much smaller doses of GHRP-6 (a few animals responded to as little as 8 ng GHRP-6 injected icv), whereas these doses were completely without effect when given iv, and comparable GH responses required injections of at least 1 μg GHRP-6 when given iv (not shown). In a similar experiment "585" was also shown to release GH after icv injection of 1 μg (Fig. 15.7), a dose that was ineffective when given iv in anesthetized animals (33). We interpret

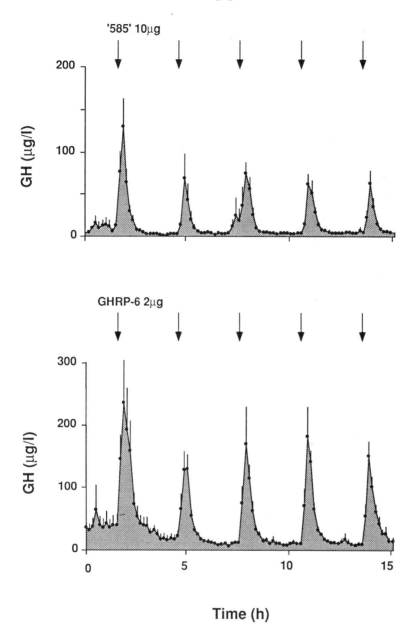

FIGURE 15.4. Regular GH release in response to 3-hourly pulses of GHRP-6 or "585." Groups of six conscious guinea pigs were given iv pulses of "585" (10 μg top panel) or GHRP-6 (2 μg, bottom panel) at 3-hourly intervals (arrows). Reproduced by permission of the Journal of Endocrinology Ltd. (22).

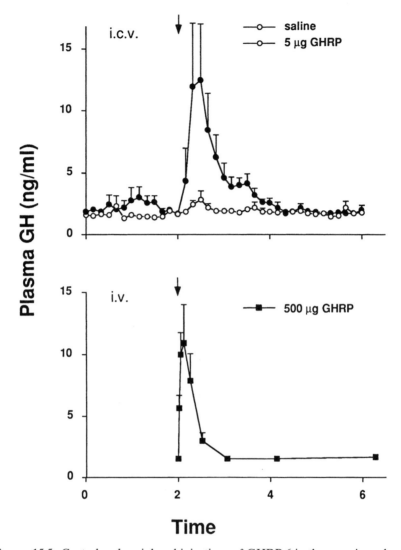

FIGURE 15.5. Central and peripheral injections of GHRP-6 in the conscious sheep. On different occasions in the same group of six ewes, GHRP-6 was given icv (5 μg) or iv (500 μg) and blood samples withdrawn and assayed for ovine GH as previously described (55).

these experiments to show that GHRP-6 and "585," but not GHRF, release GH when given icv in relatively small doses, probably reflecting their action on a central target, and is not simply due to the secretagogue leaking to the bloodstream and activating the pituitary directly.

There are many possible targets for centrally injected GH secretagogues, but we and others have accumulated evidence suggesting that at least one effect is to increase the release of GHRF from arcuate neurons (18, 29, 30). One way to test the involvement of GHRP-6–induced GHRF release would be to attempt to inhibit GHRF neurons. Electrophysiologic studies have shown that arcuate neurons that project to the median eminence and

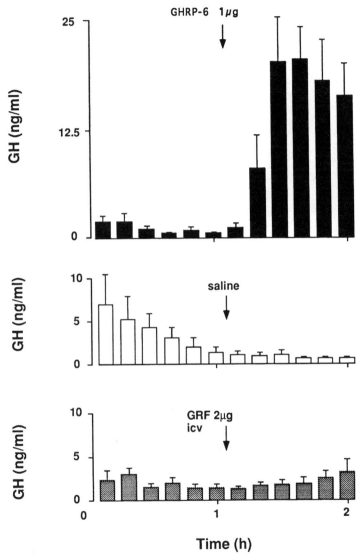

FIGURE 15.6. GH is released by GHRP-6, but not by GHRF, given icv. Blood samples were withdrawn from three groups of anesthetized guinea pigs before and after icv injection (arrows) of GHRP-6 (1 μg, top panel, $n = 5$), saline (middle panel, $n = 5$), or GHRF (2 μg, bottom panel, $n = 8$). Reproduced by permission of the Journal of Endocrinology Ltd. (33).

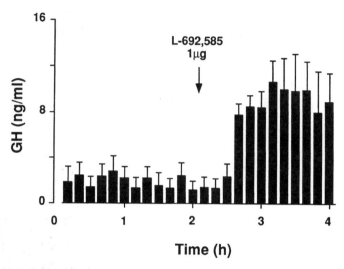

FIGURE 15.7. Serial blood samples were withdrawn from a group of four anesthetized guinea pigs before and after the icv injection (arrow) of 1 μg L-692,585, and assayed for GH. Reproduced by permission of the Journal of Endocrinology Ltd. (33).

are excited by GHRP-6 can be inhibited by periventricular stimulation (35), which may release SRIF to act on those SRIF receptors present on GHRF neurons (36). We therefore attempted to block the GH release induced by central GHRP-6 by prior administration of the long-acting SRIF agonist SMS 202-995. Figure 15.8 illustrates the results from these experiments. GHRP-6 alone (200 ng icv, $n = 15$) released GH ($p < .01$), whereas pretreatment with SMS (10 μg) icv blocked the GH release induced by GHRP-6 in 10 further animals. We were concerned to exclude the possibility that SMS was blocking the response simply by leaking from the ventricle to block the pituitary. However, central SMS injections did not block the pituitary response to 10 μg GHRF given iv 20 min later (Fig. 15.8).

Discussion

The first aim of these experiments was to compare the GH responses to repeated GHRP-6 exposure with those to the nonpeptide GH secretagogue, L-629,585 (21), a more potent analogue of the prototype compound L-629,429, (9, 27). We chose to use guinea pigs rather than rats since they respond well to GHRP-6 (24, 25), releasing more GH in response to GHRP-6 than to GHRF, which is also the case in man but not in the rat. We found that "585" behaves very similarly to GHRP-6 in vivo, although it was less potent in the guinea pig than we expected from the data on rat cells in

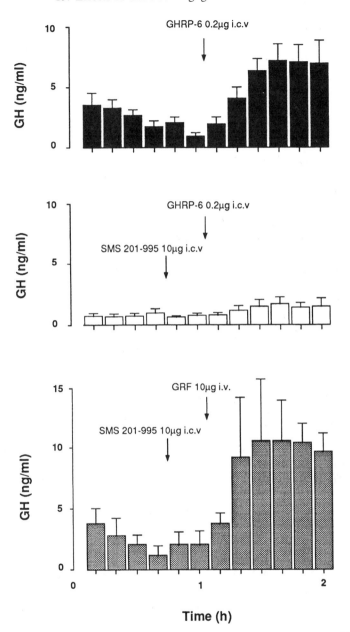

FIGURE 15.8. Central effects of GHRP-6 are blocked by pretreatment with a long-acting somatostatin agonist. Three experiments are shown. In (*a*), 15 guinea pigs were given an icv injection of 200 ng GHRP-6 (arrow) to release GH. In (*b*), 20 minutes prior to GHRP-6 injection, 10 animals received 10 µg of SMS 201-995, which blocked GH release. In (*c*) GH could still be released by an iv injection of GHRF (10 µg) after the pretreatment with SMS 201-995 (10 µg icv $n = 5$). Reproduced by permission of the Journal of Endocrinology Ltd. (33).

vitro (21). If this is not simply a species difference, it may point to a significant difference in potencies when these secretagogues are compared in vivo vs. in vitro, and caution against overreliance on in vitro screening for structure/activity studies.

Both GHRP-6 and "585" stimulated GH release, but the responses to single injections were highly variable. This probably relates to variable SRIF tone, since this peptide inhibits the actions of all the GH secretagogues (2, 27), and there is good evidence for a regular variation in endogenous SRIF tone (16, 37). Frequent serial injections of GHRF, GHRP-6, or "585" showed that intermittent GH responsiveness applies to all the GH secretagogues tested and is a feature of the intact conscious animal. This underlying variability in response should be borne in mind when evaluating GH responses in conscious individuals given single injections of these secretagogues, as in diagnostic applications.

Regular GH responses were obtained to repeated injections of either "585" or GHRP-6 given every 3 hours. This is more in line with the endogenous GH rhythm (24), and such serial injections tended to entrain GH pulses and suppress GH secretion between the peaks. Our previous studies showed that constant GHRP-6 exposure removed the variability in GHRF responsiveness in the rat (18), and we speculate that regular GHRP-6 or "585" injections may work by synchronizing activity in the neural mechanism(s) that generate the spontaneous GH secretory episodes.

The ability of GH secretagogues to enhance GH secretion with repeated administration is promising, but needs to act over longer periods than those tested here. Positive effects have been obtained with prolonged administration of GHRP-6 with (38) or without (2, 3) GHRF, suggesting that an enhanced GH release can indeed be achieved. Continuous GHRF administration amplifies endogenous GH secretion in children (39), probably having a trophic action on pituitary somatotrophs via stimulation of adenylate cyclase. Current evidence suggests that the non-GHRF secretagogues do not increase adenylate cyclase activity (8, 27) although they may enhance this effect of GHRF (40). It will be important to determine whether chronic secretagogue administration does have trophic effects on somatotrophs since this will probably be more important than acute GH-releasing potency in establishing the clinical benefits of these compounds.

In our view, the in vivo mechanism of action of these secretagogues remains unclear. The secretagogues clearly release GH from pituitary cells in vitro, so that must be one possible site of action (2, 8, 26, 27). Although GHRP-6 does not act as a simple agonist on GHRF receptors (26–28, 41–43), there is evidence that an intact GHRF receptor system is necessary (44, 45), implying some interaction between GHRP-6 and GHRF transduction mechanisms at the pituitary.

We have suggested that a central target for secretagogue action predominates in the conscious animal (18). Binding sites for GHRP-6 have been described in the hypothalamus (28, 42). These secretagogues increase neuronal firing and c-*fos* expression in the arcuate nucleus after peripheral administration (29). Although a variety of indirect studies support the direct observation that secretagogues increase GHRF release (30), this cannot be the only mechanism of action because secretagogues can release GH after GHRF responsiveness has been lost (17, 18, 46, 47).

Bowers et al. (48) concluded that both pituitary and hypothalamic mechanisms may be involved at different doses of these secretagogues. If GHRP-6 can act centrally after iv administration, central actions must be more sensitive, since the amounts reaching the hypothalamus would be lower than those reaching the pituitary gland directly. It should therefore be possible to demonstrate GH release at much lower doses of GHRP-6 when given directly into the central nervous system.

In preliminary experiments with conscious sheep we confirmed this, with equivalent GH release obtained at 100-fold lower doses of GHRP-6 when injected icv compared with iv. On the other hand, Chihara et al. (31) did not report this to be the case in the rat. Since the GH responses to GHRP-6 are highly variable in the rat, we decided to test this in more detail in the guinea pig, for which we had developed icv cannulation methods (34). Although the response was smaller in the anesthetized animal, we again observed a much more potent effect when secretagogues were given icv. Since central injections of GHRF were ineffective, this suggests that GH release was not the result of centrally administered peptide leaking out into the bloodstream. L-692,585 was also able to release GH following 1 μg icv injections that were ineffective when given iv. We conclude that small amounts of GH secretagogues release GH by a specific and sensitive central action, not shared by GHRF.

In parallel with these studies, *fos* expression has also been demonstrated in the rat arcuate nucleus following lower doses of GHRP-6 and L-692,585 injected icv than those injected iv (49, 50). This implicates activation of the arcuate nucleus, but does not differentiate between a direct or indirect effect of secretagogues on these neurons, or on other neuronal targets. Other studies have suggested that GHRP-6 may also change SRIF release or inhibit its effect (18, 48) or that there are other unknown hypothalamic factors involved (48, 51).

Since the direct pituitary effects of GHRP-6 can be inhibited by SRIF and the inhibitory potency of SRIF is reduced by GHRP-6, this secretagogue could act as a functional antagonist of SRIF at the pituitary gland (48). We therefore wondered whether SRIF could also functionally antagonize GHRP-6 actions centrally, perhaps via SRIF receptors on GHRF neurons (36) that normally respond to SRIF from either periventricular or even intraarcuate SRIF interneurons (52). The large rebound release of GH caused by abrupt removal of endogenous (53) or exogenous (54) SRIF also involves GHRF release, implying that SRIF receptors on GHRF neurons could restrain GHRF release. If one of the central effects of GHRP-6 was

232   K.M. Fairhall et al.

to stimulate GHRF neurons, this might be functionally antagonized by central SRIF administration.

Our results showed that pretreatment with SMS did block GH release in response to central GHRP-6 injection, while the pituitary remained responsive to peripheral injections of GHRF. Although SMS blocks the GHRP-6 response, this does not prove that it acts immediately on the same cellular target. GHRP-6 could activate an afferent input to the GHRF neurons (or even disinhibit an inhibitory input) and SRIF could act downstream of this to block the GHRF output. Whatever the central mechanism, our study provides further evidence that a central action of GHRP-6 is an important component of the complex in vivo response to this type of secretagogue.

Acknowledgments. We would like to thank Dr. A.L. Parlow for providing reagents for GH assays, Dr. J. Trojnar (Ferring AB) for GHRP-6 and the GHRF analog, Dr. R. Smith for L-692,585, and Sandoz for SMS 201-995.

References

1. Thomas GB, Cummins JT, Francis H, Sudbury AW, McCloud PI, Clarke IJ. Effect of restricted feeding on the relationship between hypophysial portal concentrations of growth hormone (GH)-releasing factor and somatostatin, and jugular concentrations of GH in ovariectomized ewes. Endocrinology 1991;128:1151–8.
2. Bowers CY, Momany F, Reynolds GA, Hong A. On the in vitro and in vivo activity of a new synthetic hexapeptide that acts on the pituitary to specifically release growth hormone. Endocrinology 1984;114:1537–45.
3. Bowers CY, Sartor AO, Reynolds GA, Badger TM. On the actions of the growth hormone-releasing hexapeptide, GHRP. Endocrinology 1991;128:2027–35.
4. Bowers CY, Reynolds GA, Durham D, Barrera CM, Pezzoli SS, Thorner MO. Growth-hormone (GH)-releasing peptide stimulates GH release in normal men and acts synergistically with GH-releasing hormone. J Clin Endocrinol Metab 1990;70:975–82.
5. Walker RF, Codd EE, Barone FC, Nelson AH, Goodwin T, Campbell SA. Oral activity of the growth-hormone releasing peptide His-D-Trp-Ala-Trp-D-Phe-Lys-NH₂ in rats, dogs and monkeys. Life Sci 1990;47:29–36.
6. Hartman ML, Farello G, Pezzoli SRIF, Thorner MO. Oral administration of growth-hormone (GH)-releasing peptide stimulates GH secretion in normal men. J Clin Endocrinol Metab 1992;74:1378–84.
7. Bowers CY. GH releasing peptides—structure and kinetics. J Paediatr Endocrinol 1993;6:21–31.
8. Akman MS, Girard M, O'Brien LF, Ho AK, Chik CL. Mechanisms of action of a 2nd generation growth hormone-releasing peptide (Ala-His-D-beta-Nal-Ala-

Trp-D-Phe-Lys-NH2) in rat anterior-pituitary-cells. Endocrinology 1993;132: 1286–91.

9. Smith RG, Cheng K, Schoen WR, Pong SS, Hickey G, Jacks T, et al. A nonpeptidyl growth-hormone secretagogue. Science 1993;260:1640–3.

10. Jacks T, Hickey G, Judith J, Taylor H, Chen H, Krupa D, et al. Effects of acute and repeated intravenous administration of L-692,585, a novel non-peptidyl growth hormone secretagogue, on plasma growth hormone, IGF-1, ACTH, cortisol, prolactin, insulin, and thyroxine levels in beagles. J Endocrinol 1994;143:399–406.

11. Thorner MO, Vance ML, Rogol AD, Blizzard RM, Veldhuis JD, Cauter EV, et al. Growth hormone-releasing hormone and growth hormone-releasing peptide as potential therapeutic modalities. Acta Paediatr Scand 1990;367:29–32.

12. Alster DK, Bowers CY, Jaffe CA, Ho PJ, Barkan AL. The growth hormone (GH) response to GH-releasing peptide (His-DTrp-Ala-Trp-DPhe-Lys-NH$_2$), GH-releasing hormone, and thyrotropin-releasing-hormone in acromegaly. J Clin Endocrinol Metab 1993;77:842–5.

13. Gertz BJ, Barrett JS, Eisenhandler R, Krupa DA, Wittreich JM, Seibold JR, et al. Growth-hormone response in man to L-692,429, a novel nonpeptide mimic of growth hormone-releasing peptide-6. J Clin Endocrinol Metab 1993;77: 1393–7.

14. Cordido F, Penalva A, Dieguez C, Casanueva FF. Massive growth hormone (GH) discharge in obese subjects after the combined administration of GH-releasing hormone and GHRP-6—evidence for a marked somatotroph secretory capability in obesity. J Clin Endocrinol Metab 1993;76:819–23.

15. Clark RG, Robinson ICAF. Growth hormone (GH) responses to multiple injections of a fragment of human GH-releasing factor in conscious male and female rats. J Endocrinol 1985;106:281–9.

16. Tannenbaum GS, Ling N. The interrelationship of growth hormone (GH)-releasing factor and somatostatin in the generation of the ultradian rhythm of GH secretion. Endocrinology 1984;115:1952–7.

17. McCormick GF, Millard WJ, Badger TM, Bowers CY, Martin JB. Dose-response characteristics of various peptides with growth hormone-releasing activity in the unanesthetized male rat. Endocrinology 1985;117:97–105.

18. Clark RG, Carlsson L, Trojnar J, Robinson ICAF. The effects of a growth hormone-releasing peptide and growth hormone-releasing factor in conscious and anesthetized rats. J Neuroendocrinol 1989;1:249–55.

19. Debell WK, Pezzoli SS, Thorner MO. Growth-hormone (GH) secretion during continuous infusion of GH-releasing peptide—partial response attenuation. J Clin Endocrinol Metab 1991;72:1312–6.

20. Hickey G, Jacks T, Judith F, Taylor J, Schoen WR, Krupa D, et al. Efficacy and specificity of L-692,429, a novel nonpeptidyl growth-hormone secretagogue, in beagles. Endocrinology 1994;134:695–701.

21. Schoen WR, Ok D, DeVita RJ, Pisano JM, Hodges P, Cheng K, et al. Structure-activity relationships in the amino acid side chain of L-692,429 Bioorg Med Chem Lett 1994;4:1117–22.

22. Fairhall KM, Mynett A, Smith RG, Robinson ICAF. Consistent GH responses to repeated injections of growth hormone-releasing hexapeptide (GHRP-6) and the nonpeptide GH secretagogue, L-692,585. J Endocrinol 1995; 145: 417–26.

23. Clark RG, Chambers G, Lewin J, Robinson ICAF. Automated repetitive microsampling of blood growth hormone secretion in conscious male rats. J Endocrinol 1986;111:27–35.
24. Gabrielsson BG, Fairhall KM, Robinson ICAF. Growth hormone secretion in the guinea pig. J Endocrinol 1990;124:371–80.
25. Fairhall KM, Mynett A, Robinson ICAF. Central effects of GHRP-6 on growth hormone (GH) release. J Endocrinol 1994;140(suppl):74P.
26. Goth MI, Lyons CE, Canny BJ, Thorner MO. Pituitary adenylate-cyclase activating polypeptide, growth-hormone (GH)-releasing peptide and GH-releasing hormone stimulate GH release through distinct pituitary receptors. Endocrinology 1992;130:939–44.
27. Cheng K, Chan W, Butler B, Wei LT, Schoen WR, Wyvratt MJ, et al. Stimulation of growth-hormone release from rat primary pituitary-cells by L-692,429, a novel non-peptidyl GH secretagogue. Endocrinology 1993;132:2729–31.
28. Codd EE, Shu AYL, Walker RF. Binding of a growth hormone releasing hexapeptide to specific hypothalamic and pituitary binding sites. Neuropharmacology 1989;28:1139–44.
29. Dickson SL, Leng G, Robinson ICAF. Systemic administration of growth hormone-releasing peptide activates hypothalamic arcuate neurons. Neuroscience 1993;53:303–6.
30. Guillaume V, Magnan E, Cataldi M, Dutour A, Sauze N, Renard M, et al. Growth-hormone (GH)-releasing hormone-secretion is stimulated by a new GH-releasing hexapeptide in sheep. Endocrinology 1994;135:1073–6.
31. Chihara K, Kaji H, Hayashi S, Yagi H, Takeshima Y, Mitani M, et al. Growth hormone releasing hexapeptide: basic research and clinical application. In: Bercu BB, Walker RF, eds. Growth hormone II: basic and clinical Aspects. New York: Serono Symposia, 1994:223–30.
32. Hickey GJ, Baumhover J, Faidley T, Chang C, Anderson LL, Nicolich S, et al. Effect of hypothalamo-pituitary stalk transection in the pig on GH secretory activity of L-692,585. Proceedings of the 76th Annual Meeting of the U.S. Endocrine Society, 1994;661(abstr).
33. Fairhall KM, Mynett A, Robinson ICAF. Central effects of GHRP-6 on growth hormone (GH) release are inhibited by central somatostatin action. J Endocrinol 1995; 144:555–60.
34. Robinson ICAF, Jones PM. Neurohypophysial peptides in cerebrospinal fluid: recent studies. In: Baertschi AJ, Dreifuss JJ, eds. Neuroendocrinology of vasopressin, corticoliberin and opiomelanocortins. London: Academic Press, 1983:21–31.
35. Dickson SL, Leng G, Robinson ICAF. Electrical stimulation of the paraventricular nucleues influences the activity of hypothalamic arcuate neurones. J Neuroendocrinol 1994;6:359–67.
36. Epelbaum J, Moyse E, Tannenbaum GS, Kordon C, Beaudet A. Combined autoradiographic and immunohistochemical evidence for an association of somatostatin binding sites with growth hormone-releasing factor-containing nerve cell bodies in the rat arcuate nucleus. J Neuroendocrinol 1989;1:109–15.
37. Vance ML, Kaiser DL, Evans WS, Furlanetto R, Vale W, Rivier J, et al. Pulsatile growth hormone secretion in normal man during a continuous 24-hour infusion of human growth hormone releasing factor (1-40): evidence for intermittent somatostatin secretion. J Clin Invest 1985;75:1584–90.

38. Bercu BB, Yang SW, Masuda R, Hu C-S, Walker RF. Effects of co-administered growth hormone (GH)-releasing hormone and GH-releasing hexapeptide on maladaptive aspects of obesity in Zucker rats. Endocrinology 1992;131:2800–4.
39. Brain CE, Hindmarsh PC, Brook CGD. Continuous subcutaneous GHRH (1-29)NH2 promotes growth over one year in short, slowly growing children. Clin Endocrinol 1990;32:153–63.
40. Cheng K, Chan WWS, Barreto A, Convey EM, Smith RG. The synergistic effects of His-DTrp-Ala-Trp-DPhe-LysNH2 on growth hormone (GH)-releasing factor-stimulated release and intracellular adenosine 3'5'-monophosphate accumulation in rat primary pituitary cell culture. Endocrinology 1989;124:2791–8.
41. Blake AD, Smith RG. Desensitization studies using perifused pituitary cells show that growth-hormone releasing hormone and His-D-Trp-Ala-Trp-D-Phe-Lys-NH2 stimulate growth hormone release through distinct receptor sites. J Endocrinol 1991;129:11–19.
42. Sethumadhavan K, Veeraragavan K, Bowers CY. Demonstration and characterization of the specific binding of growth hormone-releasing peptide to rat anterior-pituitary and hypothalamic membranes. Biochem Biophys Res Commun 1991;178:31–7.
43. Bitar KG, Bowers CY, Coy DH. Effect of substance-P/bombesin antagonists on the release of growth-hormone by GHRP and GHRH. Biochem Biophys Res Commun 1991;180:156–61.
44. Jansson JO, Downs TR, Beamer WG, Frohman LA. Receptor associated resistance to growth hormone-releasing hormone in dwarf "little" mice. Science 1986;232:511–3.
45. Wu D, Chen C, Katoh K, Zhang J, Clarke IJ. The effect of GH-releasing peptide-2 (GHRP-2 or KP102) on GH secretion from primary cultured ovine pituitary cells can be abolished by a specific GH-releasing factor (GHRF) receptor antagonist. J Endocrinol 1994;140:R9–R13.
46. Malozowski S, Hao EH, Ren SG, Marin G, Liu L, Southers JL, et al. Growth-hormone (GH) responses to the hexapeptide GH-releasing peptide and GH-releasing hormone (GHRH) in the cynomolgus macaque—evidence for non-GHRH-mediated responses. J Clin Endocrinol Metab 1991;73:314–7.
47. Robinson BM, DeMott-Friberg R, Bowers CY, Barkan AL. Acute growth hormone (GH) response to GH-releasing hexapeptide in humans is independent of endogenous GH-releasing hormone. J Clin Endocrinol Metab 1992;75:1121–4.
48. Bowers CY, Veeraragavan K, Sethumadhavan K. Atypical growth hormone releasing peptides. In: Bercu BB, Walker RF, eds. Growth hormone II: basic and clinical aspects. New York: Serono Symposia, 1994:203–22.
49. Dickson S. Intracerebroventricular (icv) administration of growth hormone-releasing peptide (GHRP-6) induces *fos* expression in the hypothalamic arcuate nucleus of the rat. J Physiol 1994;475:149P.
50. Dickson SL, Leng G, Dyball REJ, Smith RG. Central actions of peptide and nonpeptide growth hormone secretagogues in the rat. Neuroendocrinology 1995;61:36–43.
51. Bowers CY. GH releasing peptides—structure and kinetics. J Paediatr Endocrinol 1993;6:21–31.

52. Willoughby J, Brogan M, Kapoor R. Hypothalamic interconnections of soma-
    tostatin and growth hormone releasing factor on growth hormone secretion.
    Neuroendocrinology 1989;50:584–91.
53. Miki N, Ono M, Shizume K. Withdrawal of endogenous somatostatin induces
    secretion of growth hormone-releasing factor in rats. J Endocrinol 1988;
    117:245–52.
54. Clark RG, Carlsson LMS, Rafferty B, Robinson ICAF. The rebound release of
    growth hormone (GH) following somatostatin infusion in rats involves hypotha-
    lamic GH-releasing factor release. J Endocrinol 1988;119:397–404.
55. Thomas GB, Mercer JE, Karalis T, Rao A, Cummins JT, Clarke IJ. Effect of
    restricted feeding on the concentrations of growth hormone (GH), gonadotro-
    pins, and prolactin (PRL) in plasma, and on the amounts of messenger ribon-
    ucleic acid for GH, gonadotropin subunits, and PRL in the pituitary glands of
    adult ovariectomized ewes. Endocrinology 1990;126:1361–7.

16

Evidence for a Central Site and Mechanism of Action of Growth Hormone Releasing Peptide (GHRP-6)

SUZANNE L. DICKSON

Genesis

GHRP-6: A Centrally Acting Growth Hormone Secretagogue

Over ten years have elapsed since C.Y. Bowers and colleagues (1) first identified GHRP-6, a potent synthetic hexapeptide that selectively induces growth hormone (GH) secretion from the anterior pituitary gland. During this time, the majority of research has been directed toward understanding the pituitary site and mechanism of action of this compound. More recently, those of us attempting to unravel the neuroendocrine mechanisms controlling pulsatile GH secretion have been forced to consider the possibility that GHRP-6 also acts centrally to control the activity of GH releasing hormone (GHRF) and/or somatostatinergic neurons. By the end of the 1980s, there were various pieces of circumstantial evidence that GHRP-6 may have a direct central action, including the observation that GHRP-6 is considerably more potent when administered in vivo (1) than in vitro (2). Clark and colleagues (3) observed that, in conscious male rats, the spontaneous pulsatile pattern of GH secretion (known to be critically controlled by the GHRF and somatostatin systems) was completely disrupted by an intravenous infusion of GHRP-6. Furthermore, GHRP-6 completely disrupts the normal cyclic refractoriness in the GH response to GHRF (3, 4); since this intermittent responsiveness to GHRF has been attributed to cyclic variations in portal somatostatin concentrations (4), this would imply that GHRP-6 may act centrally to alter the pattern of somatostatin release. These authors also demonstrated that injection of anti-GHRF serum greatly attenuates the GH response to GHRP-6 (3), indicating either that the pituitary actions of GHRP-6 are dependent on GHRF release into portal blood or that one of the central actions of GHRP-6 may be to stimulate GHRF secretion.

We have used a multidisciplinary approach, employing both electrophysiologic techniques and the detection of immediate early genes as markers of neuronal activation, to investigate whether GHRP-6 has a central site and mechanism of action. Some of these studies employ novel potent nonpeptide GH secretagogues, including L-692,429 and L-692,585, which mimic the pituitary actions of GHRP-6 (5, 6). Intravenous injection of these compounds elicits a dose-related increase in plasma GH concentration and their effects appear to be selective for GH release, with only slight stimulatory effects on adrenocorticotropic hormone release (6, 7).

In the Wilderness: Electrophysiologic Studies of Neurosecretory Arcuate Neurons

GHRP-6 Induces Increased Spike Activity in Putative GHRF Neurons

Our interest in GHRP-6 arose during studies in which we were attempting to identify the GHRF neuron electrophysiologically (that is, to establish a method for distinguishing GHRF neurons from other arcuate neurons by their electrophysiologic properties). Electrophysiologic studies of releasing hormone neurons are fundamental to understanding the hypothalamic regulation of their release. The interaction between spike activity and peptide secretion is comparatively well understood for the neurohypophysial system (8), but the difficulty of identifying cells at the time of recording has to date prevented similar investigations of releasing hormone cells. Recordings in vivo of GHRF neurons are facilitated by the fact that these neurons are densely clustered in the ventral regions of the arcuate nucleus where they constitute a major cell group (9); thus, it is likely that many of the cells recorded during our electrophysiologic studies are GHRF neurons. Furthermore, the GHRF neurons are neurosecretory cells that project to the median eminence; putative neurosecretory cells can be readily identified (and distinguished from other arcuate neurons) using the test of antidromic identification (10, 11). Note that the majority of cells that are not antidromically identified as projecting to the median eminence are likely to be nonneurosecretory neurons and may therefore include populations such as the somatostatin- (12) or proopiomelanocorticotropin-containing cells (13) in this region.

To further identify putative neurosecretory neurons as GHRF neurons, we stimulated two regions of the brain that are known to influence GH secretion: electrical stimulation of the basolateral amygdala causes a large increase in GH secretion without influencing the release of any other pituitary hormone (14); electrical stimulation of the periventricular nucleus suppresses GH secretion during stimulation and this is followed by a large rebound hypersecretion following the end of stimulation (15). In electrophysiologic studies, we identified a subpopulation of arcuate neurons that (a) project to the median eminence (and are therefore likely to include neurosecretory neurons), (b) are transsynaptically excited following electrical stimulation of the basolateral amygdala, (c) are inhibited during electrical stimulation of the periventricular nucleus, and (d) show a rebound hyperactivation following the end of a period of electrical stimulation of the periventricular nucleus (11). Since it is extremely likely that the majority of arcuate neurons fulfilling these multiple criteria are GHRF neurons, we tested these cells with intravenous injection of GHRP-6. Each of three putative GHRF neurons tested showed a large increase in firing rate, occurring within 3 to 5 min following systemic injection of GHRP-6 (9). Intravenous injection of 50 μg of the nonpeptidyl GH secretagogue L-692,585 also elicited a large increase in the firing rate of putative neurosecretory neurons in the arcuate nucleus (16), and both the magnitude and the duration of the response was similar to that observed following GHRP-6 injection (Fig. 16.1). These results provided us with the first direct demonstration that GHRP-6 and the nonpeptide GH secretagogues act centrally to activate hypothalamic arcuate neurons, and the rapid time

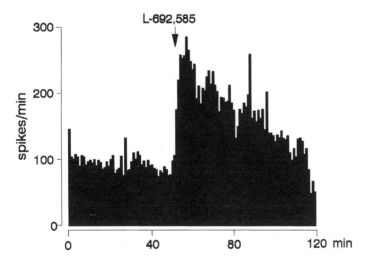

FIGURE 16.1. Extracellular recording from a single putative neurosecretory neuron in the arcuate nucleus that was excited following systemic injection of 50 μg of the GH secretagogue L-692,585. Data were collected in 1-min bins and expressed in spikes per min. The firing rate during the 15-min period following injection (averaged for 90 × 10-sec bins) was significantly higher than the firing rate before injection (averaged for 30 × 10-sec bins; $p < .001$, t-test).

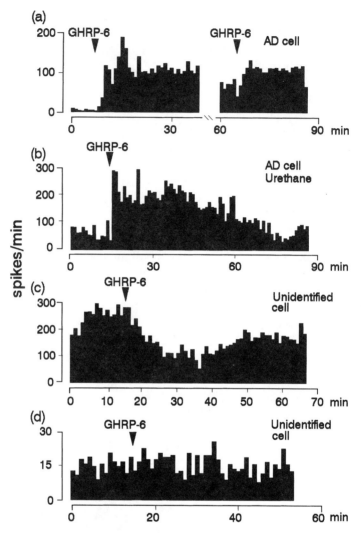

FIGURE 16.2. Extracellular recordings in vivo from four different single cells in the ventrally exposed arcuate nucleus that were tested with systemic injection of 50 µg GHRP-6. A stimulating electrode was positioned on the median eminence to anti-dromically activate cells projecting to this site; thus, putative neurosecretory neu-rons, the AD cells (*a* and *b*), were identified and distinguished from other unidentified cells in this region (*c* and *d*). Note that the unidentified cells did not fulfill criteria for antidromic identification and are therefore unlikely to be neuro-secretory neurons. Recordings (*a*) and (*b*) illustrate excitatory responses following GHRP-6 injection in two different putative neurosecretory neurons in male rats that were anesthetized with pentobarbitone or urethane respectively. In (*a*) the cell was tested with a second injection of 50 µg GHRP-6 (one hour after the first injection—notice the break in the record), to demonstrate that these cells are able to respond to repeated injections of this compound. Recordings (*c*) and (*d*) are from two

course of the response is consistent with a direct central action of these compounds.

Characterization of the Electrophysiologic Responses to GHRP-6

In pentobarbitone-anesthetized rats, we have characterized the responses recorded at the cell bodies of putative neurosecretory arcuate neurons and other unidentified cells in this region (16). The predominant response recorded at the cells bodies of the putative neurosecretory neurons was excitatory: of the 16 cells tested, 8 were excited (Fig. 16.2a), 1 was inhibited, and 7 did not respond. For the unidentified cell group the predominant response was inhibitory: of the 16 cells tested, 8 were inhibited (Fig. 16.2c), 3 were excited, and 5 did not respond (Fig. 16.2d). Thus, we have found evidence to suggest that, in addition to having an excitatory action on putative neurosecretory neurons, GHRP-6 also inhibits a subpopulation of arcuate neurons, the majority of which are likely to be nonneurosecretory. Neurosecretory arcuate neurons also show excitatory responses under urethane anesthesia (Fig. 16.2b), which is interesting in view of some of the early electrophysiologic studies suggesting that arcuate neurons are relatively quiescent in urethane-anaesthetized rats (17, 18).

Manna from Heaven: c-*fos*, a Marker of Neuronal Activation

Immediate Early Gene Expression

Recently the expression of immediate early genes and their protein products have been exploited for functional neuroanatomical studies as markers of neuronal activation (19–21). For example, the magnocellular oxytocin neurons of the hypothalamic supraoptic nucleus express c-*fos* in a number of physiologic and experimental situations (22–24) and in each case the expression of Fos protein followed the increase in firing rate of these cells (25). The induction of Fos protein in these neurons appears to be a consequence of synaptic activation, rather than being directly related to an increase in spike activity (26). C-*fos* is rapidly activated and this is followed by

---

unidentified cells in this region; in (*c*) the cell was inhibited following GHRP-6 injection, whereas in (*d*), no change in firing rate was observed. The data were collected in 1-min bins and expressed as spikes/min. In each case, the firing rate during the 15-min period following injection (averaged for 90×10-sec bins) was significantly different from the firing rate before injection (averaged for 30×10-sec bins; $p < .001$, t-test).

the synthesis of Fos protein, which has a half-life of approximately 2 hours (26). Fos protein can be detected in cell nuclei using immunocytochemical techniques; since Fos protein is more stable than its encoding message, the immunocytochemical detection of nuclear Fos protein can be used to indicate neuronal activation (21). We employed the immunocytochemical detection of Fos protein to investigate the central site and mechanism of action of GHRP-6.

*Induction of c-*fos *Following Systemic Injection of GHRP-6*

Fos-like immunoreactivity was observed in the ventral arcuate nucleus of conscious male rats injected systemically with GHRP-6, but not in saline-treated controls (Figs. 16.3, 16.4, 16.5a) (10). The distribution of arcuate neurons activated by GHRP-6 was very distinctive. Fos-positive nuclei were scattered throughout the entire arcuate nucleus, extending forward to where the third ventricle comes into contact with the median eminence and posteriorly to the ventral regions of the posterior periventricular nucleus (Fig. 16.4). The highest numbers of Fos-positive nuclei were located in

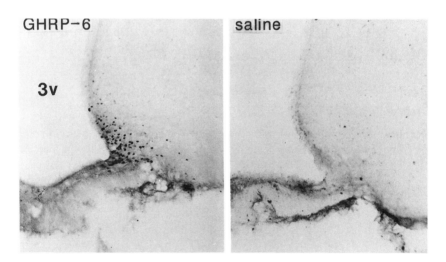

FIGURE 16.3. Fos-like immunoreactivity in the arcuate nucleus following systemic injection of 100 µg GHRP-6 (left) or an equal volume of saline vehicle (right). Ninety minutes following injection rats were terminally anesthetized and perfused with fixative (4% paraformaldehyde in 0.1 M phosphate buffer). Coronal sections (40 µm) were processed for the immunocytochemical detection of Fos protein (9). Dense nuclear staining for Fos protein was observed in the arcuate nucleus of rats injected with GHRP-6 but not in saline-treated controls. Scale bar = 0.2 mm. 3V, third ventricle.

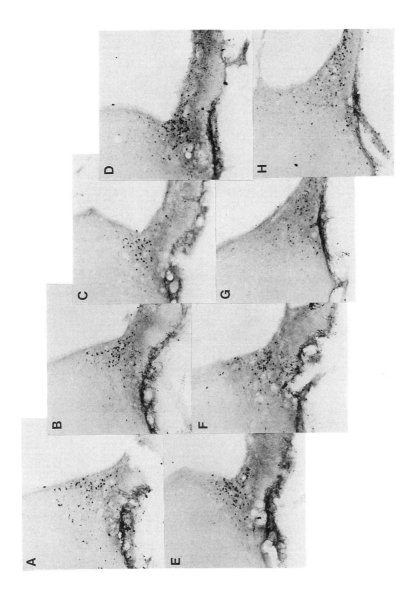

FIGURE 16.4. Serial coronal sections of the rat arcuate nucleus following systemic injection of 100 μg GHRP-6. *A* to *H* represent every third section taken throughout the arcuate nucleus extending anteriorly to where the third ventricle joins the median eminence and posteriorly to the posterior periventricular nucleus. Section thickness = 40 μm.

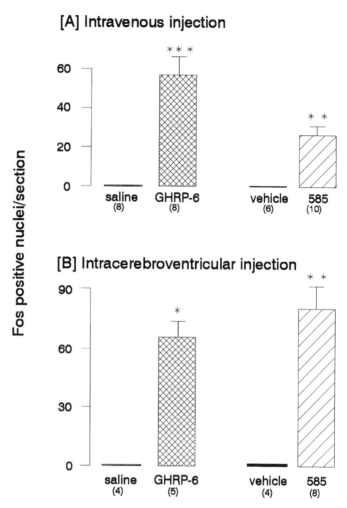

FIGURE 16.5. Mean number of Fos-positive nuclei in the arcuate nucleus of rats injected (*A*) intravenously or (*B*) intracerebroventricularly with either GHRP-6 or L-692,585 ("585"). Vehicle-treated controls were injected with saline for intravenous studies and artificial cerebrospinal fluid for intracerebroventricular studies. For each rat, the mean number of nuclei per section was calculated for every third section throughout the entire arcuate nucleus, and the bars show mean ± SE of these individual means. The numbers in parentheses indicate the number of rats in each experimental group. *$p < .05$, **$p < .01$, ***$p < .001$ (Mann-Whitney U-test, compared with vehicle-treated controls).

the more ventral regions of the arcuate nucleus including a notable medial cluster toward the ventral tip of the third ventricle. The biochemical identity of the cells activated by GHRP-6 is not known and does not correspond to the distribution of any single identified cell group in this region. In view of the overlap in the distribution of GHRF neurons with the cells activated by GHRP-6 (particularly in the more ventral portions of the arcuate nucleus), it is tempting to consider the possibility that at least a subpopulation of the arcuate neurons activated by GHRP-6 are GHRF neurons.

The Parting of the Red Sea: Does GHRP-6 Cross the Blood-Brain Barrier?

Several important questions have arisen as a result of our initial observation that systemically administered GHRP-6 activates hypothalamic arcuate neurons. Since many peptides and other blood-borne factors are not able to gain access to the central nervous system, it is important to establish whether the activation of arcuate neurons observed following GHRP-6 injection is due to a direct central action of this compound. Alternatively, the central actions of GHRP-6 could be mediated by GH (since GHRP-6 induces GH secretion from the pituitary). Indeed, previous studies have demonstrated that, in hypophysectomized rats, intravenous injection of very high doses of GH induces Fos expression in the arcuate nucleus and the periventricular nucleus (28).

Intracerebroventricular Injection of GHRP-6 and the Nonpeptide GH Secretagogue L-692,585

We have demonstrated that intracerebroventricular injection of as little as 0.1 to 1 µg GHRP-6 induces Fos protein expression in the arcuate nucleus of conscious male rats (16) in an identical manner to that described previously for intravenous injection of 50 µg of this compound (Fig. 16.5b) (9). The intracerebroventricular dose required to induce Fos expression in the arcuate nucleus was much lower than the minimum dose required by iv injection to elicit GH release in conscious male rats (10 µg/kg) (29). For all experimental groups, Fos-like immunoreactivity was examined on coronal sections in all regions of the brain that extend the entire length of the hypothalamus, and no increase in Fos expression was observed in any other structure studied, including the neighboring ventromedial nucleus, the supraoptic nucleus, and the periventricular nucleus. Thus, whether administered by intravenous or intracerebroventricular injection, the central effects of GHRP-6 were selective for a subpopulation of arcuate neurons and both the magnitude of the response (that is, the number of Fos-positive nuclei) and distribution of cells activated were identical. There are several important conclusions that can be drawn from this study: (a) the induction of Fos protein in the arcuate nucleus by GHRP-6 must be the consequence of a direct central site of action of this compound, (b) systemically administered

GHRP-6 appears to be able to gain access to central sites, and (c) its central actions are not mediated by GH secretion from the anterior pituitary gland. This latter conclusion is further supported by studies with GH-deficient animals (see below).

The hypothalamic response to the nonpeptide GH secretagogue L-692,585 mirrors that of GHRP-6. When administered by intravenous or intracerebroventricular injection L-692,585 elicited a large increase in the number of cells expressing Fos protein in the arcuate nucleus (Fig. 16.5b); both the magnitude of the response and distribution of the cells activated was identical to that described for GHRP-6 (10, 16).

Systemic or Intracerebroventricular Injection of GHRF

To investigate further the question of whether the central actions of GHRP-6 are mediated by increased GH release from the pituitary gland, rats were injected systemically with 2 μg GHRF, a dose that can elicit a larger increase in plasma GH concentration than that elicited by 50 μg GHRP-6 (29). No increase in Fos expression was observed in the arcuate nucleus (or in any other hypothalamic structure studied) following systemic injection of GHRF (16), indicating that the central actions of GHRP-6 are not mediated by GHRF itself or by increased GH secretion from the pituitary gland. Even when GHRF was administered by intra-cerebroventricular injection, no increase in Fos immunoreactivity was observed in the arcuate nucleus or in any other hypothalamic structure studied. These findings indicate that the central actions of GHRP-6 are not mediated by GHRF and that GHRF and GHRP-6 have a different central site and mechanism of action. This study also provided the first evidence to suggest that the arcuate Fos response following GHRP-6 administration is not mediated by GH, because plasma levels of GH would be elevated in rats injected intravenously with 2 μg GHRF, yet there was no induction of Fos in GHRF-injected rats.

Studies on GH-Deficient Animals

The genetically transmitted dwarfism in the mouse, referred to as "*lit/lit*," has been characterized recently and is the result of a point mutation in the extracellular binding domain of the GHRF receptor (30). These mice provide an excellent experimental model, not only for studying possible mechanisms of central feedback regulation of GH secretion, since basal GH secretion is undetectable, but also for studying the central site and mechanism of action of GHRP-6, since these animals appear to be completely unresponsive to GHRF (31) or GHRP-6 (32). It is somewhat surprising that a GHRF receptor deficiency renders the somatotrophs unresponsive to

GHRP-6 since GHRP-6 binds to a different receptor site to GHRF (33). Collectively, these findings suggest that the pituitary actions of GHRP-6 are dependent upon the presence of a functionally intact GHRF receptor. We have demonstrated that intraperitoneal injection of GHRP-6 (300 μg/kg) elicited a large increase in the number of Fos-positive nuclei in the arcuate nucleus of the *lit/lit* mice (unpublished observation). Thus, even in animals that are completely GH deficient, GHRP-6 activates hypothalamic arcuate neurons.

Similar results were obtained in studies using a GH-deficient strain of dwarf *dw/dw* rats, which have a different pituitary deficiency from that of the *lit/lit* mice. The deficiency in these rats has not been fully characterized but appears to result in a reduction both in the number of somatotrophs and in the ability of remaining somatotrophs to produce GH (34, 35). Unlike the *lit/lit* mice, the *dw/dw* rats have low but detectable levels of GH and are able to elicit very tiny GH responses to GHRF (35) and GHRP-6 (Iain C.A.F. Robinson, personal communication). Systemic injection of 50 μg GHRP-6 results in a large increase in the number of cells expressing nuclear Fos protein in the arcuate nucleus of these animals.

To investigate a possible central site of action of GH, Fos-immunoreactivity was examined in the hypothalamus of *dw/dw* rats injected intracerebroventricularly with recombinant bovine GH, using a concentration that has previously been shown to inhibit GH secretion (15 μg), when administered by this route (36). Since insulin-like growth factor-1 (IGF-1) may be involved in some of the central inhibitory feedback mechanisms controlling GH secretion (37, 38) an additional group of rats was injected intracerebroventricularly with 1 μg recombinant human insulin-like growth factor-1 (rhIGF-1). No increase in Fos-immunoreactivity was observed in the arcuate nucleus (or in any other hypothalamic structure studied) of rats injected intracerebroventricularly with rbGH or rhIGF-1 (unpublished observation). Taken together, these studies suggest that the central actions of GHRP-6 are not mediated by GH because (a) the hypothalamic response to GHRP-6 remains intact in GH-deficient animals, and (b) direct central administration of GH or IGF-1 into the lateral ventricle does not induce Fos expression in the arcuate nucleus.

Revelation

Although it is now generally accepted that GHRP-6 has a central site of action, this idea was met originally with considerable skepticism. This may reflect disbelief in the suggestion that GHRP-6 is able to gain access to central sites or the difficulties in localizing GHRP-6 binding sites in the central nervous system. Using immediate early gene expression, coupled with electrophysiologic studies, we have identified a population of cells in the arcuate nucleus that are activated by GHRP-6.

Induction of the c-*fos* gene cannot be used ubiquitously as a marker for neuronal activation since, for example, neural systems may express immediate genes other than c-*fos* when activated. Thus, although the central effects of GHRP-6 on Fos protein expression were selective for a subpopulation of arcuate neurons, it is not possible to determine whether these are the only neurons activated by this compound. Also, it remains to be determined whether GHRP-6 acts directly on the cells expressing Fos protein or indirectly, either by activating an afferent stimulatory pathway or by disinhibiting an afferent inhibitory pathway.

A great deal can be learned about the central site and mechanism of actions of GHRP-6 by considering the distribution of the cells activated by GHRP-6. In particular the central effects on Fos expression appear to be restricted to the arcuate nucleus, since there was no increase in Fos-immunoreactivity in any other structure studied. GHRP does not indiscriminately activate cells in all regions of the brain that have access to blood-borne factors as demonstrated, for instance, by the lack of Fos-immunoreactivity in the subfornical organ and the supraoptic nucleus. It is also interesting to notice the lack of Fos-immunoreactivity in the periventricular nucleus following GHRP-6 injection because this nucleus is believed to be the site of origin of the adenohypophysiotropic somatostatin neurons (39), which are involved in the inhibition of GH secretion (11, 15, 40).

It is now looking increasingly likely that the central actions of GHRP-6 include the activation of GHRF neurons. This was first suggested by Clark and colleagues (3), who demonstrated that administration of anti-GHRF serum greatly attenuates the GH response to GHRP-6. This premise has been substantiated by the rather elegant demonstration that systemic injection of hexarelin (a synthetic peptide analogue of GHRP-6) induces GHRF secretion into portal blood of the sheep (41). Our results are consistent with this hypothesis because the distribution of cells activated by GHRP-6 shows considerable overlap with the distribution of GHRF cells in this region. Also, in electrophysiologic studies, a subpopulation of the arcuate neurons activated by GHRP-6 fulfilled multiple criteria for identification as GHRF neurons (10). It is unlikely that the sole action of GHRP-6 in the central nervous system is to activate GHRF neurons, because this would not explain the large synergy between GHRP-6 and GHRF for GH secretion (1, 2). Indeed, in electrophysiologic studies, we observed inhibitory responses in putative nonneurosecretory arcuate neurons, and so it is likely that the central effects of GHRP-6 are not solely mediated by a single population of cells in this region.

In conclusion, we have found evidence for a direct central action of GHRP-6 (and the nonpeptidyl GH secretagogue L-692,585) to activate hypothalamic arcuate neurons, from both electrophysiologic studies and from immunocytochemical studies employing the detection of Fos protein as a marker of neuronal activation. The recognition of GHRP-6 as a centrally acting GH secretagogue raises the question of the possible involvement of an as yet unidentified GHRP-like ligand and receptor in the neuroendocrine regulation of pulsatile GH secretion.

Acknowledgments. I would like to thank Prof. Gareth Leng, Dr. Richard E.J. Dyball, Dr. Roy G. Smith, Dr. K.T. O'Byrne, and Dr. Clive Coen for commenting on the manuscript, Dr. Odile Doutrelant-Viltart for assistance, and Iain C.A.F. Robinson for helpful discussion. Research supported by Merck & Co. Research Laboratories (New Jersey) and the Medical Research Council (London).

References

1. Bowers CY, Momany FA, Reynolds GA, Hong A. On the in vitro and in vivo activity of a new synthetic hexapeptide that acts on the pituitary to specifically release growth hormone. Endocrinology 1984;114:1537–45.
2. Cheng K, Chan WWS, Barreto A, Convey EM, Smith RG. The synergistic effects of His-D-Trp-Ala-Trp-D-Phe-Lys-NH$_2$ on intracellular adenosine 3',5'-monophosphate accumulation in rat primary pituitary cell culture. Endocrinology 1989;124:2791–8.
3. Clark RG, Carlsson LMS, Trojnar J, Robinson ICAF. The effects of a growth hormone-releasing peptide and growth hormone-releasing factor in conscious and anaesthetized rats. J Neuroendocrinol 1989;1:249–55.
4. Clark RG, Robinson ICAF. Growth hormone responses to multiple injections of a fragment of human growth hormone releasing factor in conscious male and female rats. J Endocrinol 1985;106:281–9.
5. Smith RG, Cheng K, Schoen W, Pong S-S, Hickey G, Jacks T, Butler B, Chan WW-S, Chaung L-YP, Judith F, Taylor J, Wyvratt MJ, Fisher MH. A non-peptidyl growth hormone secretagogue. Science 1993;260:1640–3.
6. Jacks T, Hickey G, Judith F, Taylor J, Chen H, Krupa D, Feeney W, Schoen W, Ok D, Fisher D, Wyvratt M, Smith R. Effects of acute and repeated intravenous administration of L-692,585, a novel non-peptidyl growth hormone secretagogue, on plasma growth hormone, IGF-1, ACTH, cortisol, prolactin, insulin, and thyroxine levels in beagles. J Endocrinology 1994;143:399–406.
7. Cheng K, Chan WW-S, Butler B, Wei L, Trumbauer ME, Chen H, Schoen WR, Wyratt MJ, Fisher MH, Smith RG. Mechanism of action of L-692,429, a novel non-peptidyl GH secretagogue, on rat growth hormone release both in vitro and in vivo. Proceedings of the 75th Annual Meeting of the Endocrine Society, Las Vegas, 1993;988.
8. Poulain DA, Wakerley JB. Electrophysiology of hypothalamic magnocellular neurones secreting oxytocin and vasopressin. Neuroscience 1982;7:773.
9. Jacobowitz DM, Schulte H, Chrousos GP, Loriaux DL. Localization of GRF-like immunoreactive neurones in the rat brain. Peptides 1983;4:521–4.
10. Dickson SL, Leng G, Robinson ICAF. Systemic administration of growth hormone-releasing peptide (GHRP-6) activates hypothalamic arcuate neurons. Neuroscience 1993;53:303–6.

11. Dickson SL, Leng G, Robinson ICAF. Electrical stimulation of the rat periventricular nucleus influences the activity of hypothalamic neurones. J Neuroendocrinol 1994;6:359–67.
12. Dierickx K, Vandesande F. Immunocytochemical localization of somatostatin containing neurons in the rat hypothalamus. Cell Tissue Res 1979;201:349–59.
13. Meister B, Ceccatelli S, Hokfelt T, Anden N-E, Anden M, Theodorsson E. Neurotransmitters, neuropeptides and binding sites in the rat mediobasal hypothalamus: effects of monosodium glutamate (MSG) lesions. Exp Brain Res 1989;76:343–68.
14. Koibuchi N, Kakagawa T, Suzuki M. Electrical stimulation of the basolateral amygdala (ABL) elicits only growth hormone secretion among six pituitary hormones. J Neuroendocrinol 1991;3:685–7.
15. Okada K, Wakabayashi I, Sugihara H, Minami S, Kitamura T, Yamada J. Electrical stimulation of hypothalamic periventricular nucleus is followed by a large rebound secretion of growth hormone in unanesthetized rats. Neuroendocrinology 1991;53:306–12.
16. Dickson SL, Leng G, Dyball REJ, Smith RG. Central actions of peptide and non-peptide growth hormone secretagogues in the rat. Neuroendocrinology 1995;61:36–43.
17. Makara GB, Harris MC, Spyer KM. Identification and distribution of tuberoinfundibular neurones. Brain Res 1972;40:283–90.
18. Renaud LP. Neurophysiological organization of the endocrine hypothalamus. Res Publ Assoc Res Nerv Ment Dis 1978;56:269–301.
19. Morgan JI, Curran T. The role of ion flux in the control of c-*fos* expression. Nature 1986;322:552–5.
20. Morgan JI, Curran T. Stimulus-transcription coupling in neurones: role of cellular immediate-early genes. Trends Neurosci 1989;12:459–62.
21. Hunt SP, Pini A, Evan G. Induction of c-*fos*-like protein in spinal cord neurons following sensory stimulation. Nature 1987;328:632–4.
22. Hammamura M, Leng G, Emson PC, Kiyama H. Electrical activation and c-*fos* mRNA expression in rat neurosecretory neurons after systemic administration of cholecystokinin. J Physiol 1991;444:51–63.
23. Verbalis JG, Stricker EM, Robinson AG, Hoffman GE. Cholecystokinin activates c-fos expression in hypothalamic oxytocin and corticotrophin-releasing hormone neurons. J Neuroendocrinol 1991;3:205–13.
24. Luckman SM. Fos-like immunoreactivity in the brainstem of the rat following peripheral administration of cholecystokinin. J Neuroendocrinol 1992;3:149–52.
25. Leng G, Way S, Dyball REJ. Identification of oxytocin cells in the rat supraoptic nucleus by their response to cholecystokinin injection. Neurosci Lett 1991;122:159–62.
26. Luckman SM, Dyball REJ, Leng G. Induction of c-*fos* expression in hypothalamic magnocellular neurones requires synaptic activation and not simply increased spike activity. J Neurosci 1994;14:4825–30.
27. Curran T, Miller AD, Zokas L, Verma IM. Viral and cellular *Fos* proteins: a comparative analysis. Cell 1984;36:259–68.
28. Minami S, Kamegai J, Sugihara H, Hasegawa O. Systemic administration of recombinant human growth hormone induces expression of the c-*fos* gene in the hypothalamic arcuate and periventricular nuclei in hypophysectomised rats. Endocrinology 1993;131:247–53.

29. McCormick GF, Millard WJ, Badger TN, Bowers CY, Martin JB. Dose-response characteristics of various peptides with growth hormone-releasing hormone activity in the unanaesthetized male rat. Endocrinology 1985;117:97–105.

30. Lin S-C, Lin CR, Gukovsky I, Lusis AJ, Sawchenko PE, Rosenfeld MG. Molecular basis of the little mouse phenotype and implications For cell type-specific growth. Nature 1993;364:208–13.

31. Jansson J-O, Downs TR, Beamer WG, Frohman LA. Receptor associated resistance to growth hormone-releasing hormone in dwarf "little" mice. Science 1986;232:511–2.

32. Jansson J-O, Downs TR, Beamer WG, Frohman LA. The dwarf "little" (lit/lit) mouse is resistant to growth hormone (GH)-releasing peptide (GH-RP-6) as well as to GH-releasing hormone (GRH). Proceedings of the 68th Annual Meeting of the Endocrine Society, Anaheim, 1986;397.

33. Blake AD, Smith RG. Desensitization studies using perifused rat pituitary cells show that growth hormone-releasing hormone and His-D-Trp-D-Phe-Lys-NH$_2$ stimulates growth hormone release through distinct receptor sites. J Endocrinol 1991;129:11–19.

34. Charlton HM, Clark RG, Robinson ICAF, Porter Goff AE, Cox BS, Bugnon C, Bloch BA. Growth hormone-deficient dwarfism in the rat: a new mutation. J Endocrinol 1988;119:51–8.

35. Carmignac DF, Robinson ICAF. Growth hormone (GH) secretion in the dwarf rat: release, clearance and responsiveness to GH-releasing factor and somatostatin. J Endocrinol 1990;127:69–75.

36. Tannenbaum GS. Evidence for autoregulation of growth hormone secretion via the central nervous system. Endocrinology 1980;107:2117–20.

37. Berelowitz M, Szabo M, Frohman LA, Firestone S, Chu L, Hintz RL. Somatomedin-C mediates growth hormone negative feedback by effects on both the hypothalamus and the pituitary. Science 1981;212:1279–81.

38. Harel Z, Tannenbaum GS. Synergistic action between insulin-like growth factor-I and -II in central regulation of pulsatile growth hormone secretion. Endocrinology 1992;131:758–64.

39. Elde RP, Parsons JA. Immunocytochemical localization of somatostatin in cell bodies of the rat hypothalamus. Am J Anat 1975;144:541–8.

40. Kato M, Suzuki M, Kakegawa T. Inhibitory effect of hypothalamic stimulation on growth hormone (GH) release induced by GH-releasing factor in the rat. Endocrinology 1985;116:382–8.

41. Guillaume V, Magnan E, Cataldi M, Dutour A, Sauze N, Renard M, Razafindraibe H, Conte-Devolx B, Deghenghi R, Lenaerts V, Oliver C. Growth hormone (GH)-releasing hormone secretion is stimulated by a new GH-releasing hexapeptide in sheep. Endocrinology 1994;135:1073–6.

17

Animal Models for Evaluating Xenobiotic Growth Hormone Secretagogue Activity

RICHARD F. WALKER AND BARRY B. BERCU

The decision to develop a drug for commercial application after its discovery is dependent upon preclinical information derived from studies in animal models. In vitro tests are generally the first screen used to evaluate various aspects of efficacy. Primary cell cultures or tumor cell lines and membrane preparations are generally employed as primary screens to define pharmacologic effects on cell function and binding characteristics, respectively. Thereafter, the compound is administered to whole animals so as to determine and quantitate the responses to various doses as a more meaningful measure of efficacy and potency.

It is this secondary level of screening that often provides the incentive for attempting the costly and time-onsuming process of commercialization, or conversely the disincentive for such effort. Because of the pivotal role of testing in drug development, it is important to recognize that misinformation derived from an inappropriate animal model could have significant, adverse effects on the decision-making process. For example, robust responses in unusually sensitive animals might not reflect similar sensitivity in humans. However, the more usual case is that selection of a seemingly appropriate model for the human condition is in fact not a good model, is poorly responsive, and, accordingly, causes the compound not to be designated for development.

This scenario was in fact, the one that retarded scientific interest in, and delayed commercial development of a new family of growth hormone (GH) secretagogues that was discovered in the 1970s following structural alteration of met-enkephalin. The prototype of this family was the hexapeptide His-D-Trp-Ala-Trp-D-Phe-Lys-NH$_2$, named GH releasing hexapeptide or GHRP-6 (GHRP). It released growth hormone, albeit weakly in vitro, and was subsequently tested in the juvenile rat model. Although this choice of a model was logical, it was a poor one because of the peculiar requirement of GHRP for its endogenous, complementary GH secretagogue, the 44 amino acid, hypothalamic hypophysiotrophic peptide, GH releasing hormone (GHRH).

In the early studies, primary cell cultures for testing GHRP activity were prepared from the pituitary glands of preweanling rats based upon the logic that growing animals would be most sensitive to the GH secretagogue. The responses that resulted from use of these cultures were very modest (1), leading some investigators to suggest that they resulted from hypothalamic contamination with GHRH. In fact, the partially functional GH neuroendocrine of the immature, developing rat, combined with absence of GHRH from the primary cultures, were factors directly responsible for attenuation of the GHRP response. This result gave the impression that GHRP had low potency compared with GHRH.

Paradoxically, this impression was confirmed by another laboratory that chose another logical model to test the activity of GHRP in vivo. To compare the activity of GHRH and GHRP in rats, McCormick et al. (2) selected a model in which experimental animals were pretreated with diethyldithiocarbamate (DDC), an inhibitor of catecholamine synthesis. The purpose of pretreating with DDC was to eliminate the neural stimuli for GHRH secretion and thereby evaluate the effects of GHRP on the pituitary, independent of secondary stimulation resulting from endogenous GHRH. This choice seemed to have merit, because episodes of GH release occur spontaneously in male rats. These spontaneous episodes had the potential to obliterate GH episodes resulting from provocative challenges with GHRH or GHRP.

However, female rats that do not produce high amplitude, spontaneous episodes of GH secretion provided an alternative that would not have required DDC pretreatment. This choice would have had more merit than the DDC-treated model, because endogenous GHRH would not have caused episodes of GH secretion in females, and the milieu that naturally supports responsivity to pituitary stimuli would have remained intact. As the result of DDC treatment and attenuation of endogenous GHRH concentrations, the in vivo potency of GHRP was inaccurately estimated. The results that were derived from the DDC-treated model were not even representative of a "pure" GHRP response, because, as we discovered in a subsequent experiment, not all endogenous GHRH is eliminated by DDC.

As seen in Figure 17.1, the effects of passive immunization against endogenous GHRH with specific antibody nearly abolished the response to GHRP, whereas treatment with α-methyl-ρ-tyrosine, an inhibitor of catecholamine synthesis like DDC, only partially attenuated the GHRP response compared with intact controls. Therefore, stimulated GH secretion in DDC-treated rats was not an accurate estimate of in vivo GHRP activity.

It could be argued that "pure" activity was even lower than that reported by McCormick et al. (2) as suggested by our studies with GHRH antibody, and perhaps even nonexistent if total removal of endogenous GHRH was achieved. On the other hand, it might also be validly argued that the "true"

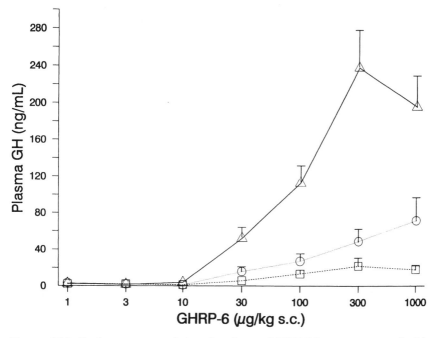

FIGURE 17.1. Peak responses to increasing doses of GHRP in rats pretreated with vehicle (triangles), α-methyl-ρ-tyrosine (circles), or GHRH antiserum (squares). Reprinted with permission from Bercu et al. (45).

activity of GHRP in vivo is a function of its relationship with endogenous GHRH and that any values derived by altering endogenous GHRH is contrived, and only a scientific curiosity. The same might be said about GHRH. Since it is a complement of GHRP, the reciprocal might be also be true, i.e., that the endogenous analogue of GHRP is the complement of GHRH and therefore is required for full expression of its potency in vivo. This concept is employed in a clinical diagnostic test for pituitary GH secretory capability that is presented by Bercu and Walker in Chapter 18 of this book. However, the complementary hypothesis does not presently lend itself to direct testing as with precipitating antibodies against GHRH, because the endogenous analogue of GHRP has not yet been discovered.

Presumably GHRP and its endogenous analogue can be likened to morphine and the endorphins. While an endogenous analogue of GHRP seems likely, it hasn't yet been identified, and therefore its complementary effects on the activity of GHRH cannot presently be evaluated. Nonetheless, the aforementioned examples show how logical but unfortunately inappropriate choices of animal models for testing GHRP caused it to be essentially ignored by the scientific community and delayed for commercial development as a therapeutic agent for nearly a decade.

This experience underscores the importance of proper selection of animal models for future evaluation of the xenobiotic GH secretagogues, which now include nonpeptidyl moieties. Accordingly, some models of GH insufficiency in which GH secretagogues such as GHRP could be tested will be briefly reviewed and critiqued. The juvenile rat has already been discussed, and its appeal as a model of physical development for testing growth acceleration was mentioned above. Although the main negative criticism of this model was its immature GH neuroendocrine axis, it is also important to recognize that in a normal animal acceleration of growth would be an abnormal process and the body would logically resist forces causing the aberration, at least initially. For example, if growth rate is optimal, then administration of GHRP would have to perturb homeostasis, shifting all those factors that regulate normal growth to a higher level. One such change would presumably have to occur in concentrations of insulin-like growth factor (IGF-1), which in young rats are already quite high, compared with older ones. In fact, plasma IGF-1 concentrations are very high in young rats, ranging from 1,000 to 1,400 ng/ml, and progressively declining to as low as 200 to 300 ng/ml in old rats. Therefore, in addition to lacking sufficient complement to support optimal GHRP activity, attempts to elevate IGF-1 as an indicator of GHRP efficacy in juvenile rats would be resisted by the model. These observations suggest that better alternatives for testing GH secretagogues in vivo would be animals with functionally compromised GH secretory mechanisms, especially those in which decremental GH secretion is associated with inadequate or inappropriate pituitary stimulation.

On the other hand, models with inherent defects in pituitary function would be less desirable. One example is the homozygous little mouse (*lit/lit*), which carries a genetic mutation causing a deleted or defective receptor for GHRH (3). We coadministered GHRH and GHRP to this model and only two of six animals released GH above the detection limits of our assay, while GH secretion was potentiated in heterozygous mice with normal growth phenotypes (mean $\pm$ S.E.M. peak plasma GH: 1.65 $\pm$ 0.7 vs. 316.3 $\pm$ 109.2 ng/ml, respectively, $p < .001$). Since coadministered GHRP and GHRH did not stimulate optimal secretion of GH in homozygous (*lit/lit*) mice compared with heterozygous controls with normal growth phenotypes, the data suggest that GHRH complements the activity of GHRP at GHRH specific sites, not by enhancing the activity of GHRP at its own site. This conclusion was supported by our subsequent findings that administration of a GHRH receptor antagonist to normal adult rats significantly blocked the response to GHRP.

As seen in Figure 17.2, administration of [N-Ac-Tyr^1, D-Arg^2]-GHRF 1-29 amide at doses up to 50 µg/kg blocked GH secretion induced by GHRP or morphine. Morphine was tested in parallel with GHRP because the opioid releases GH as part of its action, presumably by stimulating GHRH release from the hypothalamus. At 150 µg/kg, the opioid stimulated GH

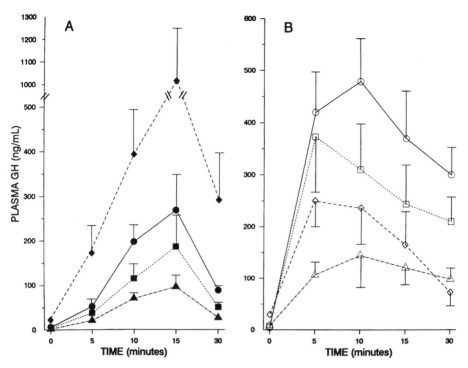

FIGURE 17.2. Effect of [N-Ac-Tyr^1, D-Arg^2]-GHRF 1-29 amide. A GH receptor agonist, on GH release in response to (A) GHRP (30 µg/kg) or (B) morphine (1.5 mg/kg). The GHRH receptor antagonist was administered in doses of 0 (circle), 5 (square), 50 (triangle), or 150 (diamond) µg/kg. Reprinted with permission from Bercu et al. (45).

release slightly more than at 50 µg/kg in morphine-treated rats, but potentiated it in GHRP-treated rats. The data presented in Table 17.1 show that the slight increase in GH secretion at the highest dose of the GHRH receptor antagonist was the result of its partial agonist activity at high dosages. This is indicated by the fact that plasma GH was significantly elevated above basal levels in rats administered the GHRH receptor antagonist alone, without GHRP or morphine. However, potentiated GH release in the GHRP-treated rats demonstrates that activation of the GHRH receptor was accompanied by amplification of the GH secretory signal initiated by activation of the GHRP receptor. Therefore, these data show that an intact GHRH receptor is requisite for realization of GHRP activity in vivo, and, conversely, that animal models with defective GHRH receptors are poor choices for testing the xenobiotic molecules.

Along this line of reasoning, animal models with other GH secretory signal transduction defects beyond the GHRH receptor should also be poor choices for testing members of the GHRP GH secretagogue family. As seen in Figure 17.3, homozygous dwarf (*dw/dw*) rats with defective adenosine

258   R.F. Walker and B.B. Bercu

TABLE 17.1. Effect of [N-Ac-Tyr¹, D-Arg²]-GHRF 1-29 amide, a GH receptor antagonist, on basal, plasma GH concentrations in female rats.

Dose of GH antagonist (μg/kg)	Plasma basal GH concentrations (ng/ml)
0	5.7 ± 0.9
5	6.9 ± 1.1
50	4.8 ± 0.8
150	18.7 ± 2.3*

*$p < .01$ compared with all other values.
Values represent mean ± S.E.M. of GH concentrations collected 20 min after administration of the GH receptor antagonist; $n = 6$ rats/dose.

3′,5′-cyclic monophosphate (cAMP) second messenger systems (4) were hyporesponsive to coadministered GHRH and GHRP-6 compared with heterozygous (*dw/?*) rats of the same strain with normal phenotypes. Similarly, obese Zucker rats with a genetic defect underlying their abnormal body weight also have poor responsivity to GH secretagogues (Fig. 17.4) (5). This defect is probably pituitary associated, at least in part, because neither GHRP nor coadministered GHRH and GHRP stimulated GH secretion comparable to nonobese Zucker rats. The data derived from

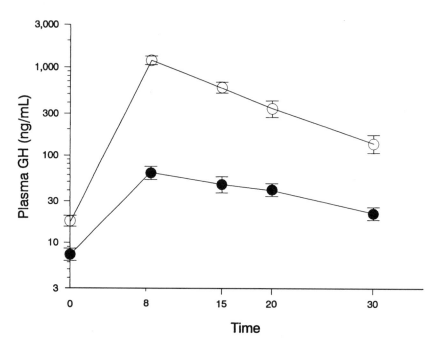

FIGURE 17.3. GH secretory responses of homozygous (*dw/dw*, closed circles) and heterozygous (*dw/?*, open circles), female Lewis dwarf rats coadministered GHRH and GHRP. Stimulated GH release in the dwarf phenotype was significantly less (*p* < .001) than in the normal statured rats (*n* = 6 rats/group).

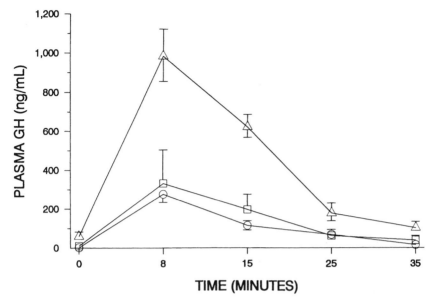

FIGURE 17.4. GH secretion in response to a provocative dose of GHRH and GHRP in lean rats (triangles) or obese rats previously administered saline (circles) or GHRH plus GHRP for 60 consecutive days. The response in lean rats was significantly greater ($p < .01$) than that in obese rats. Difference in stimulated GH secretion between the two groups of obese rats were not statistically significant ($N = 6$ rats/group). Reprinted with permission from Bercu et al. (5).

these studies with animal models whose GH deficiencies result from defects at the pituitary level show that they are clearly inappropriate for studying the potential of GH secretagogues as therapeutic entities.

A more appropriate model would be one in which GH insufficiency results from factors extrinsic to the pituitary gland, such as those resulting from decrements in hypophysiotrophic peptides. Consistent with this hypothesis, we studied the effects of GHRP and GHRH on GH secretory decline and associated sequelae in old rats. During aging, changes in the GH neuroendocrine axis are consistent with disuse atrophy at levels below the brain, and thus should respond to supplementary administration of GH secretagogues, assuming that it is accomplished before irreversible degeneration of target tissues has occurred. With advancing age there is a progressive decrease in spontaneous GH secretion that is accompanied by reduced serum concentrations of IGF-1. As the result of these decremental changes, nonendocrine systems that are influenced by GH and IGF-1 become progressively less effective. Three examples include the immune system, the hepatic lipid metabolizing systems, and the musculoskeletal system. These interactions with endogenous GH contribute to degenerative changes in form and function associated with aging. For example, the correlation between low serum GH concentrations and somatic involution/ physiologic dysfunction was recognized decades ago (6, 7). However, limited availability of cadaver-derived GH prior to development of recombinant biotechnological methods prevented valid testing of the hypothesis that GH deficiencies and aging decrements were functionally related.

When GH became available for experimentation, it was possible to show that administration of the hormone to old men significantly increased IGF-1, urinary nitrogen retention, body weight gain, lean body mass, bone density, renal function, and improved psychological attitude (8–10). Thus, while patterns of spontaneous GH secretion deteriorated and serum GH concentrations declined during aging (7, 11, 12), somatic responsiveness to GH seems to remain intact, since IGF-1 (somatomedin-C) decrements and certain physical/physiologic deficits were readily reversed by administration of exogenous GH (13).

We previously showed that it was possible to restore endogenous GH secretion to youthful values by coadministering GHRH and GHRP-6 (14), a highly potent combination of GH secretagogues (16–18). Hypothetically, reversal or prevention of certain aspects of senescence with endogenous GH would be more desirable than with exogenous GH because a more physiologic pattern of hormone secretion could be simulated by stimulating the pituitary, and more of the GH neuroendocrine axis would be reactivated. To the contrary, administration of exogenous, recombinant GH would suppress GH axial function, further promoting atrophy of the pituitary through negative feedback influence, whereas stimulation of the pituitary to increase endogenous GH output might improve pituitary function. Thus, a pharmacologic agent(s) with the potential to improve endogenous GH secretion in the elderly, opposing some of the catabolic and dysfunctional effects of aging, would have significant biomedical relevance and practical application in geriatric medicine (18).

Pituitary Gland

Prior studies demonstrated that old animals and humans were less responsive to GHRH than their young counterparts (19–26). Increased somatostatin influence was suggested as the cause of age-related hyposensitivity to GHRH (25). Also, reduced responsiveness to GHRH in old rats was accompanied by receptor and second messenger deficits as well as reduced pituitary GH concentrations (27, 28). Taken together, these findings suggest that in addition to increased somatostatin influence, deterioration of pituitary signal transduction mechanisms and/or reduced stores of pituitary GH can contribute to reduced GH secretion in old rats. However, coadministration of GHRH and GHRP to old rats produced GH hypersecretion, indicating that despite these potential impediments, the capability for stimulated GH release was preserved (14). This observation served as the basis for testing whether normal GH secretion could be achieved in old rats coadministered GHRH and GHRP, despite the negative influences of advanced age.

The pituitary gland is central to certain age-related changes that are amenable to treatment with GH secretagogues because it produces and secretes GH. If the decline in its capacity to provide the body with GH is not related to inherent defects, then it should respond positively to supplements of the secretagogues. Under these conditions, the aged rat was considered to be an appropriate model for evaluating xenobiotic GH secretagogues as therapeutic entities. To test this hypothesis, we evaluated pituitary reserve in old rats. Since pituitary GH concentrations and spontaneous as well as GHRH-stimulated GH secretion decline during aging, it was of interest to determine whether the pituitaries of old rats could sustain daily episodes of high-amplitude GH secretion under the influence of coadministered GHRH and GHRP. The data presented in Figure 17.5 demonstrate that in fact the old rats were capable of daily GH release that was quantitatively comparable to that seen in young rats. As we observed in a previous study (14), GH hypersecretion occurred in response

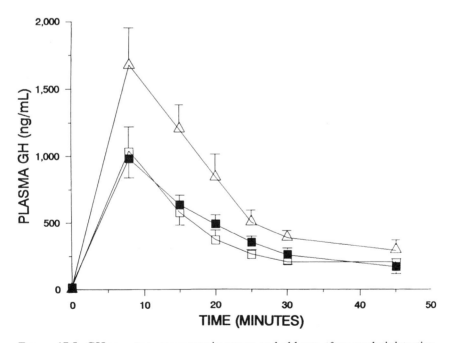

FIGURE 17.5. GH secretory responses in young and old rats after coadministration of GHRH and GHRP for 5 consecutive days. The initial response in old rats (triangles) was significantly greater ($p < .05$) than that in young rats (closed squares), whereas GH secretion in old rats administered the GH secretagogues for 5 days (open squares) and young rats was comparable.

TABLE 17.2. Effects of chronic coadministration of GHRH and GHRP on selected endocrine functions in old Fischer 344 female rats.

Parameter	Young (saline)	Old (saline)	Old (GHRH/GHRP)
Pituitary weight (mg ± S.E.M.)	11.3 ± 3.1	58 ± 11.1*	15.6 ± 6.9
Pituitary tumor incidence (%)	0	83*	17
Pituitary GH (µg/mg wet wgt)	248 ± 35	39 ± 15*	275 ± 38
Pituitary PRL (µg/mg wet wgt)	334 ± 65	169 ± 68*	422 ± 69
Plasma IGF-1 (ng/ml)	839 ± 31	542 ± 32*	780 ± 24
Plasma GH (ng/ml)	1,683 ± 213	2,956 ± 225*	1,402 ± 127
Plasma PRL (ng/ml)	22 ± 9	375 ± 121*	54 ± 8

Values represent mean ± S.E.M. for duplicate determinations from six rats per group.
*Significantly different from control ($p < .05$).

to the initial dose of GH secretagogues. However, within 5 days of treatment, stimulated GH secretory profiles in the old rats was indistinguishable from that in the young rats. As seen in Table 17.2, old rats retained the ability to release youthful quanta of GH in response to daily

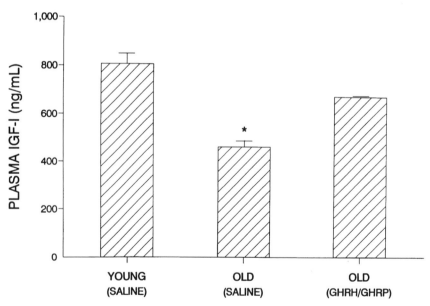

FIGURE 17.6. Increased plasma insulin-like growth factor-1 (IGF-1) in old rats administered GHRH and GHRP. The data show that plasma IGF-1 concentrations in old rats administered the GH secretagogues were significantly greater ($p < .01$) than in old rats administered saline. Thus, endogenous GH released in response to GH secretagogue administration was bioactive.

coadministration of GHRH and GHRP because peak plasma GH concentrations after treatment on the final day of study were comparable to those produced in young, naive rats. In addition to retaining adequate pituitary reserve, stimulated GH in the old rats was bioactive as indicated by its ability to increase plasma IGF-1. As seen in Figure 17.6, IGF-1 was significantly lower in old, control rats than in young control rats, whereas IGF-1 concentrations in old, GH secretagogue–treated rats were comparable to those in young control rats after only 5 days of GHRH and GHRP administration. As with GH secretion, plasma IGF-1 concentrations remained at youthful values until the end of the study (Table 17.2).

At necropsy, visual examination of the pituitary gland revealed that coadministration of GHRH and GHRP was associated with a reduced incidence of pituitary macroadenomas (Table 17.2). The reduced incidence of pituitary tumors was reflected by a low mean pituitary weight in the old, GH secretagogue–treated rats that approached the mean pituitary weight measured in young rats. The reduced incidence of macroadenomas and pituitary weight in old, treated rats was associated with increased concentrations of pituitary GH. As seen in Figure 17.7, pituitary GH concentrations were reciprocally related in young and old rats. In young rats, low

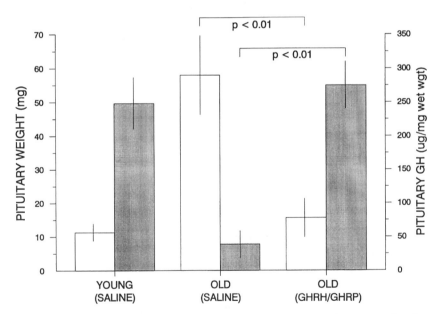

FIGURE 17.7. Effects of GHRH and GHRP coadministered for 120 consecutive days on pituitary weight (open bars) and GH concentrations (shaded bars). Pituitary weight was significantly decreased ($p < .01$) and pituitary GH concentration was significantly increased ($p < .01$) in old rats treated with the GH secretagogues. Pituitary GH concentrations and pituitary weight in young and old saline-treated rats were significantly different ($p < .01$). Reprinted with permission from Walker et al. (15).

weight was associated with high GH concentrations, while the reverse was true in old rats administered saline. On the other hand, heavy pituitary weight was decreased and low GH concentrations were increased in old rats chronically administered GHRH and GHRP. Since daily episodes of stimulated GH secretion were associated with increased GH accumulation in the pituitary gland, it was of interest to determine whether administration of the GH secretagogues activated molecular processes for GH synthesis.

To test this hypothesis, pituitary concentrations of GH messenger RNA (mRNA) and prolactin (PRL) mRNA were measured by Northern blot analysis (Fig. 17.8). Although not quantitative, the effect of GH secretagogue treatment on pituitary PRL and GH mRNAs are easily visualized in Figure 17.8. When the data were quantified, it was clear that treatment had significantly affected transcriptional mechanisms for these hormones. As

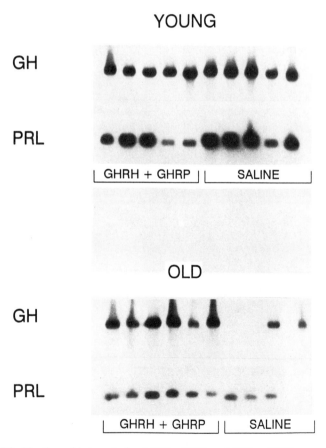

FIGURE 17.8. Northern blot analysis of GH mRNA and PRL mRNA concentrations in pituitaries of young and old rats administered GH secretagogues or saline for 120 consecutive days. Reprinted with permission from Walker et al. (15).

seen in Figure 17.9, pituitary GH mRNA concentrations increased significantly in old, treated rats to levels that actually exceeded those in young control animals. Furthermore, PRL mRNA also increased significantly. The changes in pituitary PRL mRNA were associated with increased concentrations of pituitary PRL and reduced concentrations of plasma PRL (Table 17.2). Thus, stimulation of somatotrophs with GH secretagogues

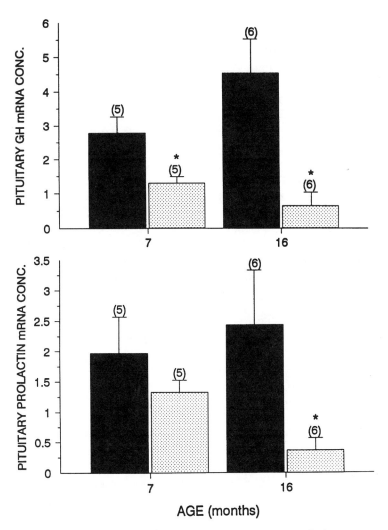

FIGURE 17.9. Effects of coadministered GHRH and GHRP on pituitary concentrations of GH mRNA and prolactin mRNA in old and young rats. GH mRNA in old rats as well as young rats administered the GH secretagogues (shaded bars) was significantly greater than in those treated with saline (stippled bars). PRL mRNA was significantly increased ($p < .05$) by GH secretagogue administration only in the older group. Reprinted with permission from Walker et al. (15).

reduced age-related hyperprolactinemia while increasing GH secretion. The results of this study show that reduced GH secretion and hyperprolactinemia in the old female rat model were not the result of intrinsic and irreversible defects in pituitary function. Rather, the data suggest that the most significant factors to perturb the GH neuroendocrine axis were external to the pituitary, involving active inhibition and/or reduced stimulation of the gland.

There are several possible explanations for the efficacy of GHRH and GHRP in old rats. The most obvious is that GHRP is a somatostatin antagonist that enhances GHRH activity passively, by blocking the inhibitory mechanism dominating GH secretion. However, the exuberant response to coadministered GHRH and GHRP so greatly exceeds that produced by other somatostatin antagonists and GHRH under optimal conditions that it is more likely that GHRP also contributes directly to GH secretion, perhaps by amplifying the GHRH signal. For example, Abribat et al. (27) showed that the high-affinity binding site for GHRH deteriorates in old rats. Since this site is presumably the one responsible for the physiologic action of GHRH, GHRP may improve GHRH binding and thus initiation of the signal transduction process. Alternatively or additionally, GHRP may markedly increase GHRH-stimulated cAMP in the pituitary of old rats, thereby overcoming age-related, GH secretory deficits. This hypothesis is supported by an in vitro study of Cheng et al. (29) in which cAMP concentrations of pituitary cells incubated with GHRH and GHRP increased to levels greater than those in cells incubated with GHRH alone, despite the fact that GHRP by itself had no effect on cellular cAMP concentrations.

Another observation was that GH secretagogue administration increased pituitary GH concentrations that were initially low in the old rats. Since GH secretion and pituitary GH concentrations both increased, it seemed likely that the GH secretagogues also activated molecular processes leading to GH synthesis. GHRH (30) but not GHRP (data not shown) increased mRNA for GH. However, in the present study, GH mRNA increased significantly in old as well as in young rats. Thus, as with cAMP, the potential for GHRH to increase pituitary GH mRNA may be amplified by GHRP. Interestingly, PRL mRNA also increased in both age groups while hyperprolactinemia was significantly reduced in the old animals. This finding suggests that transcription of GH and PRL messages is in some way linked, and that activation of synthetic pathways for both hormones may provide a regulatory influence upon PRL release. If so, then hyperprolactinemia may be more a reflection of reduced GH and PRL synthesis than of increased PRL synthesis-dependent secretion. This idea is supported to some extent by the higher concentrations of PRL and PRL mRNA in the pituitaries of young rats that had relatively low plasma concentrations of PRL. Thus, there may exist in old rats, a dysfunctional relationship between somatotrophs and lactotrophs in which decreased GH secretion is associated with increased PRL release. Accordingly, stimulation of somatotrophs with GH secretagogues may feed back upon lactotrophs to reduce PRL secretion. In fact, this relationship may be linked to phenotypic interconversion of the two cell types as previously demonstrated by Kineman et al. (31), who showed that incubation of pituitary cells with cortisol increased the percent of GH secretory cells without changing the total number of GH and PRL secretory cells that were tested. As the result of these findings, we conclude that intrinsic pituitary dysfunction is not the primary cause of diminished GH neuroendocrine activity during aging. Instead, the results strongly suggest that age-related GH secretory deficits result from insufficient or blunted pituitary stimulation by GHRH and perhaps other important GH secretagogues such as a yet unidentified endogenous analogue of GHRP.

Immune System

Immune system defects beginning relatively early in adulthood with involution of the thymus and later by T-lymphocyte dysfunction progressively increase the risk for contracting disease during aging (32, 33). Although the cause(s) of adult-onset thymic involution and T-cell dysfunction is unknown, thymus gland weight and GH secretion are maximal at puberty and decrease in parallel with advancing age (34–36). Reduced GH secretion may be causally related to age-related immune system changes because grafts of GH-secreting, GH_3 tumor cells reversed thymic involution and decrements in T-lymphocyte proliferation/differentiation and interleukin-2 (IL-2) synthesis in old rodents (37–39). Furthermore, blockade of GH synthesis with an antisense oligodeoxynucleotide complementary to GH mRNA inhibited lymphocyte proliferation (40) while administration of GH to old animals increased thymulin secretion (41), improved several other parameters of immune function, and significantly increased longevity (42). Although these reports showed that administration of exogenous, recombinant, or tumor-derived GH improved immune function, the effects of endogenous, naturally occurring GH on these parameters have not been investigated. It is possible that with advancing age, loss of endogenous GH bioactivity as well as reduced GH secretion contribute to degenerative changes in immune system structure and function. Thus, the purpose of the present study was to determine the effects of chronic administration of GH secretagogues on age-related changes in thymus and spleen weight as well as on differentiation and proliferation of lymphocytes in rats.

Although data for 32-month-old rats are shown in some of the results presented below, the numbers of animals in this group became insufficient, due to attrition during the study, to allow valid statistical analysis of the effects of GH secretagogue treatment vs. saline treatment. However, the age changes in the 18-month-old and 32-month-old groups were similarly directed, i.e., parameters either decreased or increased in all age groups from youngest to oldest, suggesting that advancing age produced quantitative but not qualitative changes.

Interim and terminal measurement of plasma hormones revealed that the GH secretagogues increased plasma GH concentrations more than saline (peaks = 1449 $\pm$ 179 ng/ml vs. 27 $\pm$ 7, respectively; $p < .01$) during secretory episodes lasting approximately 2 hours. Stimulated secretion of endogenous GH in old rats was quantitatively and temporally similar to that in young rats and correlated with elevated plasma IGF-1 concentrations (780 $\pm$ 24 ng/ml vs. 839 $\pm$ 31 ng/ml, respectively; NS) compared with that in old, saline-treated rats (542 $\pm$ 32; $p < .05$). Mean weight of thymus glands ($\pm$ standard error) in young and old saline-treated rats was 191 $\pm$ 14 mg and 140 $\pm$ 16 mg ($p < .05$), respectively, whereas mean thymus weight in old rats coadministered GHRH and GHRP (208 $\pm$ 11 mg) was significantly greater ($p < .05$) than that in old rats administered saline and not different from young rats. The GH secretagogues had no effect on thymus gland weight in younger rats or on spleen weights in rats at either age.

Figure 17.10 is a representative, fluorescence-activated cell sorting (FACS) analysis histogram showing that the distribution of thymus gland lymphocytes bearing CD4 and CD8 surface antigens was different in young and old rats. As seen in Figure 17.11A, aging was associated with a decrease

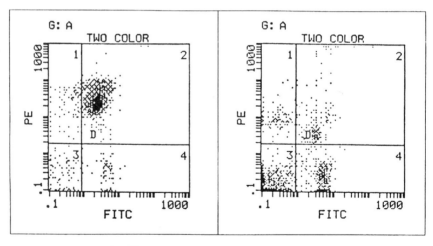

7 MONTHS   18 MONTHS

FIGURE 17.10. Representative FACS display comparing distribution of CD4 and CD8 surface antigens in populations of thymus-derived lymphocytes from 7-month-old and 18-month-old rats. CD4+CD8+ cells located in upper right quadrant. CD4−CD8+ cells in upper left quadrant. CD4−CD8− cells in lower left quadrant. CD4+CD8− cells in lower right quadrant. Note the fewer number of doubled labeled thymocytes in display from older rat. PE, phycoerythren; FITC, fluorescein-5-isothiocyanate. Reprinted with permission from Walker et al. (43).

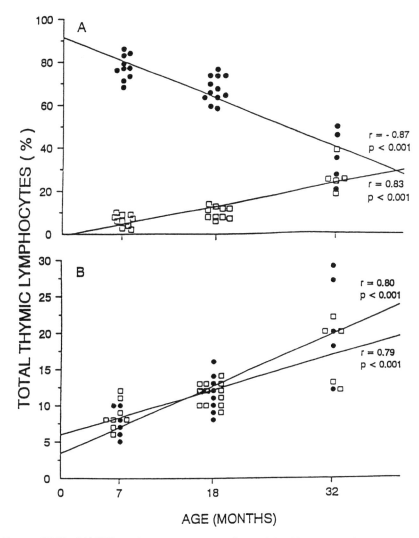

FIGURE 17.11. (A) Effect of age on concentrations of double-positive (CD4$^+$CD8$^+$) and double-negative (CD4$^-$CD8$^-$) lymphocytes in thymus glands of female rats. Linear regression analysis (cell type vs. age for individual rats) shows statistically significant decrease ($p < .05$) in double labeled cells (circles) and increase ($p < .05$) in unlabeled cells (squares). (B) Effect of age on concentrations of single positive (CD4$^+$CD8$^-$ and CD4$^-$CD8$^+$) cells in the thymus gland during aging. Linear regression analysis (cell type vs. age for individual rats) shows statistically significant increase ($p < .05$) in cells with CD4 (circles) and CD8 (squares) surface antigens. Reprinted with permission from Walker et al. (43).

in the number of thymocytes bearing both markers (CD4$^+$CD8$^+$) and an increase in cells lacking both markers (CD4$^-$CD8$^-$), respectively. Thymus cells bearing only CD4$^+$ or CD8$^+$ markers increased during aging (Fig. 17.11B). The relative concentration (percent of total labeled cells) of double-positive cells in thymus glands of control rats was 76% ± 2% at 7 months of age compared with 68% ± 1% at 18 months of age ($p < .05$). This age-related decrease was delayed in rats administered GHRH and GHRP such that the relative concentration of double-positive cells was 82% ± 2% at 7 months of age and 70% ± 2% at 18 months of age. Thus, mean concentrations of CD4$^+$CD8$^+$-labeled lymphocytes were greater in rats coadministered GHRH and GHRP than in controls administered saline. However, the differences between saline-treated and GHRH- and GHRP-treated rats reached statistical significance ($p < .05$) only in those that went from 3 to 7 months of age during the study (76% ± 2% vs. 82% ± 2%, respectively). In contrast, the age-related increase in CD4$^-$CD8$^-$ cells was not prevented by administration of GH secretagogues in either age group, although mean values were slightly reduced in the younger group, almost reaching statistical significance (7.8% ± 0.9% vs. 5.1% ± 1.8%; $p < .1$; control vs. treated group, respectively).

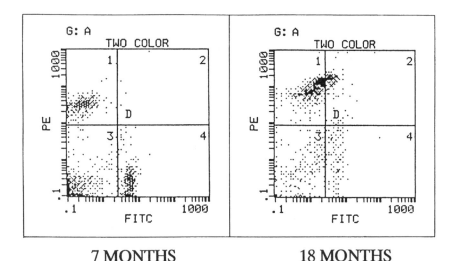

7 MONTHS   18 MONTHS

FIGURE 17.12. Representative FACS display comparing distribution of CD4 and CD8 surface antigens in populations of spleen-derived lymphocytes from 7-month-old and 18-month-old rats. CD4$^+$CD8$^+$ cells located in upper right quadrant. CD4$^-$CD8$^+$ cells in upper left quadrant. CD4$^-$CD8$^-$ cells in lower left quadrant. CD4$^+$CD8$^-$ cells in lower right quadrant. Note the greater numbers of doubled-labeled and CD4$^-$CD8$^+$ splenocytes in display from older rat. PE, phycoerythren; FITC, fluorescein-5-isothiocyanate. Reprinted with permission from Walker et al. (43).

Figure 17.12 is a representative, FACS analysis histogram showing that in the spleen, as in the thymus, age-related changes occurred in the distribution of lymphocytes bearing CD4$^+$ and CD8$^+$ surface antigens. As seen in Figure 17.13A, the major change occurred in spleen effector/suppressor cells (CD8$^+$) whose numbers significantly increased during aging. In contrast, spleen helper cells (CD4$^+$) declined slightly in number at the different ages but the trend was not statistically significant (7 months old = 28.8% $\pm$ 0.5%; 18 months old = 28.6% $\pm$ 1.4%; 32 months old = 25.2% $\pm$ 3.7%; NS). The shift in the percent population of the old rats toward the suppressor phenotype was correlated with a reduction in the CD4$^-$CD8$^-$ cells (Fig. 17.13B). The data presented in Figure 17.13B also show that aging was also associated with a slight but significant increase in double-labeled cells (CD4$^+$CD8$^+$) in the spleen. None of the age-related changes in the spleen were altered by daily coadministration of GHRH and GHRP.

As seen in Figure 17.14, proliferation of spleen cells was significantly reduced ($p < .05$) as a function of age. In contrast, proliferation of thymus cell was significantly greater in older rats than in younger rats (Fig. 17.15). GH secretagogues had no effect on proliferation of total lymphocyte populations from spleen or thymus at any age. However, proliferation of lymphocytes with the CD4$^+$ marker was significantly increased in old rats treated with GHRH and GHRP compared with those administered saline. The greatest effect occurred in the thymocytes from old rats in which statistical significance was achieved upon exposure to concanavalin A (Con A) or PHA (Fig. 17.16). Proliferation of CD4$^+$ splenocytes was significantly increased by GHRH and GHRP coadministration only in response to Con A (Fig. 17.17). GH secretagogues had no effect on proliferation of lymphocytes from spleen or thymus glands from the younger rats.

The results of this study describe certain age-related changes in immune system structure and function and also show that some of the changes can be reversed, at least in part, by administration of GH secretagogues. Important age-related changes included statistically significant reductions in thymus gland weight, thymic populations of CD4$^+$CD8$^+$ lymphocytes, splenic lymphocyte proliferation, and an increase in splenic CD8$^+$CD4$^+$ lymphocytes. Although involution of the thymus has been reported by many laboratories over the years, the reduction in CD4$^+$CD8$^+$ thymic lymphocytes was reported only within the past few years (39). Our data confirm and expand those observations to include splenic lymphocytes.

As previously reported, old rats responded to coadministration of GHRH and GHRP with episodes of GH secretion that were temporally and quantitatively identical to those occurring in similarly treated young rats (44). GHRH has been localized in lymphocytes and therefore may directly contribute at least in part, to the immune system effects seen in this study. In contrast, there are no reports documenting direct effects of GHRP on immune system structure and function, and GHRP was not mitogenic when tested in vitro with a broad range of cell lines (data not shown). Thus, the effects of chronic coadministration of these GH secretagogues on

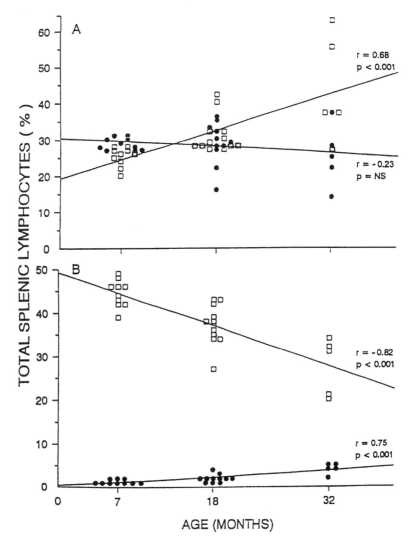

FIGURE 17.13. (A) Effect of age on concentrations of single-positive (CD4+CD8−
and CD4−CD8+) cells in the spleen during aging. Linear regression analysis (cell
type vs. age for individual rats) shows statistically significant increase ($p < .05$) in
CD8+ cells (squares) and no change in CD4+ cells. (B) Effect of age on concentrations of double-positive (CD4+CD8+) and double-negative (CD4−CD8−) lymphocytes in spleens of female rats. Linear regression analysis (cell type vs. age for
individual rats) shows statistically significant decrease ($p < .05$) in unlabeled cells
(squares) and increase ($p < .05$) in double labeled cells (circles). Reprinted with
permission from Walker et al. (43).

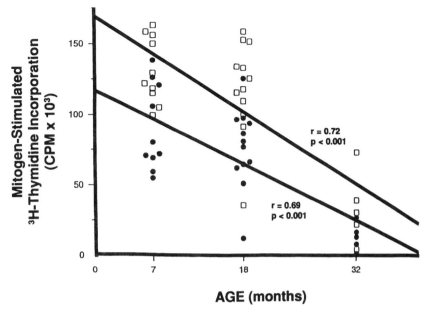

FIGURE 17.14. Mitogen-induced proliferation of spleen cells as a function of age in female rats. Linear regression analysis shows a statistically significant decrease ($p <$ .05) in the response to either Con A (circles) or PHA (squares) with advancing age. Reprinted with permission from Walker et al. (43).

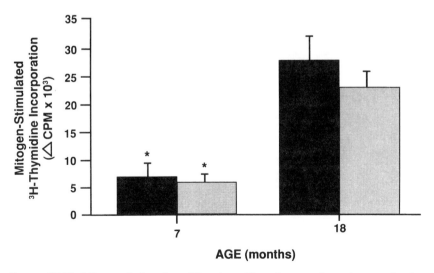

FIGURE 17.15. Mitogen induced proliferation of lymphocytes from thymus gland as a function of age in female rats. ^{3}H-thymidine accumulation was significantly greater (*$p <$ .05) in thymic lymphocytes from 18-month-old rats than in 7-month-old rats in response to Con A (solid bars) or PHA (stippled bars). Values represent means ± S.E.M. of triplicate determinations from six rats per age group. Reprinted with permission from Walker et al. (43).

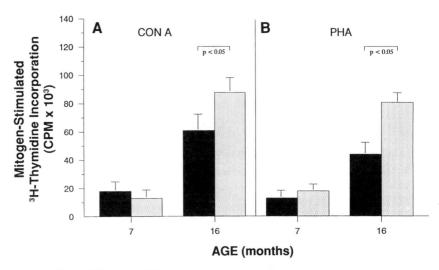

FIGURE 17.16. Effect of GH secretagogues on mitogen-induced proliferation of CD4+ lymphocytes from thymus glands of female rats at different ages. Values represent means ± S.E.M. for triplicate determinations from six rats per treatment group. ^{3}H-thymidine accumulation is compared in lymphocytes from saline treated (solid bars) or GH secretagogue–treated (stippled bars) rats following exposure to Con A (*A*) or PHA (*B*). Reprinted with permission from Walker et al. (43).

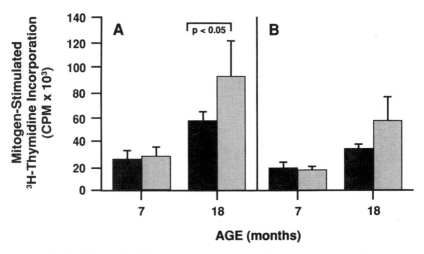

FIGURE 17.17. Effect of GH secretagogues on mitogen-induced proliferation of CD4+ lymphocytes from spleens of female rats at different ages. Values represent means ± S.E.M. for triplicate determinations from six rats per treatment group. ^{3}H-thymidine accumulation is compared in lymphocytes from saline treated (solid bars) or GH secretagogue–treated (stippled bars) rats following exposure to Con A (*A*) or PHA (*B*). Reprinted with permission from Walker et al. (43).

thymic weight and lymphocyte physiology reported in this study are ascribed more to endogenous GH than to direct effects of the peptides. This conclusion is consistent with similar observations in rats administered GH or GH secreting cell lines (37–39). The GH secretagogues were administered only once each day producing single daily episodes of GH release lasting approximately 90 min each (45). The effect of this transient exposure to endogenous GH completely reversed thymic involution associated with aging, and thus was as effective as continuous exposure to GH resulting from administration of the exogenous hormone (38). On the other hand, stimulation of endogenous GH release had a lesser effect on lymphocyte differentiation and proliferation than that resulting from administration of exogenous GH. The differential responses suggest that certain components of the immune system are more sensitive to GH than others, or that the effect of GH on thymic weight is indirect, being mediated by another factor such as IGF-1.

This second hypothesis is based upon the fact that plasma IGF-1 concentrations were fully restored to youthful values in old rats administered the GH secretagogues. Restored IGF-1 concentrations correlated with full recovery of thymic weight ($r = .91; p < .05$) in the old rats, whereas they did not correlate with lymphocyte differentiation and proliferation, which was more affected by exogenous, recombinant, or tumor-derived GH than the GH secretagogues. Thus, lymphocyte function may be more influenced by serum GH concentrations than by IGF-1. Additional support for this idea derives from the report of Weigent et al. (40), in which blockade of GH synthesis with an antisense oligodeoxynucleotide complementary to GH mRNA inhibited lymphocyte proliferation.

The conclusion to be drawn from this animal model study is that the age-related decline in GH release may contribute to progressive involution of the thymus and certain aspects of lymphocyte dysfunction. Stimulated GH secretion opposed some, but not all, the changes that were observed in the lymphocytes. Since greater efficacy has been reported for exogenous GH, the effect of GH secretagogues on immunoreconstitution in old animals may be further optimized by altering the dose and/or frequency of administration.

Cholesterol Metabolism

Elevated serum cholesterol is a concomitant of aging in many animal species. Causes of hypercholesterolemia in the old are unknown, although determinants of cholesterol accumulation in the young are fairly well understood. Cholesterol biosynthesis is controlled by 3-hydroxy-3-methylglutaryl coenzyme A (HMG-CoA) reductase whose activity is regulated by serum cholesterol via negative feedback (46). Possible defects in this feedback mechanism could contribute to age-related hypercholesterolemia by increasing cholesterol biosynthesis. However, the activity of HMG-CoA reductase is low in old rats (47), suggesting that increased synthesis of cholesterol during aging is unlikely. On the other hand, age-related defects in plasma low-density lipoprotein (LDL) clearance have been conclusively demonstrated (48, 49), making this determinant of hypercholesterolemia more relevant for study. Factors influencing cholesterol clearance include cholesterol 7α-hydroxylase, an enzyme that catalyzes the rate-limiting step in bile acid synthesis, and hepatic LDL receptors (46, 50, 51), which bind and reduce plasma cholesterol. The expression of both these proteins is hormonally regulated (51–53).

GH administration decreased plasma cholesterol in vivo (51, 54–56), while experimental hypocholesterolemia was abolished by hypophysectomy and restored by exogenous GH (51), suggesting that the pituitary hormone is a physiologic regulator of circulating cholesterol concentrations. A possible mechanism for this effect of GH on plasma cholesterol was recently identified as involving LDL receptors whose numbers increased in relation to hepatic LDL mRNA concentrations (51). Since GH secretion declines over the life span (7, 12), deficits in this pituitary hormone may contribute to cholesterol accumulation during aging. Thus, the present study was designed to characterize age-related changes in plasma cholesterol as well as hepatic mRNAs for cholesterol 7α-hydroxylase and LDL receptors in rats, and to determine how they might be affected by administration of GH secretagogues.

We first determined the effects of aging on circulating cholesterol concentrations. This objective was met by measuring plasma cholesterol in tail vein blood from rats ranging in age from 3 to 24 months. As seen in Figure 17.18, there was a progressive, age-related increase in plasma cholesterol concentrations beginning as early as 7 months of age. This trend continued throughout life without reaching plateau values.

It was then of interest to determine the influence of endogenous GH whose secretion declines during aging, upon age-related hypercholesterolemia. To investigate the functional implications of this relationship, young adult and middle-aged rats were administered GHRH and GHRP each day for 120 days so as to stimulate daily episodes of GH secretion lasting for approximately 90 minutes (15, 45). The data in Figure 17.19, show that in the young adult group, the difference between plasma cholesterol concentrations at the beginning and end of the study was greater ($p < .05$) in saline-treated (40.7 ± 2.5 mg/dl) than in GH secretagogue–treated (26.0 ± 1.7 mg/dl) rats. Similarly in the older group, plasma cholesterol was greater ($p < .05$) in saline treated (30.0 ± 3.6 mg/dl) than in GH secretagogue-treated (18.7 ± 2.7 mg/dl) rats. Since administration of GHRH and GHRP was associated with reduced plasma cholesterol concentrations in both age groups, the data suggest that daily episodes of

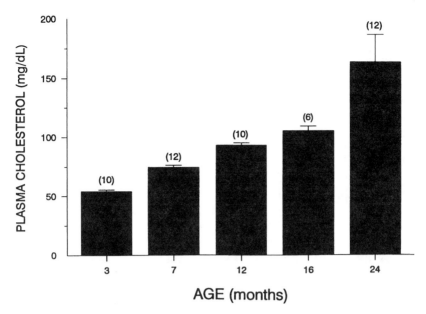

FIGURE 17.18. Age-related increase in plasma cholesterol concentrations in female rats. Measurements were made in untreated animals upon receipt from the breeder after 2 weeks' acclimation to the local facility. Values represent mean ± S.E.M. for rats in each age group (n = number in parenthesis). Differences between each age group are statistically significant ($p < .05$). Reprinted with permission from Walker et al. (58).

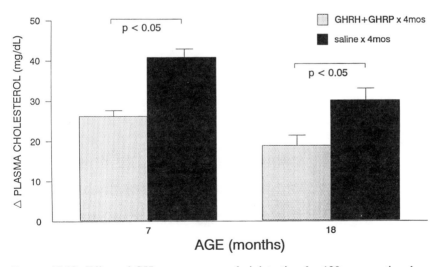

FIGURE 17.19. Effect of GH secretagogue administration for 120 consecutive days on the age-related increase in plasma cholesterol. Values represent the differences in plasma cholesterol concentrations between the first and last days of treatment (mean ± S.E.M.) for six rats/group. Dark symbols represent values in GH secretagogue–treated rats. Light symbols represent values in saline-treated rats. Reprinted with permission from Walker et al. (58).

stimulated GH secretion reduced age-related accumulation of the lipid. However, despite the effect of GH, plasma concentrations of cholesterol in the groups after 4 months of treatment (92.6 ± 3.2 mg/dl and 80.5 ± 1.7 mg/dl; older rats and younger rats, respectively) were not restored to their previous values (73.8 ± 1.8 mg/dl and 54.5 ± 0.7 mg/dl; older rats and younger rats, respectively), further suggesting that factors other than GH were contributing to the age-related increase in cholesterol.

Reduced cholesterol accumulation in rats administered GHRH and GHRP was associated in the 18-month-old group but not in the 7-month-old group with a significant increase in hepatic LDL receptor mRNA (Fig. 17.20) but not cholesterol 7α-hydroxylase mRNA (Fig. 17.21). The data showing this increase in mRNAs are presented in these figures as Northern blot analyses. Quantitative data presented in Table 17.3 show that higher concentrations ($p < .05$) of hepatic LDL receptor mRNA occurred in the older group treated with the GH secretagogues compared with those administered saline. There were no differences in the cholesterol 7α-hydroxylase mRNA concentrations in either group of 18-month-old rats and no

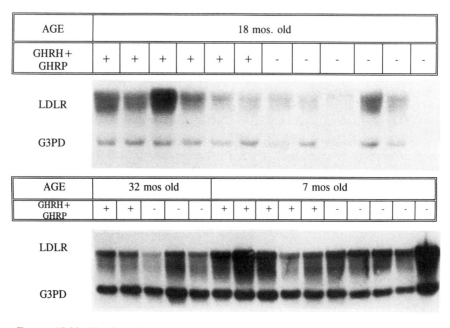

FIGURE 17.20. Northern blotting analysis of LDL receptor mRNA from 7-, 18-, and 32-month-old rats treated with GHRH and GHRP (+) or saline (−). The upper bands represent LDL receptor (LDLR) and the lower bands are glyceraldehyde 3-phosphate dehydrogenase (G3PD), which was used as an internal control. Ten micrograms of poly A⁺ RNA were applied to each lane. Although data for 32-month-old rats are shown, the numbers of GH secretagogue–treated and untreated animals were insufficient for statistical analysis. Reprinted with permission from Walker et al. (58).

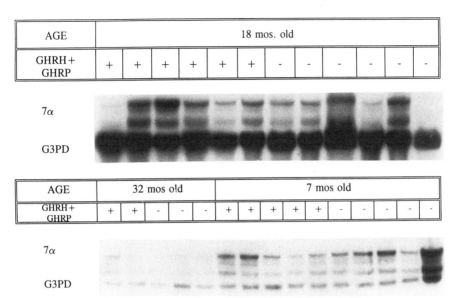

AGE	18 mos. old											
GHRH + GHRP	+	+	+	+	+	+	-	-	-	-	-	-

AGE	32 mos old					7 mos old							
GHRH + GHRP	+	+	-	-	-	+	+	+	+	+	-	-	-	-	-

FIGURE 17.21. Northern blotting analysis of hepatic cholesterol 7α-hydroxylase mRNA (7α) from 7-, 18-, and 32-month-old rats treated with GHRH and GHRP (+) or saline (−). The 4.0, 3.6, and 2.1 kb forms of the enzyme are present. The lowest band is glyceraldehyde 3-phosphate dehydrogenase (G3PD), which was used as an internal control. Ten micrograms of poly A⁺ RNA was applied to each lane. Although data for 32-month-old rats are shown, the numbers of GH secretagogue–treated and untreated animals were insufficient for statistical analysis. Reprinted with permission from Walker et al. (58).

significant differences in LDL receptor mRNA or cholesterol 7α-hydroxy-lase mRNA in either of the 7-month-old groups (Tables 17.3 and 17.4). Although mean values for hepatic LDL receptor mRNA were elevated in secretagogue-treated, 18-month-old rats but not in the 7-month-old rats,

TABLE 17.3. Effects of GH releasing peptides on hepatic LDL receptor mRNA concentrations in young and old rats.

Age group	GHRH and GHRP treated	Saline treated
7 months old	1.86 ± 0.32 (5)	2.20 ± 0.16 (5)
18 months old	1.27 ± 0.36* (6)	0.41 ± 0.10 (6)

Hepatic LDL receptor mRNA concentrations are expressed relative to a glyceraldehyde 3-phosphate dehydrogenase internal control. Values are presented as mean ± S.E.M. Numbers of rats per group are presented in parentheses. * = significantly different ($p < .05$) than same age group receiving opposite treatment.

TABLE 17.4. Lack of effect of GH releasing peptides on hepatic cholesterol 7α-hydroxylase mRNA concentrations in young and old rats.

Age group	GHRH and GHRP treated	Saline treated
7 months old	0.25 ± 0.07 (5)	0.48 ± 0.13 (5)
18 months old	1.00 ± 0.23 (6)	0.60 ± 0.09 (6)

Cholesterol 7α-hydroxylase mRNA concentrations are expressed relative to a glyceraldehyde 3-phosphate dehydrogenase internal control. Values are presented as mean ± S.E.M. Numbers of rats per group are presented in parentheses.

there was a significant change in the relationship of hepatic LDL receptor mRNA and plasma cholesterol for the individual younger animals.

As seen in Figure 17.22A, the two parameters were reciprocally related ($r = -.89$; $p < .05$) in the 7-month-old rats administered GHRH and GHRP but not in those administered saline. Although statistical significance was not reached in the 18-month-old group administered GHRH and GHRP, a similar trend was observed such that higher plasma cholesterol values were generally associated with lower hepatic LDL receptor mRNA concentrations (Fig. 17.22B). Similar correlations between plasma cholesterol and hepatic cholesterol 7α-hydroxylase mRNA concentrations were not evident in rats from either age group or treatment (Fig. 17.23).

The results of this study showed that aging in the animal model was correlated with hypercholesterolemia. Since blood cholesterol concentrations were previously shown to be reduced by administration of exogenous GH (51, 55, 56), we tried to determine whether administration of GH secretagogues would oppose age-related hypercholesterolemia. In fact, stimulated secretion of endogenous GH was effective in that regard. GHRH or GHRP administered alone to old rats was ineffective in releasing endogenous GH and reducing plasma cholesterol concentrations (data not shown), suggesting that the effects of coadministered GH secretagogues in the present study did not result from a direct effect on the liver, but instead by stimulating episodes of pituitary GH secretion. Thus, cholesterol accumulation during aging may be related, at least in part, to GH insufficiency. Apparently the influence of GH insufficiency on cholesterol occurs relatively early in life because GH secretagogues were effective in reducing plasma cholesterol concentrations in rats as early as 7 months of age.

Presumably, the effect of stimulated GH secretion on plasma cholesterol concentrations was mediated by changes in LDL receptor expression since hepatic LDL receptor mRNA concentrations were significantly higher in old rats treated with GHRH and GHRP compared with old rats adminis-

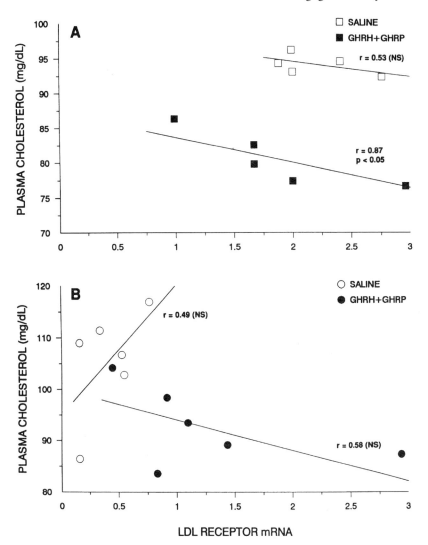

FIGURE 17.22. Effect of GH secretagogue administration on correlation between plasma cholesterol and hepatic LDL receptor mRNA in young rats (*A*) and older rats (*B*). Open symbols represent values for saline-treated rats and closed symbols represent values for GH secretagogue–treated rats. Reprinted with permission from Walker et al. (58).

tered vehicle. Although LDL receptors were not measured in the present study, hepatic LDL receptor mRNA concentrations were previously shown to be associated with increased expression of LDL receptors and reduced plasma cholesterol (51, 57). However, unlike old rats, cholesterol accumulation in young rats administered GH secretagogues in the present study

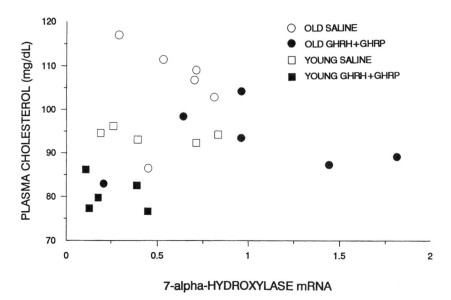

FIGURE 17.23. Lack of effect of GH secretagogue administration on correlation between plasma cholesterol and hepatic cholesterol 7α-hydroxylase. Open circles and open squares = saline-treated old and young rats, respectively. Closed circles and closed squares = GH secretagogue–treated old and young rats, respectively. Reprinted with permission from Walker et al. (58).

was significantly reduced without a concomitant increase in hepatic LDL receptor mRNA concentrations. Presumably, mean LDL mRNA concentration in the younger group of rats was at maximal value and could not be further increased by GH stimulation. However, upon using linear regression analysis of data for individual rats it became apparent that administration of GH secretagogues in the younger animals actually improved the relationship between hepatic LDL mRNA and plasma cholesterol as indicated by the fact that the negative correlation between the two variables reached statistical significance. A similar, reciprocal trend between plasma cholesterol and hepatic LDL receptor mRNA concentrations was observed in old rats administered GH secretagogues, although the relationship was not statistically significant. Perhaps a longer period of GH secretagogue administration will be required to achieve significance in the older, more GH deficient rats.

A final observation drawn from studying the aged rat as a model in which to test GH secretagogues is that GH insufficiency may contribute to the decrease in plasma cholesterol clearance that occurs during aging (48, 49) through a mechanism involving hepatic LDL receptors. The absence of an effect of GH secretagogues on hepatic cholesterol 7α-hydroxylase mRNA in young and old rats suggests that this enzyme does not contribute to the hypocholesterolemic effect of endogenous GH.

Conclusion

This chapter showed how appropriate choice of an animal model for investigating the activity of novel, biologically active compounds can significantly affect scientific interest in and commercial development of them. The specific family of compounds discussed were the family of GH secretagogues, collectively referred to as GHRP. Because of their relative inactivity in the juvenile rat model, development of GHRP for clinical application was delayed for nearly 10 years after their discovery. However, once they were studied in intact adult animals and a better understanding of their pharmacology was achieved, they rapidly gained access to clinical settings and are now being developed for a broad spectrum of uses ranging from growth stimulation in children, to treating prolactin-secreting adenomas, to many geriatric conditions including frailty, osteoporosis, and perhaps certain types of cardiovascular disease. The lessons to be learned about the use of animals from the history of GHRP is that many factors are relevant in selecting the appropriate model. For GHRP the relevant issues required that the investigator be cognizant that

1. normal values, such as basal GH and IGF-1 concentrations, are resistant to change in the normal animal;
2. there is sexual heterogeneity with regard to response, which made the female capable of providing more consistent and reliable data than the male;
3. animals with genetic disorders impeding pituitary function are inappropriate for testing GHRP; and
4. animals with GH insufficiency resulting from inadequate stimulation of the pituitary gland, such as aging rats, are most appropriate for testing GHRP.

Having passed the initial phase of development where results from animal models have the biggest impact on the future of new drugs, GHRP is in a growth phase of clinical development and may soon be in use for treating a wide spectrum of disorders associated with GH insufficiency.

References

1. Sartor O, Bowers CY, Reynolds GA, Momany FA. Variables determining the growth hormone response of His-D-Trp-Ala-Trp-D-Phe-Lys-NH$_2$ in the rat. Endocrinology 1985;117:1441–7.
2. McCormick GF, Millard WJ, Badger TM, Bowers CY, Martin JB. Dose-response characteristics of various peptides with growth hormone-releasing activity in the unanesthetized male rat. Endocrinology 1985;117:97–105.
3. Jansson J-O, Downs TR, Beamer WG, Frohman LA. The dwarf "little" (LIT/LIT) mouse is resistant to growth hormone releasing peptide (GHRP-6) as well as to GH-releasing hormone (GRH). Endocr Soc 1986;130(abstr).

4. Downs TR, Frohman LA. Evidence for a defect in growth hormone-releasing factor signal transduction in the dwarf (dw/dw) rat pituitary. Endocrinology 1991;129:58–67.
5. Bercu BB, Yang S-W, Masuda R, Hu C-S, Walker F. Effects of co-administered growth hormone (GH) releasing hormone and GH-releasing peptide on maladaptive aspects of obesity in Zucker rats. Endocrinology 1992;131:2800–4.
6. Root AW, Oski FA. Effects of human growth hormone in elderly males. J Gerontol 1969;24:97–104.
7. Rudman D, Kutner MH, Rogers CM, Lubin MF, Fleming GA, Bain RP. Impaired growth hormone secretion in the adult population. J Clin Invest 1981;67:1361–9.
8. Rudman D, Feller AG, Hoskote SN, Gergans GA, Lalitha PY, Goldberg AF, Schlenker RA, Chon L, Rudman IW, Mattson DE. Effects of human growth hormone in men over 60 years old. N Engl J Med 1990;323:1–6.
9. Kaiser FE, Silver AJ, Morley JE. The effect of recombinant human growth hormone on malnourished older individuals. J Am Geriatr Soc 1991;39:235–40.
10. Marcus R, Butterfield G, Holloway L, Gilliland L, Baylink DJ, Hintz RL, Sherman BM. Effects of short term administration of recombinant human growth hormone to elderly people. J Clin Endocrinol Metab 1990;70:519–27.
11. Finkelstein JW, Roffwarg HP, Boyar RM, Kream J, Hellman L. Age-related change in the twenty-four-hour spontaneous secretion of growth hormone. J Clin Endocrinol Metab 1972;35:665–70.
12. Ho KY, Evans WS, Blizzard RM, Veldhuis JD, Merriam GR, Samojlik E, Furlanetto R, Rogol AD, Kaiser DL, Thorner MO. Effects of sex and age on the 24-hour profile of growth hormone secretion in man: importance of endogenous estradiol concentrations. J Clin Endocrinol Metab 1987;64:51–7.
13. Pavlov EP, Harman SM, Merriam GR, Gelato MC, Blackman MR. Responses of growth hormone (GH) and somatomedin-C to GH-releasing hormone in healthy aging men. J Clin Endocrinol Metab 1986;62:595–600.
14. Walker RF, Yang S-W, Bercu BB. Robust growth hormone (GH) secretion in aged female rats co-administered GH-releasing hexapeptide (GHRP-6) or GH releasing hormone (GHRH). Life Sci 1991;49:1499–504.
15. Walker RF, Eichler DC, Bercu BB. Inadequate pituitary stimulation: a possible cause of growth hormone insufficiency and hyperprolactinemia in aged rats. Endocrine 1994;2:633–8.
16. Bowers CY, Reynolds GA, Durham D, Barrera CM, Pezzoli SS, Thorner MO. Growth hormone (GH)-releasing peptide stimulates GH release in normal men and acts synergistically with GH-releasing hormone. J Clin Endocrinol Metab 1990;70:975–82.
17. Ilson BE, Jorkasky DK, Curnow RT, Stote RM. Effect of a new synthetic hexapeptide to selectively stimulate growth hormone release in healthy human subjects. J Clin Endocrinol Metab 1989;69:212–4.
18. Thorner MO, Vance ML, Rogol AD, Blizzard RM, Veldhuis JD, Cauter EV, Copinschi G, Bowers CY. Growth hormone releasing hormone and growth hormone releasing peptide as potential therapeutic modalities. Acta Paediatr Scand Suppl 1990;327:29–30.
19. Ceda GP, Valenti G, Butturini U, Hoffman AR. Diminished pituitary responsiveness to growth hormone-releasing factor in aging male rats. Endocrinology 1986;118:2109–14.

20. Ghigo E, Goffi S, Arvat E, Nicolosi M, Procopio M, Bellone J, Imperiale E, Mazza E, Baracchi G, Camanni F. Pyridostigmine partially restores the GH responsiveness to GHRH in normal aging. Acta Endocrinol 1990;123:169–74.

21. Iovino M, Monteleone P, Steardo L. Repetitive growth hormone-releasing hormone administration restores the attenuated growth hormone (GH) response to GH-releasing hormone testing in normal aging. J Clin Endocrinol Metab 1989;69:910–3.

22. Lang I, Kruz R, Geyer G, Tragl KH. The influence of age on human pancreatic growth hormone releasing hormone stimulated growth hormone secretion. Horm Metabol Res 1988;20:574–8.

23. Lang I, Schernthaner G, Pietschmann P, Kurz R, Stephenson JM, Templ H. Effects of sex and age on growth hormone response to growth hormone-releasing hormone in healthy individuals. J Clin Endocrinol Metab 1987;65:535–40.

24. Shibasaki T, Shizume K, Nakahara M, Masuda A, Jibiki K, Demura H, Wakabayashi I, Ling N. Age-related changes in plasma growth hormone response to growth hormone-releasing factor in man. J Clin Endocrinol Metab 1984;58:212–4.

25. Sonntag WE, Gough MA. Growth hormone releasing hormone induced release of growth hormone in aging male rats: dependence on pharmacological manipulation of endogenous somatostatin release. Neuroendocrinology 1988;47:482–8.

26. Sonntag WE, Hylka VW, Meites J. Impaired ability of old male rats to secrete growth hormone in vivo but not in vitro in response to hpGRF(1-44). Endocrinology 1983;113:2305–7.

27. Abribat T, Desiauriers N, Brazeau P, Gaudreau P. Alterations of pituitary growth hormone-releasing factor binding sites in aging rats. Endocrinology 1991;128:633–5.

28. Parenti M, Dall'ara A, Rusconi L, Coochi D, Muller EE. Different regulation of growth hormone-releasing factor-sensitive adenylate cyclase in the anterior pituitary of young and aged rats. Endocrinology 1987;121:1649–53.

29. Cheng K, Chan WW-S, Barreta A, Convey DM, Smith RG. The synergistic effects of HIS-D-TRP-ALA-TRP-D-PHE-LYS-NH$_2$ on growth hormone (GH)-releasing factor-stimulated GH release and intracellular adenosine 3′,5′-monophosphate accumulation in rat primary pituitary cell culture. Endocrinology 1989;124:2791–8.

30. Barinaga M, Yamonoto G, Rivier C, Vale W, Evans R, Rosenfeld MG. Transcriptional regulation of growth hormone by growth hormone-releasing factor. Nature 1983;306:84–5.

31. Kineman RD, Faught WJ, Frawley LS. Steroids can modulate trans-differentiation of prolactin and growth hormone cells in bovine pituitary cultures. Endocrinology 1992;130:3289–94.

32. Cheney KE, Walford RL. Immune function and dysfunction in relation to aging. Life Sci 1974;14:2075–84.

33. Kay MMB. Immunological aspects of aging: early changes in thymic activity. Mech Ageing Dev 1984;28:193–218.

34. Denckla WD. Systems analysis of possible mechanisms of mammalian aging. Mech Ageing Dev 1977;6:143–53.

35. Gil-Ad I, Gurewitz R, Marcovici O, Rosenfeld J, Laron Z. Effect of aging on human plasma growth hormone response to clonidine. Mech Ageing Dev 1984;27:97–100.

36. Weksler ME. The immune system and the aging process in man. Proc Soc Exp Biol Med 1980;165:200–35.
37. Davila DR, Brief S, Simon J, Hammer RE, Brinster RL, Kelly KW. Role of growth hormone in regulating T-dependent immune events in aged, nude, and transgenic rodents. J Neurosci Res 1988;18:108–16.
38. Kelley KW, Brief S, Westly JH, Novakofski J, Bechtel PJ, Simon J, Walker EB. GH₃ pituitary adenoma cells can reverse thymic aging in rats. Proc Natl Acad Sci USA 1986;83:5663–7.
39. Li YM, Brunke DL, Dantzer R, Kelley KW. Pituitary epithelial cell implants reverse the accumulation of CD4⁻CD8⁻ lymphocytes in thymus glands of aged rats. Endocrinology 1992;130:2703–9.
40. Weigent DA, Blalock JE, LeBoeuf RD. An antisense oligodeoxynucleotide to growth hormone messenger ribonucleic acid inhibits lymphocyte proliferation. Endocrinology 1991;128:2053–7.
41. Goff B, Roth JA, Arp LH, Incefy GS. Growth hormone treatment stimulates thymulin production in aged dogs. Clin Exp Immunol 1987;68:580–7.
42. Khansari DN, Gustad T. Effects of long-term, low dose growth hormone therapy on immune function and life expectancy of mice. Mech Ageing Dev 1991;57:87–100.
43. Walker RF, Engelman RW, Pross S, Bercu BB. Effects of growth hormone secretagogues on age-related changes in the rat immune system. Endocrine 1994;2:857–62.
44. Walker RF, Yang S-W, Masuda R, Hu C-S, Bercu BB. Effects of growth hormone releasing peptides on stimulated growth hormone secretion in old rats. In: Bercu BB, Walker RF, eds. Growth hormone II: basic and clinical aspects. New York: Springer-Verlag, 1994:167–92.
45. Bercu BB, Yang S-W, Masuda R, Walker RF. Role of selected endogenous peptides in growth hormone releasing hexapeptide (GHRP) activity: analysis of GHRH, TRH and GnRH. Endocrinology 1992;130:2579–86.
46. Brown MS, Goldstein JL. A receptor-mediated pathway for cholesterol homeostasis. Science 1986;232:34–47.
47. Ness GC, Miller JP, Moffler MH, Woods LH, Harris HB. Perinatal development of 3-hydroxy-3-methylglutaryl coenzyme A reductase activity in rat lung, liver and brain. Lipids 1979;14:447–50.
48. Grundy SM, Vega GL, Bilheimer DW. Kinetic mechanisms determining variability in low density lipoprotein levels and rise with age. Arteriosclerosis 1985;5:623–30.
49. Ericsson E, Eriksson M, Vitols S, Einarsson K, Berglund K, Angelin L. Influence of age on the metabolism of plasma low density lipoproteins in healthy males. J Clin Invest 1991;87:591–6.
50. Myant NG. Cholesterol metabolism, LDL, and the LDL receptor. San Diego: Academic Press, 1990.
51. Rudling M, Norstedt G, Olivecrona H, Reihner E, Gustafsson J-A, Angelin B. Importance of growth hormone for the induction of hepatic low density lipoprotein receptors. Proc Natl Acad Sci USA 1992;89:6983–7.
52. Kovanen PT, Brown MS, Goldstein JL. Increased binding of low density lipoprotein to liver membranes from rats treated with 17-α-ethinyl estradiol. J Biol Chem 1979;254:11367–73.

53. Ness GC, Pendleton LC, Li YC, Chiang JYL. Effect of thyroid hormone on hepatic cholesterol 7α hydroxylase, LDL receptor, HMG-CoA reductase, farnesyl pyrophosphate synthetase and apolipoprotein A-I mRNA levels in hypophysectomized rats. Biochem Biophys Res Commun 1990;172:1150–6.
54. Steinberg M, Tolksdorf S, Gordon AS. Relation of the adrenal and pituitary to the hypocholesterolemic effect of estrogen in rats. Endocrinology 1967;81: 340–4.
55. Friedman M, Byers SO, Elek SR. Pituitary growth hormone essential for regulation of serum cholesterol. Nature (Lond) 1970;255:464–7.
56. Friedman M, Byers SO, Rosenman RH, Li CH, Neumann R. Effect of subacute administration of human growth hormone levels of hypercholesterolemic and normocholesterolemic subjects. Metabolism 1974;23:905–12.
57. Ma PT, Yamamoto T, Goldstein JL, Brown MS. Increased mRNA for low density lipoprotein receptor in livers of rabbits treated with 17 alpha-ethinyl estradiol. Proc Natl Acad Sci USA 1986;83:792–6.
58. Walker RF, Ness GC, Zhao Z, Bercu BB. Effects of stimulated GH secretion of age-related changes in plasma cholesterol and hepatic low density lipoprotein messenger RNA concentrations. Mech Ageing Dev 1994;75:215–26.

18

A Diagnostic Test Employing Growth Hormone Secretagogues for Evaluating Pituitary Function in the Elderly

Barry B. Bercu and Richard F. Walker

There is a progressive decrease in growth hormone (GH) secretion during aging that could result from several factors. Two of the most obvious are inadequate stimulation of the pituitary gland by hypothalamic hypophysiotropic factors, and/or degeneration of pituitary-based mechanisms for GH production/secretion. Of the two, it is most likely that decremental changes in the relationship of the pituitary with GH regulatory hormones of hypothalamic origin are the primary etiologic factors in age-related decline in GH secretion. However, this view is seemingly inconsistent with many reports that state the pituitary becomes refractory to stimulation during aging. For example, except for two reports (1, 2), there is general consensus that GH secretion in response to growth hormone releasing hormone (GHRH) administration in vivo declines with advancing age (3–11). Assuming that GHRH is the only stimulatory agent controlling GH secretion, this progressive decrement could result from several factors. One factor could be that chronic reduction in pituitary stimulation by GHRH causes desensitization to the GH secretagogue because hormones often induce their own receptors. Support for reduced stimulation of the pituitary by GHRH derives from the fact that available and/or appropriate GH secretagogues seem to decline during aging (12–14). However, GHRH responsiveness in aging rats was immediately restored by coadministration with growth hormone releasing peptide (GHRP) (15), suggesting that age-related, reduced efficacy of GHRH may be due to the absence or reduced concentrations of other as yet undefined endogenous co-secretagogue(s). This hypothesis is supported by the finding that "at least three factors operating via distinct receptors, are able to increase GH secretion" and that the data "ascribe a potential physiological role for hitherto putative hypophysiotrophic factors" (16).

Deficits in pituitary GH concentrations observed in animals may also contribute to age-related insensitivity to GHRH and subsequent GH deficiency (17, 18). However, depletion of pituitary GH may not occur in humans because the only published report states that pituitary GH concentrations did not vary significantly with age when measured at autopsy (19). However, this report has not been corroborated by other investigators, and in old laboratory animals pituitary GH concentrations are significantly lower than in young animals of the same species. However, the low concentrations of pituitary GH in old animals are probably not due to inherent defects in pituitary function, but more likely to reduced stimulation of the gland. Support for this view derives from the fact that pituitary GH concentrations as well as pituitary GH messenger RNA (mRNA) were significantly increased in old rats chronically administered GHRH and GHRP (18).

Another factor having the potential to reduce GHRH efficacy during aging is deterioration of its pituitary binding sites. GHRH binding sites with high and low affinities have been identified on rat pituitary membranes (20). Since decreased capacity and loss of the high-affinity site correlated with reduced GH secretion during aging, it was suggested that the high-affinity binding site mediates GHRH action in somatotrophs (21). Prior to this GHRH binding study, we (22) reported that GHRP has two binding sites on pituitary and hypothalamic membranes. GHRH did not compete for these GHRP binding sites, suggesting that the two peptides bind different entities. These differential binding sites may be functionally related, since somatotrophs lacking GHRH receptors are unresponsive to GHRP administered alone (23) or in combination with GHRH (17). Functional linkage has been demonstrated many times in the ability of GHRP to enhance GHRH efficacy (24, 25), and we recently showed that passive immunization against GHRH reduced GHRP activity approximately 90% (17). The ability of GHRP to increase GHRH activity in aging rats (15) may result from facilitation of GHRH binding through cooperativity/allosteric interactions. Changes in intracellular transduction of GHRH has also been associated with aging. GHRH-activated adenylate cyclase activity and adenosine $3',5'$-cyclic monophosphate (cAMP) concentrations were lower in pituitaries from old rats than from young ones (3, 26). However, this apparent deficit was immediately reversed by coadministration of GHRH and GHRP in vivo (15), again suggesting cooperativity between the peptides, and deficiency or loss of a co-secretagogue(s) for GHRH during aging. Similarly, gene defects involving transcription of GH mRNA could account for reduced synthesis and secretion of GH in the elderly, and we recently showed that GHRH and GHRP administered together positively affect the molecular biology of GH endocrinology in old rodents (18).

Finally, age-related loss of GHRH efficacy could result from increased exposure of the pituitary gland to somatostatin [somatotropin release inhibiting factor (SRIF)] (8–10), a peptide that inhibits GH secretion. Passive immunization with SRIF antiserum increased GH release more in old rats than in young rats (9, 27), and reduced pulsatile GH secretion in aging female rats (10) correlated with SRIF hypersecretion (28). In a clinical study, administration of the acetylcholinesterase inhibitor pyridostigmine to block endogenous SRIF release partially restored GH responsiveness to GHRH in elderly subjects (4). Direct measurement of hormones released from perifused hypothalamic explants confirmed increased SRIF secretion from tissues of old rats (29). It has been proposed that the activity of GHRP is greater in vivo than in vitro because the hexapeptide is a functional antagonist of SRIF, perhaps acting at the pituitary to prevent SRIF hyperpolarization of the somatotroph (16, 30, 31). These possibilities are consistent with our finding that coadministration of the complementary GH secretagogues GHRH and GHRP immediately restores youthful patterns of GH secretion in old animals (15, 18, 32).

The relevance of these findings to the present discussion is that they support the hypothesis that GHRH and the endogenous analogue of GHRP are complementary GH secretagogues that together provide appropriate stimulation of the pituitary gland to sustain normal GH production and release. If this were true, inadequate endogenous concentrations of GHRH or GHRP analogue would cause blunting of the response to its exogenous complement, GHRP or GHRH, respectively. Similarly, inadequate endogenous GHRH and GHRP would blunt responses to GHRH or GHRP but would not affect the response to coadministered GHRH and GHRP. On the other hand, lack of GH secretion to coadministered GHRH and GHRP would indicate a pituitary defect(s) in GH production and/or release mechanisms. By this logic, GHRH and GHRP could be used alone and in combination to diagnose pituitary-based causes of inadequate GH secretion during aging.

Support for using GHRH and GHRP to test for an extrapituitary etiology of GH insufficiency derives from the following experimental and clinical findings.

Animal Studies

At the core of the proposed diagnostic application are data from a pilot study in which we assessed the contribution of endogenous GHRH to GHRP activity. This study was based upon GHRP's ability to potentiate GHRH efficacy in vivo and in vitro (24, 25, 33). Since GHRH-efficacy is reduced 50% to 75% in old rats (11), we tested the effect of GHRP alone and in combination with GHRH on GH release in old rats. Peak plasma GH concentrations resulting from GHRP administration in old rats were approximately 60% lower than in young rats. In contrast, peak plasma GH concentrations were greater in old rats than in young rats administered GHRP and GHRH. Since target organs sometimes become hyperresponsive when tonic and/or phasic stimulation decreases, then one would expect exaggerated responses to provoke exogenous stimuli under experimental conditions. GH hypersecretion observed in naive, old rats administered a single bolus of GHRH and GHRP (15) provides support for the hypothesis that deficits in stimulated GH secretion in aged rats were due to insufficient signals or inappropriately transduced GH releasing stimuli.

These two possibilities were tested in preliminary studies and serve as the scientific basis for designing the diagnostic test. The first possible cause of age-related decrements in GH secretion that we considered was that a progressive reduction in pituitary stimulation occurs during aging. This reduction could be attributed to GHRH and/or to an endogenous ligand for GHRP. Although the search has begun, a purported endogenous ligand for GHRP has not yet been identified (17). Nonetheless, we were able to demonstrate the functional dependence of GHRP upon endogenous GHRH by causing deficits in GHRH that in turn attenuated responses to GHRP (17). Young rats were passively immunized against endogenous GHRH or administered α-methyl-ρ-tyrosine. If GHRH and the endogenous ligand of GHRP are physiologic coagonists of GH secretion in the young rat, then removal of one or the other should be expressed as attenuated activity of its coagonist when administered alone. Passive immunization, which inactivates GHRH with neutralizing antibodies, or α-methyl-ρ-tyrosine, which blocks stimulation of hypothalamic GHRH

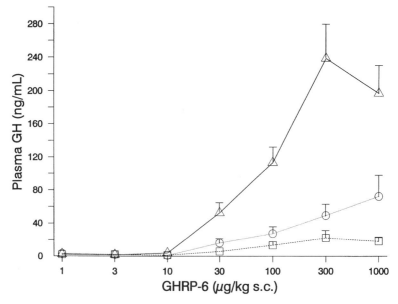

FIGURE 18.1. Peak responses to increasing doses of GHRP in rats pretreated with vehicle (triangles), α-methyl-ρ-tyrosine (circles), or GHRH antiserum (squares). Reprinted with permission from Bercu et al. (17).

neurons, was used to remove or reduce concentrations of endogenous GHRH in the young experimental animals. These treatments were intended to simulate the aged condition in which concentrations of GHRH and GHRH mRNA are low (12, 13). Afterward, GHRP was administered to test and compare its efficacy with that in old rats (15).

As seen in Figure 18.1, GHRP activity was significantly attenuated in young female rats administered GHRH antiserum or α-methyl-p-tyrosine to reduce endogenous GHRH concentrations. Presumably, naturally occurring decrements in endogenous GHRH during aging contributed to the blunted response to GHRP that we previously observed (15). If GHRH activity is also dependent upon an as yet unidentified endogenous cosecretagogue whose concentrations decline during aging, then our preliminary data support the hypothesis that extrinsic pituitary deficits contribute at least in part, to attenuated GH secretory responses to administered GHRH.

The second possible cause of age-related decrements in GH secretion we considered was that deterioration in GH secretagogue signal transduction at the pituitary level contributes to the progressive insensitivity to provocative stimuli. In a preliminary test of this hypothesis and in continuing analysis of the interactions between GHRH and GHRP as discussed above, we examined GH secretory responses to GHRP in young rats administered [N-Ac-Tyr[1], D-Arg[2]]-GHRF 1-29 amide, an antagonist to the pituitary GHRH receptor. As seen in Figure 18.2, preadministration of 5 to 50 g/kg of [N-Ac-Tyr[1], D-Arg[2]]-GHRF 1-29 amide, which blocks pituitary GHRH receptors, also attenuated GHRP activity. However, 150 g/kg of [N-Ac-Tyr[1], D-Arg[2]]-GHRF 1-29 amide potentiated GHRP activity, presumably due to partial agonist activity of the GHRH receptor agonist at the higher dose (17). These data suggest that activation of pituitary GHRH receptors contributes to full expression of GHRP activity in vivo.

In conclusion, age-related pituitary GH hyposecretion probably reflects decrements in the combined influence of several peptides, including SRIF on the pituitary, requiring analysis of responses to coadministered substances as provocative, diagnostic challenges. This hypothesis is in agreement with a recent report in which GH regulation was shown to be complex, involving several secretagogues acting in concert with GHRH to provide tightly controlled, temporal and quantitative secretion of GH (16). Accordingly, we devised a model for direct testing of pituitary growth hormone secretory capability (Fig. 18.3). The model assumes that a "normal" response to administration of GHRP or GHRH requires the presence of its endogenous analogue, i.e., GHRH or GHRP, respectively. Blunted responses to either exogenous GH secretagogue is interpreted as indicating a deficiency of its endogenous complement. Blunted responses to both exogenous GH secretagogues administered sequentially implies deficiencies of both endogenous complements. This condition can be differentiated from inherent pituitary problems such as those involving receptor or second

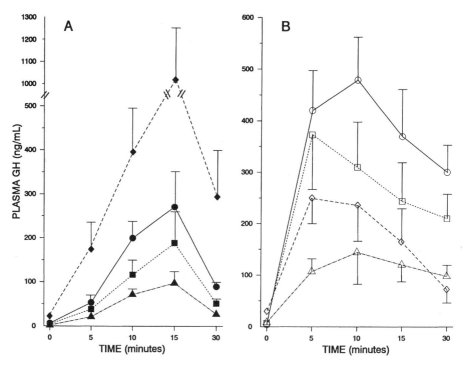

FIGURE 18.2. Effect of [N-Ac-Tyr¹, D-Arg²]-GHRF 1-29 amide, a GH receptor agonist, on GH release in response to (*A*) GHRP (30 µg/kg) and (*B*) morphine (1.5 mg/kg). The GHRH receptor antagonist was administered in doses of 0 (circle), 5 (square), 50 (triangle), or 150 (diamond) µg/kg. Reprinted with permission from Bercu et al. (17).

messenger deficits by a "normal" response to GHRH and GHRP coadministration. A blunted response following coadministration of both GH secretagogues would indicate inherent pituitary dysfunction rather than inadequate endogenous stimuli.

Human Studies

A pilot study in children was performed and preliminary data collected to support the clinical project. The results of this study support the basic concept of GH secretagogue complementarity as a useful diagnostic tool for evaluating pituitary-based GH deficiency.

The protocol used in the pediatric study from which the data presented below were collected involved the sequential administration of exogenous GHRP and exogenous GHRH as described in Figure 18.3. Briefly, each subject began testing at approximately 0900 hours, one half hour after placement of an intravenous catheter. Two blood samples for basal concen-

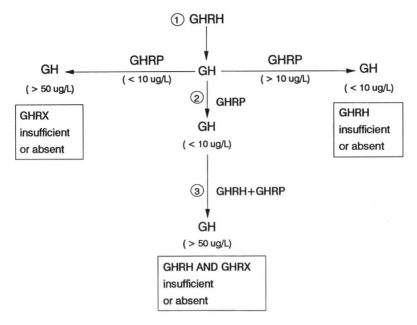

FIGURE 18.3. A pituitary function test using responses to sequentially administered GH secretagogues as diagnostic parameters. GHRX, peptidyl and nonpeptidyl forms of GHRP.

trations of serum GH were drawn, and GHRP (1 g/kg) was administered as a bolus. Blood samples were then drawn at 5-min intervals for 20 min during which the GH secretory response occurred, and at longer intervals for about 2 hours thereafter. GHRH (1 g/kg) was then administered as a bolus, when basal GH concentrations were again established. Blood samples were similarly collected after GHRH administration as they were following GHRP administration. GH concentrations in each serum sample was then determined by radioimmunoassay. The study population was composed of children with normal GH secretory function and those with GH secretory dysfunction of various clinical diagnoses. The purpose of evaluating different etiologies of GH secretory dysfunction was an attempt to validate the hypothesis that complementary endogenous GH secretagogues must be present for optimal expression of GHRP or GHRH stimulatory potential.

As seen in Figure 18.4A, peak concentrations of serum GH were reached at the 15- or 20-min time point following stimulation with both secretagogues. In all five normal children tested, peak GH concentrations following GHRP administration were greater than those following GHRH administration. The greater response to the xenobiotic GH secretagogue was not due to its being administered first, because reversal of the sequence produced the same differential response. Similar responses have also been observed in other laboratories. Peak concentrations of serum GH following

296   B.B. Bercu and R.F. Walker

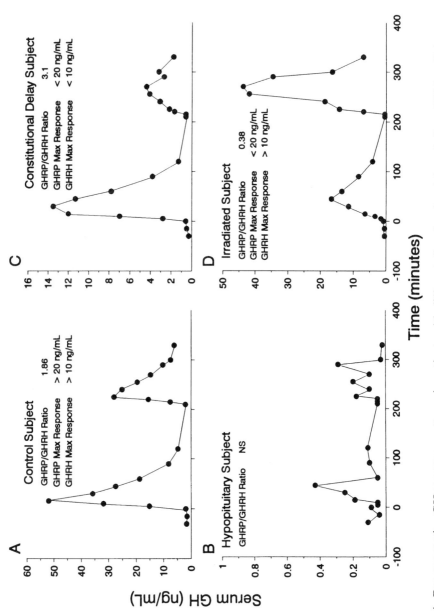

FIGURE 18.4. Representative GH secretory patterns in patients with different disorders involving pituitary GH insufficiency.

GHRP administration in normal children ranged between 42 and 66 ng/ml, whereas peak concentrations of serum GH following GHRH administration ranged between 18 and 36 ng/ml. Thus, the ratio of GHRP to GHRH in the normal children ranged between 1.4 and 2.3, reflecting the greater responses to GHRP in these subjects.

Three primary assumptions about these data were used to compare with the data collected from children with GH secretory dysfunction:

1. Peak concentrations of serum GH following administration of exogenous GHRP or GHRP reflected not only the dose of GH secretagogue administered, but also the "normal" concentration of endogenous complementary GH secretagogue present in the subject.
2. GHRP:GHRH ratios between 1.4 and 2.3 reflected the "normal" concentrations of endogenous GH secretagogues contributing to the peak concentrations measured in normal individuals.
3. Peak concentrations of serum GH in response to exogenous GHRP or GHRH should occur within 15 or 20 min of their administration.

With these considerations in mind we tested children with various causes of GH secretory dysfunction ranging from hypopituitarism to that resulting from radiation therapy. The results of these trials substantiated the basic premises of the test hypothesis. For example, the data presented in Figure 18.4B show that children with inherently dysfunctional pituitary glands, did not respond to any combination of exogenous GH secretagogue administration. This representative subject had an atrophic pituitary gland secondary to autoimmune hypophysitis, which responded to neither sequentially administered GHRH and GHRP nor to the coadministered peptides.

Radiation of the head for treatment of cancer in children retards growth, presumably by reducing the secretion of endogenous GHRH from hypothalamic neurosecretory neurons. In the preliminary clinical study, children that were presumably deficient in endogenous GHRH responded poorly to GHRP but exuberantly to GHRH, resulting in a GHRP:GHRH ratio significantly <1. As seen in Figure 18.4D, a representative subject from this group produced a peak response of 18.9 ng/ml serum GH in response to GHRP, whereas the response to GHRH was 49.4 ng/ml, resulting in a GHRP to GHRH ratio of 0.38. The data were interpreted to mean that the poor response to GHRP resulted from a paucity of endogenous GHRH after radiation. The reduced availability of this GHRP complement therefore attenuated the response to a provocative dose of exogenous GHRP. Conversely, the data also suggest that radiation did not alter the availability of GHRP because the response to a provocative dose of GHRH was amplified. This finding could be explained by enhanced sensitivity of the pituitary to GHRH resulting either from increased exposure to the endogenous analogue of GHRP and/or by a mechanism involving low-level exposure to GHRH that in turn increased the pituitary response to a substantial bolus of the secretagogue. The actual mechanism of this change in responsivity is unknown, but data collected from other patients suggest that the reciprocal effect also occurs.

For example, as seen in Figure 18.4C, some children responded poorly to both secretagogues but much better to GHRP than to GHRH. This pattern of response suggests a greater deficiency in endogenous GHRP than in GHRH, giving a GHRP:GHRH ratio >3. As seen in this subject, a peak response of 13.5 ng/ml serum GH occurred after administration of GHRP, whereas a peak response of only 4.4 ng/ml occurred after administration of GHRH. The GHRP to GHRH ratio for this subject was 3 to 1. Interestingly, in all individuals with GHRP to GHRH ratios greater than 3, the quantitative responses to both sequentially administered exogenous secretagogues were significantly lower than those observed in normal subjects. These data suggest that GHRP deficiency may be more tightly linked to GHRH deficiency than the reciprocal, because as shown above, quantitatively normal and even supranormal responses to GHRH can occur in the presence of poor responses to GHRP. This observation suggests that endogenous GHRH deficiency is not necessarily accompanied by GHRP deficiency. However, since we have not observed quantitatively normal responses to GHRP in the presence of poor responses to GHRH, a deficit in endogenous GHRP may always be accompanied by a deficit in GHRH.

An interesting and logical extension of these observations is to differentiate the responses of individuals with GHRP to GHRH ratios of <1 from those with ratios >3, because in GH-deficient animals, which provide a more homogeneous population than humans, coadministration of GHRH and GHRP produced supranormal GH secretory responses (15). In limited studies with GH-deficient elderly human subjects, Bowers et al. (personal communication) observed attenuated mean (population) responses to the coadministered GH secretagogues. However, they did not screen the subjects for individual responses to GHRP and GHRH, so that the variance in their data could have resulted from differences in the endogenous secretagogues, i.e., robust responders may have had GHRP to GHRH ratios >3 while poor responders had ratios <1. This hypothesis is under investigation and will in fact be evaluated in the proposed research. Another, alternative explanation for poor responses to both secretagogues is that under this condition, SRIF inhibitory influence over GH secretion is particularly strong. This possibility can be tested by using a combination of the GH secretagogues, administered under the influence of SRIF inhibitors.

Study Design for Diagnosis of Age-Related GH Insufficiency

The purpose of the proposed diagnostic test is to determine whether growth hormone deficiencies in elderly subjects result from factors that are intrinsic and/or extrinsic to the pituitary and if in fact different GH secretory responses can be identified. As in children, GH secretory deficits could presumably result from intrinsic defects in pituitary function and/or from extrinsic factors such as decreased or aberrant pituitary signaling by GH secretagogues, excessive SRIF secretion, and/or increased pituitary SRIF sensitivity. Based on these possibilities, individuals between the ages of 50 and 80 years will be requested to participate in the following study.

Drugs and Chemicals

Three substances should be employed in the diagnostic test. These include a peptidyl or nonpeptidyl representative of the xenobiotic family of GH secretagogues (GHRP) and one of the GHRH molecules presently available. These compounds will be used in the provocative tests to determine response capacity of the pituitary gland. Arginine hydrochloride or pyridostigmine can be used as a pretreatment when the results of provocative testing suggest that blunted GH secretory responses are due to excessive somatostatin influence.

Participants

People in generally good health, relative to their age group, should be sought for this study. They should have normal lifestyles, be within 20% of their ideal body weight, and preferably not be taking medication except if it does not alter GHRH or GHRP efficacy. Informed consent should be obtained from each person participating in the study. Subjects should be divided equally between the sexes, with postmenopausal women selected for participation. Young individuals between the ages of 20 and 30 years should be asked to provide control data for comparison with those from the elderly participants. Females in the young group should be scheduled for study during their menstrual cycles, when serum concentrations of gonadal steroids are relatively low. Medical histories on the subjects should be taken before testing with particular attention given to sex hormone replacement therapy in the women. Concern for estrogen replacement is appropriate because the hormone alters receptors and other cellular elements affecting GH secretion. Preferably, this confounding element will be avoided by selection of postmenopausal volunteers who are not being treated with estrogen or progesterone. Serum samples from each participant should be screened for insulin-like growth factor-1 (IGF-1) concentrations as an indicator of functional status of the GH neuroendocrine axis. IGF-1 concentrations should not necessarily be used as an inclusion or exclusion criterion, but instead be used in retrospective analysis of the data to determine whether characteristics of stimulated GH secretion are in some way correlated with previous age-related changes in serum IGF-1.

Protocol

After completing the required consent forms and receiving final counseling on the nature of the study, each participant should fast for the duration of the test (6 hours). Drinking water will be allowed. After completion of a medical history form, the patients should be seated in a comfortable reclining chair, and provided with magazines, television, or other appropriate forms of entertainment. An intravenous catheter kept patent by a slow infusion of isotonic saline should be fitted and two blood samples drawn for basal concentrations of growth hormone and IGF-1 so as to determine the

SEQUENTIAL GHRP-2/GHRH STUDY

Name: _____

Date of Birth: _____

Date: _____   Weight: _____

Co-administered GHRH (GFR 1-29) + GHRP-2 Protocol

I. H_2O only after midnight

II.   IV infusion of normal saline with 3 way stopcock keep vein open rate

```
         GH
 -30     X
 -15     X
   0     X
GHRP (1  g/kg) i.v. bolus
  +5     X
 +10     X
 +15     X
 +30     X
 +45     X
 +90     X
+120     X
```

Minimum Blood Requirements:

GH 1.5 ml clot; total = 15 mL

When testing for somatostatin influence, administer arginine (0.5 gm/kg) for 30 minutes preceding co-administration of GHRH and GRHP.

FIGURE 18.5. Protocol for pituitary function test involving sequential administration of GHRP and GHRF 1-29 amide.

GH neuroendocrine status of each participant before administration of the
secretagogues. Basal blood samples should be drawn at 15-min intervals.
Then GHRP (1 g/kg) and GHRH (1 g/kg) are administered and blood
samples drawn according to the schedule provided below (Fig. 18.5). Subjects should return to the clinic for coadministration of the GH secretagogues if the results of the first diagnostic test suggest that they are deficient
in both endogenous GHRH as well as endogenous GHRP.

COMBINED GHRH/GHRP-2 STUDY

Name: _____

Date of Birth: _____

Date: _____   Weight: _____

GHRP-2 followed by GHRH (GRF 1-29) protocol

I.   H₂0 only after midnight

II.   IV infusion of normal saline with 3 way stopcock keep vein open

```
        GH
-30     X
-15     X
  0     X
GHRP-2 1.0 mcg/kg i.v. bolus
 +5     X
+10     X
+15     X
+30     X
+45     X
+60     X
+90     X
+120    X
GRF 1-29 1.0 mcg/kg i.v. bolus
+210    X
+215    X
+220    X
+225    X
+240    X
+255    X
+270    X
+300    X
+330    X
```

Minimum Blood Requirements:

GH 1.5 ml clot; total 30 mL

FIGURE 18.6. Protocol for pituitary function test involving coadminstration of
GHRP and GHFR 1-29 alone, or preceded by arginine.

If a subject responds poorly to both sequentially administered GH secretagogues, he/she should be rescheduled for testing with coadministered GHRH and GHRP. All conditions will be the same as during the first, sequential test, except that this test will take only half the time previously required. The GH secretagogues should be coadministered in the morning. If after radioimmunoassay (RIA) analysis of serum GH concentrations stimulated GH secretion remains poor, the subject will be asked to return for a third and final test to determine whether blunted GH secretion can be attributed to increased SRIF tone or to intrinsic defects in the pituitary GH secretory mechanism. During this final session, arginine will be infused at a concentration of 0.5g/kg for 30min preceding coadministration of GHRH and GHRP. This concentration of arginine has been reported to enhance exogenous GHRH efficacy by inhibiting SRIF secretion (34). A summary of the secondary and tertiary protocols are presented in Figure 18.6.

It will be of specific interest to compare the GH secretory responses in youthful subjects with those of the elderly that have IGF-1 parameters within the youthful range of values. The object is to determine whether secretagogue stimulated GH release is correlated with physiologic state of the GH neuroendocrine axis, especially IGF that directly participates in many of the physical and performance effects that are hoped to be restored by treatment of the elderly with GH or GH secretagogues.

Conclusion

We have presented a novel diagnostic test for evaluating pituitary function during aging. The objective of the test is to determine whether it is feasible to use GHRP and GHRH as diagnostic tools to investigate the etiology of GH deficiency in the elderly. Ideally, the data resulting from application of principles upon which the diagnostic test is based will allow appropriate selection of therapeutic entities, ranging from GHRH or GHRP administered separately or in combination, or alternatively, recombinant GH for those aged individuals lacking their own pituitary mechanism for GH production and/or secretion.

References

1. Wehrenberg WB, Ling N. The absence of an age-related change in the pituitary response to growth hormone-releasing factor in rats. Neuroendocrinology 1983;37:463–6.
2. Pavlov EP, Harman SM, Merriam GR, Gelato MC, Blackman MR. Responses of growth hormone (GH) and somatomedin-C to GH-releasing hormone in healthy aging men. J Clin Endocrinol Metab 1986;62:595–600.

3. Ceda GP, Valenti G, Butturini U, Hoffman AR. Diminished pituitary responsiveness to growth hormone-releasing factor in aging male rats. Endocrinology 1986;118:2109–14.

4. Ghigo E, Goffi S, Arvat E, Nicolosi M, Procopio M, Bellone J, Imperiale E, Mazza E, Baracchi G, Camanni F. Pyridostigmine partially restores the GH responsiveness to GHRH in normal aging. Acta Endocrinol 1990;123:169–74.

5. Iovino M, Monteleone P, Steardo L. Repetitive growth hormone-releasing hormone administration restores the attenuated growth hormone (GH) response to GH-releasing hormone testing in normal aging. J Clin Endocrinol Metab 1989;69:910–3.

6. Lang I, Schernthaner G, Pietschmann P, Kurz R, Stephenson JM, Templ H. Effects of sex and age on growth hormone response to growth hormone-releasing hormone in healthy individuals. J Clin Endocrinol Metab 1987;65:535–40.

7. Lang I, Kruz R, Geyer G, Tragl KH. The influence of age on human pancreatic growth hormone releasing hormone stimulated growth hormone secretion. Horm Metab Res 1988;20:574–8.

8. Shibasaki T, Shizume K, Nakahara M, Masuda A, Jibiki K, Demura H, Wakabayashi I, Ling N. Age-related changes in plasma growth hormone response to growth hormone-releasing factor in man. J Clin Endocrinol Metab 1984;58:212–4.

9. Sonntag WE, Gough MA. Growth hormone releasing hormone induced release of growth hormone in aging rats: dependence on pharmacological manipulation of endogenous somatostatin release. Neuroendocrinology 1988;47:482–8.

10. Sonntag WE, Steger RW, Forman LJ, Meites J. Decreased pulsatile release of growth hormone in old male rats. Endocrinology 1980;107:1875–9.

11. Sonntag WE, Hylka VW, Meites J. Impaired ability of old male rats to secrete growth hormone in vivo but not in vitro in response to hpGRF(1-44). Endocrinology 1983;113:2305–7.

12. De Gennaro Colonna V, Zoli M, Cocchi D, Maggi A, Marrama P, Agnati LF, Muller EE. Reduced growth hormone releasing factor (GHRF)-like immunoreactivity and GHRF gene expression in the hypothalamus of aged rats. Peptides 1989;10:705–8.

13. Morimoto N, Kawakami F, Makino S, Chihara K, Hasegawa M, Ibata Y. Age-related changes in growth hormone releasing factor and somatostatin in the rat hypothalamus. Neuroendocrinology 1988;47:459–64.

14. Ono M, Miki N, Shizume K. Release of immunoreactive growth hormone-releasing factor (GRF) and somatostatin from incubated hypothalamus in young and old male rats. Neuroendocrinology 1986;43:111(abstr).

15. Walker RF, Yang S-W, Bercu BB. Robust growth hormone (GH) secretion in aged female rats co-administered GH-releasing hexapeptide (GHRP-6) or GH releasing hormone (GHRH). Life Sci 1991;49:1499–504.

16. Goth MI, Lyons CE, Canny BJ, Thorner MO. Pituitary adenylate cyclase activating polypeptide, growth hormone (GH)-releasing peptide and GH-releasing hormone stimulate GH release through distinct pituitary receptors. Endocrinology 1992;130:939–44.

17. Bercu BB, Yang S-W, Masuda R, Walker RF. Role of selected endogenous peptides in growth hormone releasing hexapeptide (GHRP-6) activity: analysis of GHRH, TRH, and GnRH. Endocrinology 1992;130:2579–86.

18. Walker RF, Eichler DC, Bercu BB. Inadequate pituitary stimulation: a possible cause of growth hormone insufficiency and hyperprolactinemia in aged rats. Endocrine 1994;2:633–8.
19. Gershberg H. Growth hormone content and metabolic actions of human pituitary glands. Endocrinology 1957;61:160–5.
20. Abribat T, Desiauriers N, Brazeau P, Gaudreau P. Alterations of pituitary growth hormone-releasing factor binding sites in aging rats. Endocrinology 1991;128:633–5.
21. Abribat T, Boulanger L, Gaudreau P. Characterization of [^{125}I-Tyr^{10}]human growth hormone-releasing factor (1-44) amide binding to rat pituitary: evidence for high and low affinity classes of sites. Brain Res 1990;528:291–9.
22. Codd EE, Yellin T, Walker RF. Binding of growth hormone-releasing hormones and enkephalin-derived growth hormone-releasing peptides to mu and delta opioid receptors in forebrain of rat. Neuropharmacology 1988;27:1019–25.
23. Jansson J-O, Downs TR, Beamer WG, Frohman LA. The dwarf "little" (LIT/LIT) mouse is resistant to growth hormone releasing peptide (GHRP-6) as well as to GH-releasing hormone (GRH). Endocr Soc 1986;130(abstr).
24. Cheng K, Chan WW-S, Barreta A, Convey DM, Smith RG. The synergistic effects of HIS-D-TRP-ALA-TRP-D-PHE-LYS-NH$_2$ on growth hormone (GH)-releasing factor-stimulated GH release and intracellular adenosine 3',5'-monophosphate accumulation in rat primary pituitary cell culture. Endocrinology 1989;124:2791–8.
25. Bowers CY, Reynolds GA, Durham D, Barrera CM, Pezzoli SS, Thorner MO. Growth hormone (GH)-releasing peptide stimulates GH release in normal men and acts synergistically with GH-releasing hormone. J Clin Endocrinol Metab 1990;70:975–82.
26. Parenti M, Dall'ara A, Rusconi L, Coochi D, Muller EE. Different regulation of growth hormone-releasing factor-sensitive adenylate cyclase in the anterior pituitary of young and aged rats. Endocrinology 1987;121:1649–53.
27. Sonntag WE, Forman LJ, Miki N, Steger RW, Ramos T, Arimura A, Meites J. Effects of CNS active drugs and somatostatin antiserum on growth hormone release in young and old male rats. Neuroendocrinology 1981;33:73–8.
28. Takahashi S, Gottshall PE, Quigley KL, Goya RG, Meites J. Growth hormone secretory patterns in young, middle-aged and old female rats. Neuroendocrinology 1987;46:137–42.
29. Ge F, Tsagarakis S, Rees LH, Besser GM, Grossman A. Relationship between growth hormone-releasing hormone and somatostatin in the rat: effects of age and sex on content and in-vitro release from hypothalamic explants. J Endocrinol 1989;123:53–8.
30. DeBell WK, Pezzoli SS, Thorner MO. Growth hormone (GH) secretion during continuous infusion of GH-releasing peptide: partial response attenuation. J Clin Endocrinol Metab 1991;72:1312–6.
31. Jaffe CA, Ho PJ, DeMott, Friberg R, Bowers CY, Barkan AL. Effect of prolonged growth hormone-releasing peptide infusion on GH secretion in normal men. Endocr Soc 1992;169(abstr).
32. Walker RF, Ness GC, Zhihong Z, Bercu BB. Effects of stimulated growth hormone secretion on age related-changes in plasma concentrations and hepatic low density lipoprotein messenger RNA concentrations. Mech Ageing Dev 1994;75:215–26.

33. Blake AD, Smith RG. Desensitization studies using perifused pituitary cells show that growth hormone-releasing hormone and His-D-Trp-Ala-Trp-D-Phe-Lys-NH$_2$ stimulate growth hormone release through distinct receptor sites. J Endocrinol 1991;129:11–19.
34. Alba-Roth J, Muller OA, Schopohl J, VonWerder K. Arginine stimulates growth hormone secretion by suppressing endogenous somatostatin secretion. J Clin Endocrinol Metab 1988;67:1186–9.

Part V

Target Tissues and Applications for Growth Hormone Secretagogues

19

Transdifferentiation of Growth Hormone and Prolactin Secreting Cells

L. STEPHEN FRAWLEY

Until a decade ago, it was widely held that the growth hormone (GH)-producing somatotrophs and the prolactin (PRL)-secreting mammotrophs were separate and distinct cell types. This view began to change in 1985 with the publication of two papers by our group. In one of these, we characterized the ontogeny of GH- and PRL-secreting cells in fetal and neonatal rats and found that the appearance of GH secretors preceded that of PRL cells by about a week (1). In addition, we developed and utilized the sequential plaque assay to demonstrate that virtually all of the initial PRL secretors also released GH. This suggested that PRL secretors initially arose from a subset of traditional somatotrophs, and provided the first direct evidence that GH and PRL secretors could functionally interconvert.

A developmental and functional link between GH and PRL secretors was subsequently confirmed for mice by transgene studies (2, 3) and for humans (4, 5) and cattle (6) by the use of plaque assays and immunocytochemistry (ICC), either alone or in combination. In the same year we published another paper (7) showing that bihormonal mammosomatotrophs were a normal constituent cell type of the adult rat pituitary, although their proportions were obviously lower than those found in developing animals. Once again this observation was subsequently confirmed by others for rats (8–11) and extended (by use of similar techniques) to several other species including humans (12), mice (13), cattle (14–16), shrews (11), and bats (17). Yet despite the wealth of evidence that at least a subset of pituitary cells can express both these genes even in adults, the notion that GH and PRL cells are functionally interconvertible is only recently gaining universal acceptance.

Previous reluctance by some to embrace the concept of acidophilic plasticity is difficult to reconcile given the knowledge that cells such as hepatocytes exhibit dramatic swings in the types and amounts of proteins they secrete, or, to use an example closer to home, that the ratio of mono- to bihormonal gonatotropes fluctuates greatly during the course of the rat estrous cycle (18). Another relevant example was just published by Talamantes' group, which used a sequential plaque assay to show that placental cells that secrete placental lactogen (PL)-I convert to PL-II secretors via a bihormonal intermediate (19). Accordingly, this chapter outlines the argument that GH- and PRL-secreting cells are functionally interconvertible. Such a possibility would provide a mechanism whereby the long-term needs of an animal for GH or PRL could be met without a dramatic increase in the absolute numbers of pituitary cells.

In the 20 years since the term *transdifferentiation* was introduced by Selman and Kafatos (20), it has been used very loosely to describe a broad spectrum of phenomena. We subscribe to the view that transdifferentiation is a change in cell phenotype that occurs in the absence of cell division (21, 22). This process is generally referred to as direct transdifferentiation, and there are now numerous examples of phenotypic interconversions that occur without mitosis (21, 22). Consistent with the constraints of this definition, transdifferentiation of pituitary acidophils can be viewed simply as the initiation of hormone secretion (i.e., synthesis and release) in a cell not previously committed to the secretion of that particular hormone. Inasmuch as the mammosomatotroph appears to be a logical intermediate in this process, it is reasonable to propose that secretion of GH would be initiated in a cell previously dedicated to the release of PRL exclusively, or vice versa. The immediate result in both cases would be the appearance of a bihormonal mammosomatotroph from a more traditional unihormonal acidophil. Presumably, this entire process is completed, from a unihormonal cell of one type to a unihormonal cell of the other type, by subsequent cessation in the production of the initial hormone. A reasonable variation on this theme is that all acidophils may be mammosomatotrophs as was previously suggested by the ICC studies of Hashimoto et al. (15), who showed that although only 26% of acidophils in cattle pituitaries were conspicuously mammosomatotropic, even acidophils that stored predominantly GH or PRL contained significant amounts of the minority hormone, as determined by objective, quantitative criteria. In this plausible scenario, monohormonal cells might simply be mammosomatotrophs that express GH or PRL at too low a level for detection. Thus, the ability to produce GH and PRL, either alone or in combination, would be in the normal repertoire of most acidophils, and manifestation of one phenotype or another would not require "differentiation" in its strictest sense. The cell-specific requirement for Pit-1/growth hormone factor (GHF)-1 for expression of both GH and PRL genes is entirely consistent with this view.

Two primary lines of evidence are supportive of the possibility that GH and PRL cells are functionally interconvertible. First, the physiologic state of an animal can influence the proportions of single and dual hormone secretors. For example, mammosomatotrophs are common in pituitaries of pregnant and lactating musk shrews, but are not abundant in those of virgin females (9). Likewise, mammosomatotrophs are present during reproductive cycles of female bats, but become hypertrophied during pregnancy and account for almost all PRL-secreting cells found during states of reproductive quiescence such as pre- and midhibernation and intervening periods of arousal (23). Studies with rats also support the idea of acidophilic plasticity. For example, our group has shown that the proportions of cells that release GH or PRL undergo reciprocal shifts during the physiologic transition from nonpregnancy through lactation (24). The temporal characteristics of these fluctuations in single and dual hormone secretors are entirely consistent with the possibility that classical somatotrophs and mammotrophs are functionally interconvertible through the mammosomatotroph as an intermediate cell type. Interestingly, these shifts in acidophilic cell types are completely reversed within 4 days of weaning, suggesting that this phenomenon is bidirectional (25). In addition, our group conducted a conceptually similar study in cattle and observed that the percentage of all pituitary cells that released PRL in plaque assays fluctuated during the course of the bovine estrous cycle (26). These changes could not be explained by variations in the relative abundance of PRL-only cells (which remained constant) but rather were attributable to reciprocal shifts in the proportions of GH-only and mammosomatotropic cells. This relationship between ratios of acidophilic subpopulations and stage of reproductive cycle indicates that ovarian steroids may regulate this phenomenon.

More direct evidence favoring "plasticity" among single and dual hormone secretors is provided by in vitro studies from our laboratory and others. Knowing that estradiol stimulates PRL release by a direct pituitary action in rats, we treated pituitary cultures derived from male rats with the steroid and then measured the proportions of acidophilic types by plaque assay. We found that chronic (6-day) exposure to estradiol caused a decrease in the fraction of cells that released only GH, a commensurate increase in that which secreted both GH and PRL, and had no effect on the proportion of classic mammotrophs (27). Thus, the steroid evoked a shift in ratios that is strongly suggestive of a functional interconversion among acidophilic types. In other studies, our laboratory showed that steroid hormones were also capable of causing transdifferentiation of acidophilic subtypes in bovine pituitary cultures (28). Here, cortisol was found to induce reciprocal shifts in the ratio of GH to PRL secretors, and it did so by increasing the proportions of cells that released GH alone or both hormones while decreasing the fraction of PRL-only cells. These effects could not be explained by selective cell proliferation because the mitotic index was extremely low (~1%) and complete inhibition of cell division by treatment with cytosine arabinoside (an inhibitor of DNA synthesis) did not affect the response to cortisol. In addition to steroids, hypothalamic regulatory agents can also influence the proportions of GH and PRL secretors by a direct pituitary action. For example, we showed that treatment of pituitary cells from rats with GnRH also evoked a reciprocal change in the ratio of GH to PRL secretors (29, 30). The magnitude of this response was again not influenced by the presence of cytosine arabinoside, indicating that this shift was not mitosis dependent and therefore was probably the consequence of hormone induction within existing cells (30). Finally, there is a report (31) showing that mice transgenic for human growth hormone releasing factor (GHRF) exhibit a massive increase in both the number and proportion of mammosomatotrophs, indicating yet another time that the hypothalamic hypophysiotropic factors as well as steroids can influence the ratios of acidophil types.

Studies with pituitary cell lines also lend credence to the view that direct transdifferentiation can occur between GH and PRL secretors. For example, cultures of GH_3 cells are composed of about 20% mammosomatotrophs and about 30% GH-only cells; the remainder secrete neither hormone (32). When these cultures were exposed to estradiol and retested, 50% of the cells were mammosomatotrophs and monohormonal secretors disappeared, suggesting a conversion of GH-only cells to bihormonal acidophils (33). Similarly, we treated another pituitary cell line (Po) that contained negligible PRL secretors with estradiol for just 3 days and induced PRL release by 30% of the cells (34). Inasmuch as virtually all the cells released GH before and after treatment, this finding provides compelling evidence for direct transdifferentiation of traditional somatotrophs to mammosomatotrophs. Conceptually similar effects were recently observed by others when a newly established GH-secreting cell line (MtT/S) was treated with insulin or insulin-like growth factor-1 (IGF-1) (35), or when GH_3 cells were exposed to nerve growth factor (NGF) (36).

In light of these considerations, we are beginning to envision cells that release GH, PRL, or both hormones, not as a class of interrelated cells, but as a single cell type suspended as a functional pendulum (Fig. 19.1) (37). As the pendulum swings one way or the other, it becomes primarily a mammotroph or a somatotroph, whereas suspension at an intermediate point is characterized by the bihormonal (i.e., mammosomatotropic) phenotype. In contrast to a true pendulum, whose motion depends on gravitational forces, our hypothetical pendulum is driven by a series of gears, the most proximate of which represent transcriptional regulatory factors. These gears in turn are propelled by another series representing input by hypothalamic hypophysiotropic agents, growth factors, and steroid hormones. To date our efforts have focused on establishing the relationship between the most superficial layer of gears and the phenotypic expression of GH and/or PRL cells. Our future emphasis must be on the intermediate gears.

Armed with the suitable model systems described above, we are now in a position to undertake more mechanistic studies aimed at elucidating the cellular and molecular processes that govern transdifferentiation. It has become increasingly clear that the key to gaining such an understanding will

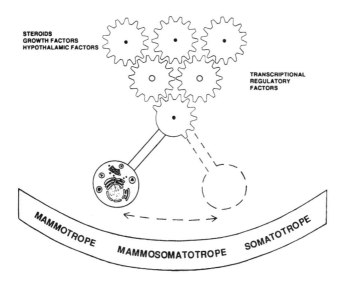

FIGURE 19.1. A pituitary cell is shown suspended as a pendulum to illustrate the "plasticity" that exists among cells that secrete GH, PRL, or both hormones. Reprinted with permission from Frawley and Boockfor (37).

reside in our capacity to repeatedly monitor the initial and terminal events of these processes (gene transcription and hormone release, respectively) in "real time," while experimentally manipulating putative regulatory molecules. The availability of the plaque assay in its various forms satisfies the latter requirement. However, a dynamic system for monitoring molecular events in single cells has been lacking until very recently. In an attempt to develop such a system, we followed the lead of our colleagues working in plant biology (38–42) and modified the standard luciferase reporter system for real-time measurement of gene expression in living endocrine cells.

As is widely known, the conceptual basis for this strategy is that a luciferase structural gene is fused to a gene promoter sequence of interest and this construct, in turn, is incorporated into a plasmid. When placed in an appropriate cellular environment, activation of the promoter under study will induce an increased rate of luciferase messenger RNA (mRNA) production by the fused luciferase structural gene. When this message in turn is translated, the amount of luciferase enzyme produced can serve as a "reporter" of the degree to which the attached promoter was activated. Quantification of luciferase activity is accomplished by measuring the amount of light emitted when the luciferase is exposed to its substrate luciferin in the presence of other intercellular agents such as adenosine triphosphate (ATP) and oxygen. Thus, in this scenario the amount of light emitted will be proportional to the degree of promoter activation. It is noteworthy that the relationship of luciferase concentration

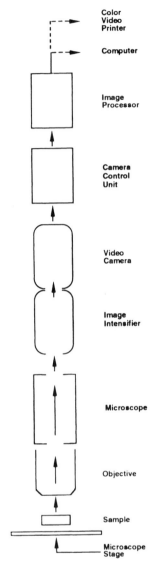

FIGURE 19.2. Flow diagram of the photon imaging system used for making "real-time" measurements of gene expression in living endocrine cells.

to light emission has been shown to be linear over several orders of magnitude.

The experimental protocol used to generate and quantify light from luciferase-producing cells was detailed in a recent publication (43) and will not be repeated here. Suffice it to say that a line of GH-secreting GC cells stably transfected with a plasmid containing the first 237 bp of the rat GH promoter fused to the firefly luciferase structural gene was exposed to dimethylsulfoxide (DMSO) (1%) for 3 hours before an experiment. This step was taken to make the cells permeable to luciferin, which was added to the cells under study just after a bright field image was acquired. The room was then made completely dark, and photon collection was commenced for a 15-min period. A block diagram showing the arrangement of equipment used for photon imaging is presented in Figure 19.2. As can be seen, light emitted from the cells (which hang upside down in a Sykes-Moore chamber) passes through a 40× objective attached to a Zeiss Axioskope to an

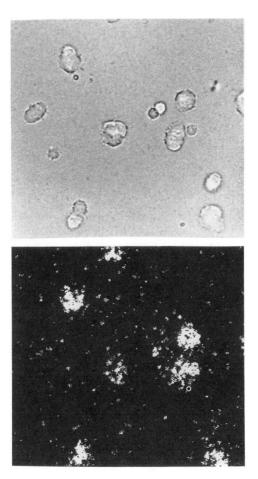

FIGURE 19.3. Photomicrographs of GC cells stably transfected with a plasmid containing a portion of the rat GH promoter fused to luciferase structural genes, which were subsequently subjected to photon imaging. Upper panel: bright field image. Lower panel: amplified photonic signals emanating from the individual cells shown above. See the accompanying text for additional details.

image intensifier/video camera mounted on the microscope. Great care was taken to maximize the efficiency of light transmission to the image intensifier. For example, augmentation of light transmission was achieved by selecting an objective with a high numerical aperture to magnification ratio along with a reduced number of lenses. Likewise, we utilized a microscope unit that was devoid of prisms and that had no optics in the coupling to the intensifier/camera unit. In fact, the entire microscope had no lenses other than the reduced number present in the objective and the tube lens. After exiting the microscope, the light activates the intensifier/camera unit and the amplified signal is output to the camera control unit and image processor. Therefore, the processed information is sent to a computer for further processing and/or storage. Alternately, the information can be sent to a video monitor for display or to a color video printer for production of a hard copy of the image obtained.

The results of a typical experiment are illustrated in Figure 19.3, where amplified photon emissions can be localized to individual cells transfected with the GH promoter/luciferase construct. It is noteworthy that this strategy can provide quantitative as well as qualitative information about promoter activation. Indeed, we have performed quantitative photon counting on cells of this type after treatment with T_3 (which activates the promoter under study) and found that the increment (2.5-fold) in gene expression measured in living cells by photon emissions compared favorably with that measured (3.1-fold) when extracts of entire cultures of companion cells were evaluated by conventional luciferase assay (43). Taken together, these results demonstrate the utility, validity, and power of the modified luciferase reporter technology for real-time measurement of gene expression in living endocrine cells. With this powerful tool along with the plaque assay still at our disposal, we are now in a position to investigate directly the mechanisms underlying the transdifferentiation of GH and PRL secreting cells.

Acknowledgments. These studies were supported by grants from the National Institutes of Health (DK 38215) and the United States Department of Agriculture (USDA 9202846).

References

1. Hoeffler JP, Boockfor FR, Frawley LS. Ontogeny of prolactin cells in neonatal rats: initial prolactin secretors also release growth hormone. Endocrinology 1985;117:187.
2. Behringer RR, Mathews LS, Palmiter RD, Brinster RL. Dwarf mice produced by genetic ablation of growth hormone-expressing cells. Genes Dev 1988;2:453.
3. Borrelli E, Heyman RA, Arias C, Sawchenko PE, Evans RM. Transgenic mice with inducible dwarfism. Nature 1989;339:538.

4. Mulchahey JJ, Jaffe RB. Detection of a potential progenitor cell in the human fetal pituitary that secretes both growth hormone and prolactin. J Clin Endocrinol Metab 1987;66:24.

5. Asa SL, Kovacs K, Horvath E, Losinski NE, Laszlo FA, Domokos I, Hallid‚y WC. Human fetal adenohypophysis. Neuroendocrinology 1988;48:423.

6. Kineman RD, Faught WJ, Frawley LS. The ontogenic and functional relationships between growth hormone- and prolactin-releasing cells during the development of the bovine pituitary. J Endocrinol 1992;134:91–6.

7. Frawley LS, Boockfor FR, Hoeffler JP. Identification by plaque assays of a pituitary cell type that secretes both growth hormone and prolactin. Endocrinology 1985;116:734.

8. Lloyd RV, Coleman K, Fields K, Nath V. Analysis of prolactin and growth hormone production in hyperplastic and neoplastic rat pituitary tissues by the hemolytic plaque assays. Cancer Res 1987;47:1087.

9. Leong DA, Lau SK, Sinha YN, Kaiser DL, Thorner MO. Enumeration of lactotropes and somatotropes among male and female pituitary cells in culture: evidence in favor of a mammosomatotrope subpopulation in the rat. Endocrinology 1985;116:1371.

10. Nikitovitch-Winer MB, Atkin J, Maley BE. Colocalization of prolactin and growth hormone within specific adenohypophyseal cells in male, female, and lactating female rats. Endocrinology 1987;121:625.

11. Ishibashi T, Shiino M. Co-localization pattern of growth hormone (GH) and prolactin (PRL) within the anterior pituitary cells in the female rat and female musk shrew. Anat Rec 1989;223:185.

12. Lloyd RV, Anagnostou D, Cano M, Barkan AL, Chandler WF. Analysis of mammosomatotropic cells in normal and neoplastic human pituitary tissues by the reverse hemolytic plaque assay and immunocytochemistry. J Clin Endocrinol Metab 1988;66:1103.

13. Sasaki F, Iwama Y. Two types of mammosomatotropes in mouse adenohypophysis. Cell Tissue Res 1989;256:645.

14. Fumagalli G, Zanini A. In cow anterior pituitary, growth hormone and prolactin can be packed in separate granules of the same cell. J Cell Biol 1985;100:2019.

15. Hashimoto S, Fumagalli G, Zanini A, Meldolesi J. Sorting of three secretory proteins to distinct secretory granules in acidophilic cells of cow anterior pituitary. J Cell Biol 1987;105:1579.

16. Kineman RD, Faught WJ, Frawley LS. Mammosomatotropes are abundant in bovine pituitaries: influence of gonadal status. Endocrinology 1991;128:2229.

17. Ishibashi T, Shiino M. Subcellular localization of prolactin in the anterior pituitary cells of the female Japanese house bat, Pipistrellus abramus. Endocrinology 1989;124:1056.

18. Childs GV, Unabia G, Lee BL, Lloyd J. Maturation of follicle stimulating hormone gonadotropes during the rat estrous cycle. Endocrinology 1992;131:29–36.

19. Yamaguchi M, Ogren L, Endo H, Thordarson G, Bigsby RM, Talamantes F. Production of mouse placental lactogen-I and placental lactogen-II by the same giant cell. Endocrinology 1992;131:1595–602.

20. Selman K, Kafatos FC. Transdifferentiation in the labial gland of silk moths: is DNA synthesis required for cellular metamorphosis? Cell Differ 1974;3: 81–94.

21. Beresford WA. Direct transdifferentiation: can cells change their phenotype without cell division? Cell Differ Dev 1990;29:81–93.
22. Okada TS. Transdifferentiation: flexibility in cell differentiation. Oxford: Clarendon Press, 1991.
23. Ishibashi T, Shiino M. Subcellular localization of PRL in the anterior pituitary cells of the female Japanese house bat. Endocrinology 1989;124:1056.
24. Porter TE, Hill JB, Wiles CD, Frawley LS. Is the mammosomatotrope a transitional cell for the functional interconversion of growth hormone- and prolactin-secreting cells? Suggestive evidence from virgin, gestating, and lactating rats. Endocrinology 1990;127:2789.
25. Porter TE, Wiles CD. Evidence for bidirectional interconversion of mammotropes and somatotropes: rapid reversion of acidophilic cell types to pregestational proportions after weaning. Endocrinology 1991;129:1215.
26. Kineman RD, Henricks DM, Faught WJ, Frawley LS. Fluctuations in the proportions of growth hormone- and prolactin-secreting cells during the bovine estrous cycle. Endocrinology 1991;129:1221–5.
27. Boockfor FR, Hoeffler JP, Frawley LS. Estradiol induces a shift in cultured cells that release prolactin or growth hormone. Am J Physiol 1986;250:E103.
28. Kineman RD, Faught WJ, Frawley LS. Steroids can modulate transdifferentiation of prolactin and growth hormone cells in bovine pituitary cultures. Endocrinology 1992;130:3289–94
29. Hoeffler JP, Frawley LS. Hypothalamic factors differentially affect the proportions of cells that secrete growth hormone or prolactin. Endocrinology 1987;120:791.
30. Frawley LS, Hoeffler JP. Hypothalamic peptides affect the ratios of GH and PRL cells: role of cell division. Peptides 1988;9:825.
31. Stefaneanu L, Kovacs K, Hovrath E, Asa SL, Losinski NE, Billestrup N, Price J, Vale W. Adenohypophysial changes in mice transgenic for human growth hormone-releasing factor: a histological, immunocytochemical, and electron microscopic investigation. Endocrinology 1989;125:2710.
32. Boockfor FR, Hoeffler JP, Frawley LS. Cultures of GH$_3$ cells are functionally heterogeneous: thyrotropin-releasing hormone, estradiol and cortisol cause reciprocal shifts in the proportions of growth hormone and prolactin secretors. Endocrinology 1985;117:418.
33. Boockfor FR, Schwarz LK. Cultures of GH$_3$ cells contain both single and dual hormone secretors. Endocrinology 1987;122:762.
34. Kineman RD, Frawley LS. Secretory characteristics and phenotypic plasticity of growth hormone and prolactin secreting cell lines. J Endocrinol 1994;140:455–63.
35. Inoue K, Sakai T. Conversion of growth hormone-secreting cells into prolactin-secreting cells and its promotion by insulin and insulin-like growth factor-1 in vitro. Exp Cell Res 1991;195:53–8.
36. Boroni F, Sigala S, Dal Toso R, Balsari A, Missale C, Spano PF. Nerve growth factor dictates the acquisition of the lactotrope phenotype in GH$_3$ cells. Program, 3rd International Pituitary Congress, 1993; Abstract MP-12.
37. Frawley LS, Boockfor FR. Mammosomatotropes: presence and functions in normal and neoplastic pituitary tissue. Endocr Rev 1991;12:337–55.
38. Ow DW, Wood KV, DeLuca M, de Wet JR, Helsinki DR, Howell SH. Transient and stable expression of the firefly luciferase gene in plant cells and transgenic plants. Science 1986;234:856–9.

39. Wick RA. Photon counting imaging: applications in biomedical research. BioTechniques 1989;7:262–8.
40. Barnes WM. Variable patterns of expression of luciferase in transgenic tobacco leaves. Proc Natl Acad Sci USA 1990;87:9183–7.
41. Kay SA, Millar AJ. Circadian regulated *Cab* gene transcription in higher plants. In: Young M, ed. The molecular biology of circadian rhythms. New York: Marcel Dekker, 1992:73–89.
42. Millar AJ, Short SR, Chua N-H, Kay SA. A novel circadian phenotype based on firefly luciferase expression in transgenic plants. Plant Cell 1992;4:1075–87.
43. Frawley LS, Faught WJ, Nicholson J, Moomaw B. Real time measurement of gene expression in living endocrine cells. Endocrinology 1994;135:468–71.

20

Growth Hormone Releasing Hormone: Behavioral Evidence for Direct Central Actions

Franco J. Vaccarino and Sidney H. Kennedy

Growth hormone releasing hormone (GHRH) peptides have been isolated and characterized from human pancreatic tumor (1, 2) and from hypothalami of several species including rat and human (3–6). GHRH peptides have strong homology with peptides of the glucagon, vasoactive intestinal polypeptide, and PHI-27 family. Many of the GHRH neurons have their cell bodies in the arcuate nucleus of the hypothalamus with fibers extending into the median eminence (7–9). GHRH-containing terminals in the median eminence are located proximal to capillaries that drain into the hypophysial portal blood system. It is through this system that GHRH stimulates the release of growth hormone from cells in the anterior pituitary gland (10). The characterization and synthesis of GHRH represents a significant development and has opened important new avenues for the direct study of GHRH function.

One important consequence of the characterization and synthesis of GHRH has been the development of immunohistochemical work aimed at characterizing the distribution of GHRH-containing neurons in the brain. In addition to the presence of terminals in the median eminence, immunohistochemical results indicate that GHRH terminals are present in intra- and extrahypothalamic sites not directly associated with the median eminence and hypophysial portal blood system (7, 8). The latter results, together with electrophysiologic findings demonstrating that iontophoretically applied GHRH can influence neuronal membrane excitability (11), suggest that, in addition to its hormonal actions, GHRH has neurotransmitter and/or neuromodulatory actions. This notion is consistent with the functional significance of other peptides found in the brain (12), and raises the distinct possibility that GHRH has direct central behavioral effects. This chapter reviews evidence derived from our laboratory and others that indicates that GHRH has direct central actions on neural systems involved in feeding and related behaviors.

Effects of Intracerebroventricular GHRH Treatment on Feeding

In addition to its potent endocrine action on the release of growth hormone, evidence from our laboratory has shown that centrally administered GHRH has stimulatory effects on feeding behavior in rats. We have found that acute intracerebroventricular (icv) injections of rat hypothalamic (rh) or human pancreatic (hp) GHRH, in doses ranging from 0.2 to 40.0 pmol enhances food intake by 25% to 75% in hungry and non–food-deprived rats (13–16). Supporting the notion that the facilitatory effects of GHRH on feeding are centrally mediated are the following findings. First, intraperitoneal (ip) administration of rhGHRH (in doses ranging from 0.2 to 200.0 pmol) or growth hormone does not influence food intake (13). Since ip-administered GHRH is unlikely to reach the brain in behaviorally significant amounts, these results suggest that GHRH-induced feeding is not mediated by peripheral effects of GHRH or growth hormone. Second, icv administration of a structurally related but physiologically inactive peptide has no effect on food intake (13, 16), suggesting that the facilitatory effects of GHRH on feeding are not due to nonspecific effects of peptide administration. Third, icv injections of rhGHRH, in feeding-stimulatory doses, do not influence general locomotor activity (13, 15), indicating that the increased feeding observed following GHRH treatment is not due to general behavioral activating properties of GHRH. Together, these findings suggest that central GHRH plays a stimulatory role in feeding.

Consistent with our findings, Riviere and Bueno (17) investigated the effects of icv GHRH treatment in sheep and found increases in food intake in doses comparable to those used in our studies. Thus, the facilitatory effects of GHRH on feeding are generalizable to other species.

While picomole doses of icv GHRH stimulate feeding, icv treatment with higher doses has been found to suppress feeding. Imaki et al. (18) reported that icv hpGHRH treatment depressed food intake at 1- and 4-nmol doses. In an effort to replicate these findings and establish a wider dose-response curve, we tested the effects of icv GHRH in doses ranging from 0.4 pmol to 4.0 nmol (14). Consistent with our original report (13), GHRH was most effective at the 4.0-pmol dose. At the 40.0- and 100.0-pmol doses, however, GHRH was less effective, and at 1 and 4 nmol GHRH was either ineffective or had suppressive effects on feeding. Together with findings from our original report, the above findings demonstrate that the facilitatory effects of icv GHRH on feeding are most evident in doses ranging from 0.2 to 40.0 pmol, while the suppressive effects are associated with nanomole doses (i.e., 4 nmol). Of special interest here is the fact that the feeding-suppressive doses of GHRH are comparable to icv GHRH doses that stimulate growth hormone release (19). This raises the possibility that increased growth hormone release contributes to the suppressive effects of GHRH on feed-ing observed with higher doses. However, findings showing that hypophysectomy does not abolish the feeding inhibitory effects of higher doses of GHRH (18) indicate that the feeding inhibition is also centrally mediated.

Photoperiod-Dependent Differences in the Feeding Effects of GHRH

Rats display a circadian pattern of feeding in which the great majority of their feeding occurs during the dark phase of the light-dark photoperiod (20). During the light phase they are relatively inactive with respect to feeding. Thus, it seems appropriate to view light and dark photoperiods as representing significantly different feeding states in the rat. The studies described thus far have examined the effects of GHRH on feeding during the light photoperiod. Previous studies examining feeding effects of pharmacologic agents suggest that photoperiod specificity is an important variable to consider in understanding mechanisms underlying feeding be-havior (21). The following findings indicate that GHRH-induced feeding is photoperiod sensitive and raises important conceptual issues regarding the role of endogenous GHRH in normal feeding.

In contrast to its appetitive effects during the light photoperiod, low to moderate doses of GHRH have either no effect or inhibitory effects when administered during the dark photoperiod (15). Interestingly, a structurally related but inactive peptide had no effect in the dark or the light photope-riod. The latter result suggests that both the facilitation and inhibition observed during the light and dark photoperiods, respectively, are GHRH specific.

These results indicate that the central actions of GHRH may be associ-ated with two separate antagonistic feeding responses, one orexigenic (light photoperiod) and the other anorexic (dark photoperiod). To account for these light-dark differences in GHRH effects on feeding, we have hypoth-esized the existence of a natural circadian oscillation in endogenous activity of GHRH (15, 22). In the light photophase, endogenous GHRH activity may be relatively low. Addition of low doses of exogenous GHRH would result in raised internal GHRH levels, activating a facilitatory response. During the dark, however, baseline levels of endogenous GHRH activity may be naturally increased to the optimal facilitatory range. This would contribute to the high baseline consumption present during the dark photo-phase. Addition of low doses of exogenous GHRH during the dark would act to push internal levels beyond the optimal facilitatory range invoking an inhibitory response and resulting in relative suppression of feeding.

This hypothesis is consistent with the fact that the doses required to produce inhibition of feeding are higher during the light photophase than during the dark photophase (15). It should be noted that this proposed circadian oscillation in endogenous GHRH levels need only be a functional one and thus may be expressed as either an actual increase in available GHRH at the critical central site of action or an increased sensitivity to GHRH. Interestingly a similar photoperiod specificity has been observed with the feeding effects of central norepinephrine treatments, and similar explanations have been proposed (21).

Central Site of Action for GHRH-Induced Feeding

In an effort to localize the central site of action for GHRH's facilitatory effects on feeding, we have tested the effects of GHRH in a variety of hypothalamic regions known to contain GHRH terminals. Immunohistochemical studies examining the distribution of GHRH neurons indicate that GHRH neurons originating in the arcuate nucleus project to numerous hypothalamic sites other than the median eminence (7, 8). These include the anterior, periventricular, dorsomedial, paraventricular, suprachiasmatic, and premamillary hypothalamic nuclei and the medial preoptic, lateral preoptic, and lateral hypothalamic regions. These sites, then, represent possible sites of action for GHRH's facilitatory effects on feeding.

Following a number of pilot studies examining the effects of direct intrahypothalamic GHRH microinjections on feeding, it was determined that the suprachiasmatic nucleus (SCN)/medial preoptic area (MPOA) region of the hypothalamus (SCN/MPOA) was the most sensitive site for the feeding-enhancing effects of GHRH. To examine the sensitivity of this region to GHRH's appetitive effects, we tested rats for their feeding response to intra-SCN/MPOA GHRH microinjections in doses of 0.0, 0.01, 0.1, and 1.0 pmol. It was found that GHRH doses as low as 0.01 pmol were effective at increasing food intake, with the highest increase observed at the 1.0-pmol dose (23). GHRH microinjections into other regions that receive GHRH projections, including the lateral preoptic and anterior hypothalamus, were ineffective.

An important issue with the SCN/MPOA results is that this hypothalamic region is located near the third ventricle. It is possible that GHRH is spreading into the ventricular system and producing its stimulatory effects on feeding in some other brain region. Since the SCN and MPOA are by definition adjacent to the third ventricle, this possibility is difficult to control. However, the following findings argue for a local SCN/MPOA effect of GHRH. The fact that some of the extra-SCN/MPOA ineffective sites were located adjacent to the third ventricle but were nonetheless ineffective argues against ventricular spread being a critical factor. Also, intra-SCN/ MPOA microinjections of GHRH doses as low as 0.01 pmol (20–40 times lower than icv doses previously found to be effective) significantly stimulated feeding. Finally, the increased feeding observed following microinjections of 0.1 and 1.0 pmol GHRH into the SCN/MPOA is about 200% to 300% higher than that observed with similar doses injected icv. More recently, we have also found that GHRH microinjections into the paraventricular nucleus, which receives GHRH projections and is an important structure for feeding (24), had no consistent effects on food intake (23). Together, these observations support the notion that the SCN/MPOA is the important target region for growth hormone releasing factor's (GHRF's) appetitive effects. The extent to which the SCN and MPOA can be differentiated with respect to their roles in GHRF-induced feeding is not yet known.

Although our studies have been carried out exclusively in male rats, Tanaka et al. (25) have examined the effects of intrahypothalamic GHRH injections in female rats. Interestingly, they found a feeding-stimulatory effect following microinjections of GHRH into the ventromedial hypothalamus of female rats. Thus, it may be that sex differences in mechanisms mediating GHRH-induced feeding exist.

The Role of GHRH in Mechanisms Controlling Circadian Activity

The fact that the SCN, a critical site for regulation of circadian rhythms (26), is a sensitive site for GHRH's effects on feeding is intriguing in light of the previously discussed photoperiod-sensitive differences in GHRH efficacy. It is therefore tempting to suggest that GHRH may be involved in mechanisms underlying circadian activity, possibly as it relates to feeding.

Indeed, recent results from our laboratory support this notion. We have recently found that, in hamsters, intra-SCN GHRH injections during the inactive phase, cause phase advances in circadian activity patterns (27). The temporal profile of the phase-altering effect of intra-SCN GHRH resembles that of nonphotic influences on the clock reported by others (27). This effect may be feeding related such that GHRH actions in the SCN could represent a feeding-specific neural signal to the SCN circadian clock. The functional relevance of this is not yet known. However, one possibility is that by producing a phase advance the next day (i.e., getting up earlier), enhanced SCN GHRH activity associated with feeding could constitute a competitive advantage to animals discovering a food source during their inactive phase. Such a phase advance would allow the organism to gain an advantage in procuring that food source the next day.

Somatostatin and Opioid Involvement in GHRH-Induced Feeding

Somatostatin

In light of the strong functional association between somatostatin (SS) and GHRH, the possible role of SS in GHRH-induced feeding is of natural interest. Earlier studies examining the potential feeding effects of central SS were inconsistent. Increased food intake (28, 29), decreased food intake (30–32), and a biphasic effect on food intake (33) have all been reported. Analysis of these studies indicated that doses used were generally in the nanomole range. Since systemic SS treatment is known to decrease feeding (34, 35), it is possible that at least some of the central SS effects reflect leakage into the peripheral circulation.

To clarify this issue, we tested a wide range of doses following icv administration (36). The results indicate that in rats, picomole doses of SS increase food intake, while nanomole doses suppress feeding. The dose-response profile of these SS effects was very similar to that observed for GHRH-induced feeding.

In an effort to examine the possible interaction between SS- and GHRH-induced feeding mechanisms, we have further examined the effects of icv SS and GHRH antisera treatments on SS and GHRH-induced feeding (37). As expected, icv SS antiserum treatment reversed SS-induced feeding and icv GHRH antiserum treatment reversed GHRH-induced feeding. Of special interest, however, was the finding that SS antiserum also reversed GHRH-induced feeding while GHRH antiserum had only marginal effects on SS-induced feeding. These results indicate that as in the case of GH regulation, GHRH and SS interact to control feeding. More specifically, these results indicate that GHRH-induced feeding relies on the integrity of central SS activity. Since these studies were carried out following icv administration of the antisera and did not examine any specific brain region, the locus of this interaction is not yet known.

Opioids

In addition to SS, we have also found that endogenous opioid activity may contribute to GHRH-induced feeding. In an earlier report it was found that naloxone treatment attenuated the feeding facilitatory effects of GHRH in doses that did not affect baseline feeding (16). These results suggested that endogenous opioid activity is involved in mechanisms mediating GHRH-induced feeding. More recently, in an effort to examine the possible brain regions involved in mediating these opioid effects, we examined the effects of local opioid receptor blockade in the paraventricular nucleus (PVN) of the hypothalamus (38). The PVN is well known for its role in regulating feeding behavior and is known to contain opioid receptors. Moreover, the PVN receives projections from the PVN and MPOA. The results demonstrated that intra-PVN microinjections of naltrexone and methylnaltrexone (nonselective opioid antagonists) attenuated the increased food intake derived from intra-SCN/MPOA microinjections of GHRH. These results indicate that increased opioid activity in the PVN is an output signal for feeding-relevant GHRH actions in the SCN/MPOA.

Behavioral Characterization of GHRH-Induced Feeding

Meal Patterns

In an effort to characterize the behavioral nature of GHRH-induced feeding, we have analyzed meal patterns in rats receiving central GHRH microinjections (39). Results indicate that at moderately effective doses, GHRH caused increases in food intake that were characterized by increases in the length of time spent eating a meal. Interestingly, at maximally effective doses, the meal duration did not differ from baseline, but the rate of feeding was increased. Thus, it appears that the behavioral profile of GHRH-induced feeding differs as a function of dose effectiveness. These findings are consistent with those of Riviere and Bueno (17), who reported that the GHRH-induced increase in food intake in sheep is associated with increased rate of ingestion. Interestingly, we have also found that latency to meal onset is not affected by GHRH treatment (39). Taken together, these observations suggest that GHRH-induced feeding is associated with meal maintenance or postingestional factors rather than meal initiation.

Motivational Variables

To investigate the extent to which the feeding effects of GHRH reflect an increased motivational state, we have also examined GHRH's ability to increase operant responding for food reward. Initial findings are consistent with the notion that increased central GHRH increases the motivation to obtain food. Rats treated with icv GHRH show increased operant responding for food reward (40). Thus, it appears that GHRH is not simply stimulating motor outputs required for feeding behavior but is enhancing the animal's motivation to obtain food. This finding raises the possibility that central GHRH plays a role in the maintenance of feeding during increased food-related drive states such as hunger. That GHRH increases the reinforcing properties of food is also interesting in light of the previously discussed findings (16) showing that blockade of endogenous opioids, which have been implicated in reward mechanisms, attenuates GHRF-induced feeding.

Macronutrient Selectivity

Considerable evidence has emerged over the years to suggest that different brain mechanisms may subserve the regulation of feeding behavior associated with different macronutients. Indeed, there is now strong evidence that norepinephrine and NPY act in the PVN to selectively enhance carbohydrate, but not protein, intake (41). In an effort to determine the macronutrient profile of GHRH-induced feeding, we have recently completed a set of studies examining the effects of central GHRH actions on protein, carbohydrate, and fat intake. The results of these studies indicate that intra-SCN/MPOA GHRH injections selectively increase protein intake without stimulating fat or carbohydrate intake (37). Consistent with the hypothesized output role for PVN opioids in GHRH-induced feeding, antagonism of opioid activity in the PVN selectively attenuated the increased protein intake induced by these central actions of GHRH (38).

Endogenous GHRH and Feeding

The data discussed thus far indicate that exogenously applied GHRH acts centrally to stimulate feeding. Although this raises the possibility that endogenous GHRH activity contributes to ongoing feeding behavior, a direct demonstration would require showing that blockade of endogenous central GHRH activity specifically attenuates ongoing baseline feeding behavior. Indeed, results from our laboratory support this notion. Utilizing centrally administered GHRH antiserum as a functional antagonist, we found that, in rats, intra-SCN/MPOA GHRH antiserum treatment attenuated the increased feeding normally observed at the onset of the active cycle (42). More recently, we have further demonstrated that this attenuation is protein-selective, such that the central antiserum treatment attenuated the increased protein intake, but not carbohydrate intake, found at the onset of the active phase of the rats' activity cycle (43).

Together, these results show that endogenous GHRH activity contributes to mechanisms controlling normal baseline feeding and that, consistent with the effects of exogenous GHRH, the role of endogenous GHRH activity appears to be to selectively stimulate protein intake. Importantly, it should be noted that in contrast to the effects of exogenously applied GHRH, which were apparent during the inactive phase of the animals' activity cycle, the effects of endogenous GHRH blockade (GHRH antiserum treatment) were apparent during the active phase of the activity cycle. This suggests that elevated endogenous GHRH activity in the SCN/MPOA region contributes to mechanisms mediating the increased feeding observed during the active phase of the photocycle.

GHRH Abnormalities in Anorexia Nervosa

The fact that alterations in endogenous GHRH activity can affect ongoing feeding behavior raises the possibility that abnormalities in human eating behavior may involve alterations in endogenous GHRH activity. Indirect evidence supports this notion as it applies to anorexia nervosa (AN), a psychiatric condition in which patients voluntarily restrict their food intake to the extent that they are in a chronically starved state. A number of investigators (44–46) have provided evidence indicating that AN patients showed a supersensitive GH response to GHRH challenge. This finding raised the possibility that diminished levels of endogenous GHRH were present in AN, resulting in the development of supersensitive GHRH receptors.

In an effort to test the hypothesis that AN patients have diminished levels of GHRH and that these lowered GHRH levels contribute to the abnormal eating pattern characteristic of these patients, we recently tested the effects of intravenous GHRH treatment on food intake in patients diagnosed with AN and normal control subjects in a double-blind, placebo-controlled study (47). Consistent with a hypothesized role for GHRH in AN, GHRH treatment tended to normalize food intake in AN patients. More specifically, GHRH increased food intake in restricting AN patients and decreased binging in bulimic AN patients. These effects did not correlate with GH release or nutritional status. Normal control subjects showed diminished food intake following GHRH treatment.

Although the mechanisms underlying these effects are unknown, it is possible that they reflect direct central actions of GHRH. This notion is indirectly supported by the following points. First, although the blood-brain barrier presents a biologic obstacle for peptide entry into the brain, this barrier is not absolute and there is considerable evidence now to show that systemically administered peptides can enter the brain (48). Second, since fentomole doses of GHRH are sufficient to produce behavioral effects, only a very small fraction of systemically administered GHRH (n.s. in the AN study; a 1-μg/kg dose was used) need enter the brain to produce behavioral effects. Finally, since the behavioral effects of GHRH in AN patients did not correlate with GH or nutritional status, it is unlikely that the effects are related to peripheral or pituitary actions of GHRH.

Central-Peripheral Integration of Function

This chapter has outlined findings that suggest that GHRH can have direct neurotransmitter-like actions on central feeding systems. These findings, in combination with data demonstrating that systemic GHRH or growth hormone does not influence feeding in the present paradigms, suggest that

GHRH-induced feeding is not directly associated with peripheral growth hormone actions. It is interesting, however, to speculate that the effects of GHRH on feeding are functionally associated and coordinated with growth hormone actions. This possibility is based on the functional compatibility between the central effects of GHRH (i.e., to stimulate feeding in a protein-selective manner) and the peripheral effects of GHRH on growth hormone (i.e., to promote growth and protein synthesis) (49). Thus, GHRH may underlie growth processes and metabolic changes in an integrated manner by promoting feeding through central actions and stimulating growth and protein synthesis through pituitary actions. More detailed investigations of the behavioral and physiologic nature of GHRH-induced feeding will be necessary to determine the extent to which the central feeding effects of GHRH are functionally coupled with GHRH's pituitary-mediated effects on growth and protein synthesis.

Indirect support for the notion that GHRH may coordinate central behavioral and peripheral physiologic functions comes from the finding that GHRH terminals in the SCN/MPOA and the median eminence derive largely from a common anatomical source, the arcuate nucleus (7, 8). Thus, anatomical evidence is consistent with the possibility that an overlap exists between the central signals controlling GHRH actions on feeding and central signals controlling GHRH actions on growth hormone release. A report showing that the pulsatile release of growth hormone (measured in plasma) in rats is correlated with feeding behavior (independent of photoperiod) lends further support to this notion (50). Future research will be necessary to further elucidate the relationship between the present behavioral effects and growth hormone function.

Conclusion

The fact that GHRH has central actions that parallel its peripheral endocrine actions raises the possibility that physiologic and behavioral actions typically ascribed to GH activity may involve central GHRH actions. In other words, there are at least two target organs for GHRH actions: the pituitary and the brain. It will be important for future research examining the role of GH in physiologic functions to consider the possible role of central GHRH actions.

Acknowledgments. Thanks to Drs. W. Vale and J. Rivier of the Salk Institute for Biological Studies, and to Dr. W. Wassenaar of Ferring, Inc. for generously providing us with GHRH and related peptides throughout the course of this work. Research cited from our laboratory was supported by an NSERC operating grant to Franco J. Vaccarino and the Ontario Mental Health Foundation.

References

1. Guillemin R, Brazeau P, Bohlen P, Esch F, Ling N, Wehrenberg WB. Growth hormone-releasing factor from a human pancreatic tumour that caused acromegaly. Science 1982;218:585–7.
2. Rivier J, Speiss J, Thorner M, Vale W. Characterization of a growth hormone-releasing factor from a human pancreatic islet tumour. Nature 1982;300: 276–8.
3. Bohlen P, Esch F, Brazeau P, Ling N, Guillemin R. Isolation and characterization of the porcine hypothalamic growth hormone-releasing factor. Biochem Biophys Res Commun 1983;117:726–34.
4. Bohlen P, Esch F, Brazeau P, Ling N, Guillemin R. Isolation and characterization of the bovine hypothalamic growth hormone-releasing factor. Biochem Biophys Res Commun 1983;117:772–9.
5. Ling N, Esch F, Bohlen P, Brazeau P, Wehrenberg WB, Guillemin R. Isolation, primary structure and synthesis of human hypothalamic somatocrinin: growth hormone releasing factor. Proc Natl Acad Sci USA 1984;81: 4302–6.
6. Spiess J, Rivier J, Vale W. Characterization of rat hypothalamic growth hormone-releasing factor. Nature 1983;303:532–5.
7. Sawchenko PE, Swanson LW, Rivier J, Vale WW. The distribution of growth-hormone-releasing factor (GRF) immunoreactivity in the central nervous system of the rat: an immunohistochemical study using antisera directed against rat hypothalamic GRF. J Comp Neurol 1985;327:100–15.
8. Merchenthaler I, Thomas CR, Arimura A. Immunohistochemical localization of growth hormone-releasing factor (GHRF)-containing structures in the rat brain using antirat GHRF serum. Peptides 1984;5:1071–5.
9. Bloch B, Brazeau P, Ling N, Bohlen P, Esch F, Wehrenberg WB, Benoit R, Bloom F. Immunohistochemical detection of growth hormone-releasing factor in the brain. Nature 1983;263:251–7.
10. Plotsky PM, Vale W. Patterns of growth hormone-releasing factor and somatostatin secretion into the hypophysial-portal circulation of the rat. Science 1985;230:461–3.
11. Twery MJ, Moss RL. Sensitivity of rat forebrain neurons to growth hormone-releasing hormone. Peptides 1985;6:609–13.
12. Scharrer B. Neurosecretion: beginnings and new directions in neuropeptide research. Annu Rev Neurosci 1987;10:1–17.
13. Vaccarino FJ, Bloom FE, Rivier J, Vale W, Koob GF. Stimulation of food intake in rats by centrally administered hypothalamic growth hormone-releasing factor. Nature 1985;314:167–8.
14. Vaccarino FJ, Feifel D, Rivier J, Vale W, Koob GF. Centrally administered growth hormone-releasing factor stimulates food intake in free-feeding rats. Peptides 1988;9(suppl 1):35–8.
15. Feifel D, Vaccarino FJ. Feeding effects of growth hormone-releasing factor in rats are photoperiod sensitive. Behav Neurosci 1989;103:824–30.
16. Vaccarino FJ, Buckenham K. Naloxone blockade of growth hormone-releasing factor-induced feeding. Regul Pept 1987;18:165–71.
17. Riviere P, Bueno L. Influence of regimen and insulinemia on orexigenic effects of GRF (1-44) in sheep. Physiol Behav 1987;39:347–450.

18. Imaki T, Shibasaki T, Hotta M, Masuda A, Demure H, Shizume K, Ling N. The satiety effect of growth hormone-releasing factor in rats. Brain Res 1985;340:168–88.
19. Wehrenberg WB, Ehlers CL. Effects of growth hormone-releasing factor in the brain. Science 1986;232:1271–3.
20. Rosenwasser AM, Boulous Z, Terman M. Circadian organization of food intake and meal patterns in the rat. Physiol Behav 1979;27:33–9.
21. Margules DL. Hypothalamic norepinephrine: circadian rhythms and the control of feeding behavior. Science 1972;178:640–2.
22. Feifel D, Vaccarino FJ. Growth hormone-regulatory peptides (GHRH and somatostatin) and feeding: a model for the integration of central and peripheral function. Neurosci Biobehav Rev 1994;18:421–33.
23. Vaccarino FJ, Hayward M. Microinjections of growth hormone-releasing factor into the medial preoptic area/suprachiasmatic nucleus region of the hypothalamus stimulate food intake in rats. Regul Pept 1988;21:21–8.
24. Leibowitz SF. Adrenergic stimulation of the paraventricular nucleus and its effects on ingestive behavior as a function of drug dose and time of injection in the light-dark cycle. Brain Res Bull 1977;3:357–63.
25. Tanaka Y, Egawa M, Inoue S, Takamura Y. Effect of hypothalamic administration of growth hormone-releasing factor (GRF) on feeding behavior in rats. Brain Res 1991;558:273–9.
26. Rusak B, Zucker I. Neural regulation of circadian rhythms. Physiol Rev 1979;59:449–526.
27. Vaccarino FJ, Sovran P, Baird JP, Ralph M. Growth hormone-releasing hormone mediates feeding-specific feedback to the suprachiasmatic circadian clock. Peptides 1995;16:595–8.
28. Danguir J. Food intake in rats increased by intracerebroventricular infusion of the somatostatin analogue SMS 201–995 and is decreased by somatostatin antiserum. Peptides 1987;9:211–3.
29. Rezek MT, Havlicek V, Hughes R, Friesen H. Central site of action of somatostatin (SRIF): role of hippocampus. Neuropsychology 1976;15:499–504.
30. Lin MT, Chen JJ, Ho LT. Hypothalamic involvement in the hyperglycemia and satiety actions of somatostatin in rats. Neuroendocrinology 1987;45:62–7.
31. Shibasaki T, Kim YS, Yamauchi N, Masuda A, Imaki T, Hotta M, Deura H, Wakabayshi I, Ling N, Shizume K. Antagonistic effect of somatostatin on corticotropin-releasing factor induced anorexia. Life Sci 1988;42:329–34.
32. Vijayan E, McCann SM. Suppression of feeding and drinking activity in rats following intraventricular injection of thyrotropin releasing hormone (TRH). Endocrinology 1977;100:1727–30.
33. Aponte G, Leung P, Gross D, Yamada T. Effects of somatostatin on food intake in rats. Life Sci 1984;35:741–6.
34. Levine AS, Morley JE. Peripherally administered somatostatin reduced feeding by a vagally mediated mechanism. Pharmacol Biochem Behav 1982;16:897–902.
35. Lotter EC, Krinksy R, McKay JM, Treener CM, Porte D, Woods SC. Somatostatin decreases the food intake of rats and baboons. J Comp Physiol Psychol 1981;95:278–87.
36. Feifel D, Vaccarino FJ. Central somatostain: a reexamination of its effects on feeding. Brain Res 1990;535:189–94.

37. Feifel D, Vaccarino FJ, Rivier J, Vale W. Evidence for a common neural mechanism mediating GRF-induced and somatostatin-induced feeding. Neuroendocrinology 1993;57(2):299–305.
38. Dickson PR, Vaccarino FJ. GRF-induced feeding: evidence for protein selectivity and opiate involvement. Peptides 1994;15:1343–52.
39. Dickson PR, Vaccarino FJ. Characterization of feeding behavior induced by central injection of GRF. Am J Physiol 1990;259:r651–7.
40. Feifel D, Vaccarino FJ. Central injections of growth hormone-releasing factor increase operant responding for food reward. Prog Neuropsychopharmacol Biol Psychiatry 1990;14:813–20.
41. Leibowitz SF. Brain monoamines and peptides: role in the control of eating behavior. Fed Proc 1986;45:1396–403.
42. Vaccarino FJ, Feifel D, Rivier J, Vale W. Antagonism of central growth hormone-releasing factor activity selectively attenuates dark-onset feeding in rats. J Neurosci 1991;11(12):3924–7.
43. Dickson PR, Feifel D, Vaccarino FJ. Blockade of endogenous GRF at dark onset selectively suppresses protein intake. Peptides 1995;16:7–9.
44. Brambilla F, Ferrari E, Cavagnini F, et al. 2-Adrenoceptor sensitivity in anorexia nervosa: GH response to clonidine or GHRH stimulation. Biol Psychiatry 1989;25:256–64.
45. Rolla M, Andreoni D, Belliti D, Ferdeghini M, Ferrannini E. Failure of glucose infusion to suppress the exaggerated GH response to GHRH in patients with anorexia nervosa. Biol Psychiatry 1990;27:215–22.
46. Tamai H, Komai G, Matsubayashi S, et al. Effects of cholinergic muscarinic receptor blockade on human growth hormone (GH)-releasing hormone-(1-44)-induced GH secretion in anorexia nervosa. J Clin Endocrinol Metab 1990;70(3):738–41.
47. Vaccarino FJ, Kennedy SH, Ralevski E, Black R. The effects of growth hormone-releasing factor on food consumption in anorexia nervosa patients and normals. Biol Psychiatry 1994;35:446–51.
48. Kastin AA, Banks WA, Zadina JE. A decade of changing perceptions about neuropeptides. Ann NY Acad Sci 1990;579:1–7.
49. Woods SC, Decke E, Vasselli JR. Metabolic hormones and regulation of body weight. Psychol Rev 1974;81:26–43.
50. Even P, Danguir J, Nicolaidis S, Rougeiot C, Dray F. Pulsatile secretion of growth hormone and insulin in relation to feeding in rats. Am J Physiol 1987;253:R772–8.

21

Growth Hormone (GH) Releasing Hormone– and Thyrotropin Releasing Hormone–Induced GH Release in the Acute Phase of Trauma

Antonio Mancini, Domenico Valle, Gianluigi Conte, Michele Perrelli, Edoardo Menini, Vittorio Mignani, Paolo Carducci, Francesco Della Corte, and Laura De Marinis

A great number of metabolic and hormonal derangements have been described in traumatized patients. Plasma concentrations of many substrates, enzymes, and metabolites rise and/or fall, and indices of organ and tissue function change without unequivocal overall patterns. Endocrine variations can mediate a lot of these metabolic pathway dysfunctions; in general two different secretory patterns can be seen in anterior pituitary hormones: an increase of adrenocorticotropic hormone (ACTH), prolactin (PRL), and growth hormone (GH) and a decrease of other hormones [thyroid stimulating hormone (TSH), luteinizing hormone (LH), follicle stimulating hormone (FSH)]. Neurohypophysial hormones can be involved depending on the kind of trauma. Catecholamines, insulin, and glucagon levels also change, and are of fundamental importance in the regulation of intermediate metabolism in such patients.

However, despite the mass of studies on single parameters, little is reported in the literature of trauma research on a possible unifying concept that can account for a global interpretation of such multiple data, even if a central control mechanism has been suggested, for example, as part of the "fight or flight" response (1). A great difficulty resides in the fact that studies are performed in small heterogeneous groups of patients at different stages in their response. Another question concerns, in fact, the possibility of a phase-dependent specificity of endocrine dysregulation.

Moreover the study of the different endocrine responses has great importance not only from a pathophysiologic standpoint but also for the possible prognostic implications of hormonal evaluation. Finally, individual endocrine disturbances that can induce a cascade of unfavorable metabolic events have obvious therapeutic implications.

This chapter briefly summarizes experimental observations of metabolic and endocrine changes in patients who suffered a severe injury, focusing on pituitary hormones, and presents data on the GH dynamics in such patients. These is a growing interest in the metabolic action of this hormone and in the introduction of trauma among the nonconventional indications for a therapeutic use of growth hormone in adult humans (2–6).

Metabolic Alterations in Traumatized Patients

The scientific study of the metabolic responses to trauma is often considered to have begun with Cuthbertson's (7) observation of the marked rise in urinary nitrogen excretion after long-bone fractures. The response to injury may begin before the injury itself, with awareness of approaching danger activating the hypothalamic defense area. This "stress" response is reinforced by afferent stimuli arising when the injury has occurred: nociceptive afferents from damaged areas, baro- and volume-receptor inputs responding to hypovolemia and hypotension, and stimuli from osmo-, gluco-, and hormone receptors responding to changes in the composition of blood (8).

In metabolic terms, this phase is characterized by a rapid mobilization of the glycogen and triglyceride fuel stores, although this massive mobilization does appear as an exaggeration of a response appropriate for more minor degrees of stress. In fact this early phase of the response is also characterized by a lack of sensitivity to the environment and a variation of metabolic rate that is very low in comparison to the extent expected from the degree of fuel availability (9). This phase has been termed "ebb phase" (10) and lasts typically around 12 to 24 hours, depending on the kind and severity of the trauma and the type of treatment. Glycogenolysis and lipolysis are quickly activated, especially for the sympathoadrenal discharge, reinforced by pituitary stimulation of cortisol secretion. The threshold levels necessary for producing metabolic changes have been studied in experimental work: a plasma adrenaline concentration above 0.5 nmol/L will promote lipolysis with elevation of glycerol and free fatty acid (FFA) concentrations in plasma; above 1.0 nmol/L will raise glucose and lactate concentrations through hepatic and muscle glycogenolysis, and above 2.2 nmol/L will inhibit insulin secretion. Hyperglycemia after injury is prolonged by a stimulation of hepatic gluconeogenesis that is not suppressed, as it would normally be, by the elevated plasma glucose concentrations. However, it also reflects an inhibition of the peripheral glucose utilization; glucose oxidation, in particular, is inhibited. It seems that the body is now trying to conserve the fuels mobilized in the initial stressful situation.

The ebb phase merges in a more prolonged period characterized by an increase in metabolic rate and breakdown in body tissue, the catabolic or "flow phase" (about 7 to 10 days after injury). There are two main metabolic adjustments: increased metabolic rate, and increase of markers suggestive of muscle protein breakdown—increased urinary excretion of nitrogen, 3-methylhistidine, zinc, creatine, and creatinine (11). The hypermetabolism is initially due primarily to an increased rate of fat oxidation (12).

Finally, the "anabolic" or convalescent phase follows in the next 2 to 4 weeks, and many factors can influence this predictable course of the disease.

Endocrine Alterations in Traumatized Patients

As discussed above, the mobilization of energy substrates is mainly mediated by an activation of sympathetic nervous system and adrenomedullary responses. The literature reports many studies on this topic (13). Space does not allow us to discuss the important studies on insulin and glucagon secretion (reviewed in refs. 8 and 13) or the recently acquired knowledge on insulin resistance in this condition, which can account for different metabolic characteristics of trauma (14).

We focus our attention on the pituitary response and the correlations between pituitary and target peripheral glands. The best described of these responses is the increase in the adrenocortical activity, induced by the release of ACTH from the anterior pituitary. The relationship of this response with the severity of trauma is complex. Although plasma cortisol increases acutely after injuries of minor and moderate severity, its levels are lower after severe injuries (15). The increase in ACTH and cortisol subsides quickly after injury, so that in most cases they return to normal by 3 to 5 days. Among the different ACTH secretagogues [norepinephrine, 5-hydroxytryptamine (5-HT), oxytocin, vasopressin, interleukin-1 (IL-1)], the main physiologic mediator appears to be corticotropin releasing factor (CRF), which also represents the "final common pathway" for many other ACTH secretagogues (16). CRF has been regarded as a possible mediator involved in the coordination of a whole-body response to trauma (13). In fact it has an appropriate neuroanatomical distribution to be a neurotransmitter involved in the regulation of endocrine and CNS response to tissue damage; it is released in response to classical stimuli of injury (nociceptive stimulation and hypovolemia) and mediates many of the responses to various cytokines (IL-1 and IL-6), whose role has been recently underlined (17–19). Cardiovascular effects of CRF have also been described (20).

Despite cortisol increase, it has poor correlation with a composite of the duration of coma and the type of neurologic defect (21), even if this is not uniformly accepted (22). Moreover, the diurnal rhythm in ACTH and cortisol release is abolished by severe injury (23); plasma cortisol response to hypoglycemia has been reported to be abnormal (24). Finally, abnormalities in cortisol secretory dynamics may persist for many months (25).

Circulating GH and PRL increase after trauma (26–28), although one study indicates low GH levels during the catabolic phase of severe injury (29). The iv administration of glucose results in a "paradoxical" increase in GH, greater in patients with worse neurologic dysfunction. Despite this increase, no corresponding increase in somatomedin activity has been found (23). In thermally injured patients an absent GH response to insulin-induced hypoglycemia (30, 31) or to arginine (31) has been reported. The data on the correlation between injury severity and GH and PRL response to trauma are still conflicting (28, 32–34).

Numerous studies have investigated the pituitary-thyroid axis, showing the well-known picture of the so-called euthyroid sick syndrome or low triiodothyronine (T_3) syndrome. The TSH response to thyroid releasing hormone (TRH), however, seems to be quite variable (being blunted, normal, or exaggerated) (35, 36). The changes in circulating hormone concentration are rapid. Declines in thyroxine (T_4) and T_3 values are apparent within 6 to 12 hours and become maximal within 4 days; reverse T_3 peaks at 12 hours after injury and returns to normal by 2 weeks. These changes appear to be functional, but influenced by many factors, including the administration of drugs (dopamine, glucocorticoids, β-adrenergic blockade) (22).

Finally, gonadotropin levels decrease, even if the response to gonadotropin releasing hormone (GnRH) can be variable (37). Testosterone levels decrease and this change persists for the duration of coma (35, 36) and in some reports for several months after trauma (37). The magnitude of decline in testosterone concentrations correlates with the severity of the injury and with changes in catecholamine and adenosine $3',5'$-cyclic monophosphate (cAMP) concentrations. A weak but significant inverse correlation has been reported between Glasgow Coma Score (GCS) and testosterone concentration at the time of admission (37).

So, it is clear in these complex circumstances that reciprocal influences can be exerted between the central and peripheral nervous system and different hormone axis controlled by the pituitary, inducing a complex net effect on metabolic pathways. In some cases they can be useful, but in other cases dangerous for the organism, providing an opening for possible therapeutic intervention.

Materials and Methods

We have studied a group of 16 subjects (14 males and 2 females), aged 18 to 57 years, admitted into the general intensive care unit of our university hospital immediately after the trauma event (severe head injury due to traffic accidents in all patients). All patients had a GCS less than 8 (average GCS = 5.2); in 11 patients (69%) multiple trauma lesions were present (bone fractures in six patients, abdominal trauma in two patients, thoracic trauma in two patients, spinal cord injury in two patients). Patients with a history of endocrine disorders, or under steroid treatment, or receiving thyroid hormones or barbiturates were excluded from our study.

Outcome at 6 months showed good recovery in five (31%), minor disability in five (31%), severe disability in three (19%) and death in three (19%).

We have performed the following tests:

1. A TRH test (Relefact Hoechst, 200 µg iv, with blood sample collection at 0, 15, 30, and 60 min) for the determination of the GH response on the 1st and 16th day after the trauma.
2. A GHRH test (Geref, Serono, 50 µg iv, with blood sample collection at 0, 15, 30, and 60 min) on the 2nd, 7th, and 15th day after the trauma.
3. Daily blood samples were collected at 7 A.M. for the determination of basal levels of GH, PRL, TSH, T_3, and T_4 during the first week and on the 15th and 16th day after the trauma.

Blood samples were centrifuged within 2 hours after collection and separate plasma aliquots were stored until assayed at $-20°C$. Plasma GH and PRL were measured by radioimmunoassay (RIA) methods; plasma TSH by immunoradiometric assay (IRMA) methods, using kits by Radim (Pomezia, Italy). Intra- and interassay coefficients of variations were respectively, 2.5% and 5.8% for GH, 2.1% and 3.1% for PRL, and 1.6% and 3.8% for TSH. T_3 and T_4 were measured by an immunoenzymatic method (IMX) based on the fluorescence polarization immunoassay (FPIA) technology, with reagents furnished by Abbott Divisione Diagnostici (Rome, Italy). Intra- and interassay coefficients of variations were <5%.

Normal basal hormone ranges in our laboratory are GH, 0–5 ng/ml; PRL, 5–15 ng/ml; TSH, 0.5–3.5 µU/ml; T_3, 0.8–2.0 ng/ml; T_4, 45–120 ng/ml.

Statistical analysis was performed employing Student's t-test for paired data, assuming $p < .05$ as significant value.

Results

The GH response to TRH (basal and peak levels in individual patients) is shown in Figure 21.1. A "paradoxical" response of GH to TRH was observed on the day following the acute trauma (Fig. 21.1, left). The same test, performed subsequently, showed the disappearance of this response on the 16th day (Fig. 21.1, right) in the three patients who were still present in the intensive care unit. The response of GH to GHRH (mean ± S.E.M. basal and peak levels) are reported in Figure 21.2. The response to GHRH on the 7th and 15th day were significantly greater than in the very acute phase ($p < .01$ comparing 7th and 15th vs. 2nd day).

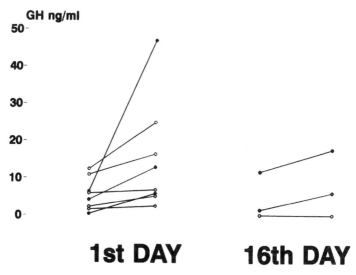

FIGURE 21.1. Individual GH response to TRH administration in traumatized patients (1st and 16th day after head injury). Basal and peak values are shown.

Table 21.1 shows the mean basal levels of the three pituitary hormones studied: basal GH and TSH showed a trend toward a simultaneous increase on the fourth day; PRL levels did not show significant variations. Moreover, T_3 and T_4 daily concentrations are reported in the same table.

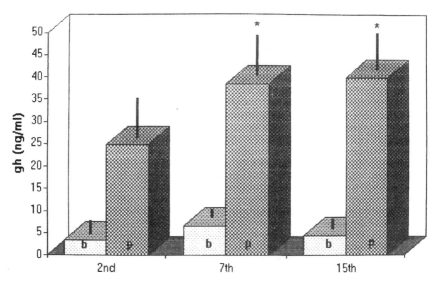

FIGURE 21.2. Mean ± S.E.M. GH response to GHRH in traumatized patients (2nd, 7th, and 15th day after injury). Basal and peak values are shown.

TABLE 21.1. Mean ± SEM daily basal levels of GH, TSH, PRL, T_3, and T_4 in our patients after head injury.

Day	GH (ng/ml)	TSH (ng/ml)	PRL (ng/ml)	T_3 (ng/ml)	T_4 (ng/ml)
1	5.7 ± 1.6	0.33 ± 0.11	5.2 ± 1.7	0.86 ± 0.07	59.2 ± 2.5
2	5.5 ± 1.4	0.27 ± 0.05	5.2 ± 1.1	0.76 ± 0.09	61.0 ± 5.5
3	9.7 ± 6.8	0.29 ± 0.09	1.0 ± 1.1	0.86 ± 0.15	59.5 ± 5.8
4	14.1 ± 6.8	0.98 ± 0.49	3.8 ± 1.3	0.70 ± 0.10	54.3 ± 5.6
5	6.4 ± 1.6	0.82 ± 0.20	2.0 ± 0.9	0.72 ± 0.13	67.4 ± 5.9
6	6.5 ± 1.3	0.61 ± 0.20	2.0 ± 1.0	0.69 ± 0.15	55.4 ± 6.6
7	1.7 ± 0.2	0.70 ± 0.20	2.5 ± 2.0	0.70 ± 0.15	51.7 ± 3.2
. . .	. . .	. . .	. . .	. . .	. . .
15	5.9 ± 2.8	0.69 ± 0.42	. . .	1.13 ± 0.13	72.4 ± 11.1
16	6.5 ± 3.1	1.15 ± 0.73	. . .	1.15 ± 0.05	76.3 ± 0.7

Discussion and Perspectives

The acute stress of a severe head injury is therefore another condition in which a paradoxical response of GH to TRH can be observed. This unexpected event in response to nonspecific stimuli was originally described in acromegalic patients (38, 39); it had been originally interpreted as an expression of receptor alteration directly related to the presence of tumor tissue (40). A GH response to TRH has been described in other nonneoplastic conditions, such as renal failure (41), anorexia nervosa (42), primary hypothyroidism (43), insulin-dependent diabetes mellitus (44), and aging (45), as well as in normal subjects pretreated with GHRH (46, 47), offering other possible interpretations for the test.

In our patients, GH response to TRH stimulation could be the first sign of an activation of the hypothalamus-pituitary axis after the trauma, pointing toward a "central" alteration in the control of GH secretion.

Other "paradoxical," GH responses have been described in traumatized subjects, for example, an increase after glucose load was reported in a group of patients with severe head injuries, while the GH response did not differ from normal subjects in patients with less-severe injuries (26). This abnormal GH elevation after glucose loading persisted for up to 2 months after injury. On the other hand, an absent GH release after insulin-induced hypoglycemia (30, 31) and arginine (31) in thermally injured patients has been reported.

While these kinds of alteration in GH dynamics could be related to an abnormal sensitivity to metabolic fuels, the paradoxical GH increase after TRH could generally indicate a blockade of tonical inhibition on GH, possibly related to an alteration of somatostatin. According to this line of thinking, we can consider that the elevation of basal GH in most studies (26, 48) and the observed elevation in TSH in our and other reports (49), was due to the known inhibition on TSH exerted by somatostatin itself.

This disinhibition appears to precede the progressive increase in GH reserve documented by the augmented GH peak in the days following the trauma. The advantage of this GH increase is obvious, since it could contribute to the decrease of the hypercatabolic response in the posttrauma period. The GH increase has also been considered to exert a delayed action on adipose tissue and to contribute to the potentiation of lipolysis (8).

However, it has also been reported in a group of patients with severe injuries, in a highly catabolic stage of multiple trauma, that plasma GH levels were decreased compared with unstressed normal subjects. Even this response has been interpreted as a defensive mechanism to facilitate the maintenance of the hypercatabolic state needed to provide amino acids for hepatic gluconeogenesis, since hyperglycemia seems to be the driving force (29).

This apparent contrast underlines the need to correlate the hormonal response with the severity of the trauma and, what is more important, with the stage of the metabolic course in which the different patients are studied.

Finally, the GH reregulation should be of fundamental importance in the anabolic phase, to prevent an excessive nitrogen loss and to promote reparation mechanisms. This opens the recently discovered field of exogenous biosynthetic GH administration in critically ill patients; different metabolic actions are exerted by GH, including significant nitrogen and phosphorus retention, increase in protein synthesis, increased levels of insulin-like growth factor-I (with further decrease in protein breakdown), weight gain, and positive mineral and water balance (2–5). The effects have potential benefits to preserve muscle function, the reduction in the period of mechanical ventilation and intensive hospitalization, and so on. Finally, different studies underline the relationship between GH and immunity (3), stressing the important link between endocrine and the immune system in the recovery phase of injured patients.

References

1. Hilton SM, Redfern WS. A search for brainstem cell groups integrating the defence reaction in the rat. J Physiol 1986;378:213–28.
2. Voerman BJ, de Boer H, Thijs G. Recombinant human growth hormone in critically ill patients. Curr Op Anaesth 1994;7:161–5.
3. Voerman HJ, Thijs LG. Administration of human growth hormone in critically ill patients. In: Vincent JL, ed. Yearbook of intensive care and emergency medicine. Berlin: Springer, 1993:160–8.
4. Ziegler TR, Young LS, Manson JM, Wilmore DW. Metabolic effects of recombinant human growth hormone in patients receiving parenteral nutrition. Ann Surg 1988;208:6–16.
5. Pichard C, Jolliet P. Biosynthetic growth hormone: impact on nitrogen metabolism and muscle function in stressed patients. In: Vincent JL, ed. Yearbook of intensive care and emergency medicine. Berlin: Springer, 1993:151–9.

6. Soroff HS, Rozin RR, Mooty J, Lister J, Raben MS. Role of human growth hormone in the response to trauma: I. Metabolic effects following burns. Ann Surg 1967;166:739–52.
7. Cuthbertson DP. The disturbance of metabolism produced by bony and non-bony injury, with notes on certain abnormal conditions of bone. Biochem J 1930;24:1244–63.
8. Frayn KN. Hormonal control of metabolism in trauma and sepsis. Clin Endocrinol 1986;24:577–99.
9. Little RA, Frayn KN, Randall PE, Stoner HB, Maycock PF. Plasma catecholamine concentrations in acute states of stress and trauma. Arch Emerg Med 1985;2:46–7.
10. Cuthbertson DP. Postshock metabolic response. Lancet 1942;1:433–7.
11. Threlfall CJ, Stoner HB, Galasko CSB. Patterns in the excretion of muscle markers after trauma and orthopaedic surgery. J Trauma 1981;21:140–7.
12. Frayn KN, Little RA, Stoner HB, Galasko CSB. Metabolic control in non-septic patients with musculoskeletal injuries. Injury 1984;16:73–9.
13. Turnbull AV, Little RA. Neuro-hormonal regulation after trauma. In: Vincent JL, ed. Yearbook of intensive care and emergency medicine. Berlin: Springer, 1992:574–81.
14. Deibert DC, De Fronzo RA. Epinephrine-induced insulin resistance in man. J Clin Invest 1980;65:717–21.
15. Barton RN, Stoner HB, Watson SM. Relationships among plasma cortisol, adrenocorticotrophin, and severity of injury in recently injured patients. J Trauma 1987;27:384–92.
16. Plotsky PM. Pathways to the secretion of adrenocorticotropin: a view from the portal. J Neuroendocrinol 1991;3:1–9.
17. Rothwell NJ. Neuroendocrine mechanisms in the thermogenic responses to diet, infection and trauma. In: Porter JC, Yezova, eds. Circulating regulatory factors and neuroendocrine function. New York: Plenum Press, 1990:371–80.
18. Rothwell NJ. Central effects of CRF on metabolism and energy balance. Neurosci Biobehav Rev 1990;14:263–71.
19. Rothwell NJ. Functions and mechanisms of interleukin-1 in the brain. TIPS 1991;12:430–6.
20. Fisher LA. Corticotropin-releasing factor: endocrine and autonomic integration of responses to stress. TIPS 1989;10:189–93.
21. King LR, McLaurin RL, Lewis HP, Knowles HC Jr. Plasma cortisol levels after head injury. Ann Surg 1970;172:975–84.
22. Woolf PD. Hormonal responses to trauma. Crit Care Med 1992;20:216–26.
23. Steinbock P, Thompson G. Serum cortisol abnormalities after craniocerebral trauma. Neurosurgery 1979;5:559–65.
24. Edwards OM, Clark JDA. Post-traumatic hypopituitarism: six cases and a review of the literature. Medicine 1986;65:281–90.
25. Jackson RD, Mysiw WJ. Abnormal cortisol dynamics after traumatic brain injury; lack of utility in predicting agitation or therapeutic response to tricyclic antidepressant. Am J Phys Med Rehabil 1989;68:18–23.
26. King LR, Knowles HC Jr, McLaurin RL, Brielmaier J, Perisutti G, Piziak VK. Pituitary hormone response to head injury. Neurosurgery 1981;9:229–35.
27. Frayn KN, Price DA, Maycock PF, Carroll SM. Plasma somatomedin activity after injury in man and its relationship to other hormonal and metabolic changes. Clin Endocrinol 1984;20:179–87.

28. Chiolero R, Lemarchand TH, Schutz Y, De Tribolet N, Felder JP, Freeman J, Jequier E. Plasma pituitary hormone levels in severe trauma with or without head injury. J Trauma 1988;28:1368–74.

29. Jeevanandam M, Ramias L, Shamos RF, Schiller WR. Decreased growth hormone levels in the catabolic phase of severe injury. Surgery 1992;111:495–501.

30. Dolecek R, Adamkova M, Sotornikova T, Zavada M, Kracmar P. Endocrine response after burn. Scand J Plast Reconstr Surg 1979;13:9–16.

31. Wilmore DW, Orcutt TW, Mason AD Jr, Pruitt EA Jr. Alterations in hypothalamic function following thermal injury. J Trauma 1975;15:697–703.

32. Matsuura H, Nakazawa S, Wakabayashi I. Thyrotropin-releasing hormone provocative release of prolactin and thyrotropin in acute head injury. Neurosurgery 1985;16:791–5.

33. Brizio-Molteni L, Molteni A, Warpeha RL, Angelatas J, Lewis N, Fors EM. Prolactin, corticotropin and gonadotropin concentrations following thermal injury in adults. J Trauma 1984;24:1–7.

34. Jennett B, Bond M. Assessment of outcome after severe brain damage: a practical scale. Lancet 1975;1:480–4.

35. Fleischer AS, Rudman DR, Payne NS, Tindall GT. Hypothalamic hypothyroidism and hypogonadism in prolonged traumatic coma. J Neurosurg 1978;49:650–7.

36. Rudman D, Fleisher AS, Kutner MH, Raggio JF. Suprahypophyseal hypogonadism and hypothyroidism during prolonged coma after head trauma. J Clin Endocrinol Metab 1977;45:747–54.

37. Clark JDA, Raggatt PR, Edwards OM. Hypothalamic hypogonadism following major head injury. Clin Endocrinol 1988;29:153–65.

38. Irie M, Tsushima T. Increase of serum growth hormone concentration following thyrotropin releasing hormone in patients with acromegaly or gigantism. J Clin Endocrinol Metab 1972;35:97–100.

39. Schalch DS, Gonzales-Barcena D, Kastin AJ, Schally AV, Lee LA. Abnormalities in the release of TSH in response to thyrotropin-releasing hormone (TRH) in patients with disorders of the pituitary, hypothalamus and basal ganglia. J Clin Endocrinol Metab 1972;35:609–13.

40. Le Dafniet M, Garnier P, Bression D, Brandia AM, Raadot J, Peillon F. Correlative studies between the presence of thyrotropin-releasing hormone receptors and the in vitro stimulation of growth hormone secretion in human GH-secreting adenomas. Horm Metab Res 1985;17:476–9.

41. Gonzales-Barcena D, Kastin AJ, Schalch DS, Torres-Zamora M, Perez-Pasten E, Kato A, Schally AV. Responses to thyrotropin-releasing hormone in patients with renal failure and after infusion in normal men. J Clin Endocrinol Metab 1973;26:117–20.

42. Maeda K, Kato Y, Yamaguchi N, Chihara K, Ohgo S, Iwasaki Y, et al. Growth hormone release following thyrotropin-releasing hormone injection into patients with anorexia nervosa. Acta Endocrinol 1976;81:1–8.

43. Collu R, Leboeuf G, Letarte J, Ducharme JR. Increase in plasma growth hormone levels following thyrotropin-releasing hormone injection in children with primary hypothyroidism. J Clin Endocrinol Metab 1977;44:743–7.

44. Dasmahapatra A, Urdanivia E, Cohen MP. Growth hormone response to thyrotropin-releasing hormone in diabetes. J Clin Endocrinol Metab 1981;52:859–62.

45. Barreca T, Franeschini R, Messina V, Bottaro L, Rolandi E. Stimulation of GH release by TRH in elderly subjects. Horm Res 1985;21:214–9.
46. Borges JL, Uskavitch DR, Kaiser DL, Cronin MJ, Evans WS, Thorner MO. Human pancreatic growth hormone-releasing factor-40 (hpGRF-40) allows stimulation of GH release by TRH. Endocrinology 1983;113:1519–21.
47. Sartorio A, Spada A, Bochicchio D, Atterrato A, Morabito F, Faglia G. Effect of TRH on GH release in normal subjects pretreated with GHRH 1-44 pulsatile administration. Neuroendocrinology 1986;44:470–4.
48. Hackl JM, Gottardis M, Wieser CH, Rumpl E, Stadler CH, Schwarz S, Monkayo R. Endocrine abnormalities in severe traumatic brain injury—a cue to prognosis in severe craniocerebral trauma? Intensive Care Med 1991;17:25–9.
49. Bacci V, Schussler GC, Kaplan TB. The relationship between serum triiodothyronine and thyrotropin during systemic illness. J Clin Endocrinol Metab 1982;54:1229–35.

22

Effects of Hexarelin on Growth Hormone Secretion in Short Normal Children, in Obese Children, and in Subjects with Growth Hormone Deficiency

Sandro Loche, Paola Cambiaso, Maria Rosaria Casini,
Bruno P. Imbimbo, Daniela Carta, Patrizia Borrelli,
and Marco Cappa

A series of small peptides [growth hormone releasing peptides (GHRPs)], analogues of enkephalin, have been recently synthesized that selectively stimulate growth hormone (GH) secretion (1). These peptides have potent GH releasing activity in all species tested so far, and are effective also after oral administration (2–4). One of these peptides, GHRP-6, has been extensively studied in vitro and in vivo. In vitro, GHRP-6 stimulates GH secretion from pituitary cells by a mechanism not mediated by either growth hormone releasing hormone (GHRH) or opioid receptors (5–9), and via signaling mechanisms distinct from those of GHRH (5, 10). In vivo, GHRP-6 stimulates GH secretion in animals (11–13) and in humans (2, 3, 14–18). Interestingly, the GH releasing activity of GHRP-6 in vivo is more potent than that observed in in vitro experiments, indicating that the peptide may also have a hypothalamic site of action, a view supported by the observation that GHRP-6 activity is enhanced when the experiments are carried out on hypothalamic-pituitary incubates (12), and by the evidence of specific hypothalamic binding sites for the peptide (7, 8). Furthermore, GHRP-6 acts synergistically with GHRH to release GH both in vitro (5, 10, 12) and in vivo (3, 15).

Hexarelin (Hex) is a new synthetic hexapeptide (His-D-2-Methyl-Trp-Ala-Trp-D-Phe-Lys-NH$_2$) analogue to GHRP-6, in which D-tryptophan has been replaced by its 2-methyl derivative (20). This new peptide is more stable to enzymatic, chemical, and oxidative degradation and is also more hydrophobic than the parent compound. These properties should theoretically increase the bioavailability of the peptide. Recently, Hex has been shown to be slightly more effective than GHRP-6 in rats (21). Recent studies have also shown that Hex is a potent GH secretagogue in both adults (4) and children (22). This chapter presents our data on the GH releasing effect of Hex in short normal children, in subjects with organic hypopituitarism, in subjects with isolated GH deficiency (GHD), and in obese children.

Short Normal Children

The study group comprised 45 children referred to our institutions for evaluation of short stature, and ultimately found to have familial short stature and/or constitutional delay of growth. Twenty-four children were prepubertal (11 boys and 13 girls, ages 5.9 to 13 years) and 21 were early pubertal (Tanner stages 2 to 3, 13 boys and 8 girls, ages 10 to 14 years). All had normal insulin-like growth factor-I (IGF-I) levels and normal thyroid function tests, and none had taken long-term medications prior to the study. All children were tested on two occasions with GHRH 1-29 (Serono, Italy), $1\mu g/kg$ iv, and with Hex (prepared and supplied by Europeptides, Argenteuil, France), $2\mu g/kg$ iv. This dosage of Hex was chosen because it has been shown to be the calculated maximal dose in men (23). Blood samples were drawn from an indwelling catheter inserted in an antecubital vein 15min before and immediately before the injection of GHRH or Hex and then every 15min for 2 hours. All experiments started between 8 and 9 A.M. after the children fasted overnight. In five boys with constitutional growth delay, ages 12.0 to 13.7 years, the GH response to Hex was reevaluated one week after priming with testosterone (testosterone enanthate, 100mg intramuscularly).

None of the subjects experienced adverse side effects after Hex administration. The GH response to GHRH and to Hex in the short normal children is shown in Figure 22.1. Mean peak GH and mean area under the curve (AUC) after GHRH were $26.6 \pm 7.1\mu g/L$ (AUC = $1,387 \pm 317\mu g.min/L$) and $23.2 \pm 5.5\mu g/L$ (AUC = $1,213 \pm 217\mu g.min.L$) in the prepubertal male and female subjects, respectively. In the pubertal subjects mean peak GH and mean AUC after GHRH were $19.0 \pm 3.9\mu g/L$ (AUC = $975 \pm 170\mu g.min.L$) in boys and $21.1 \pm 5.2\mu g/L$ (AUC = $1,164 \pm 281\mu g.min.L$) in girls. In all children Hex caused a prompt a clear-cut increase of serum GH concentrations with peaks occurring between 15 and 30min from injection and ranging between 15 and $121\mu g/L$. The GH response to Hex was significantly higher than that observed after GHRH in all groups of children evaluated both as maximum GH peak (prepubertal boys = $55.9 \pm 7.2\mu g/L$, $p < .001$ vs. GHRH; prepubertal girls = $40.9 \pm 6.1\mu g/L$, $p < .005$ vs. GHRH; pubertal boys = $59.7 \pm 6.9\mu g/L$, $p < .005$ vs. GHRH; pubertal girls = $53.9 \pm 11.5\mu g/L$, $p < .01$ vs. GHRH) or as AUC (prepubertal boys = $2,934 \pm 425\mu g.min.L$, $p < .005$ vs. GHRH; prepuber-

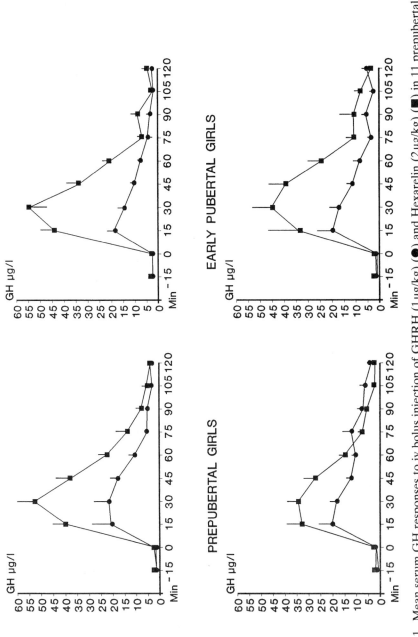

FIGURE 22.1. Mean serum GH responses to iv bolus injection of GHRH (1 μg/kg) (●) and Hexarelin (2 μg/kg) (■) in 11 prepubertal and 13 pubertal short normal boys and in 13 prepubertal and 8 pubertal short normal girls.

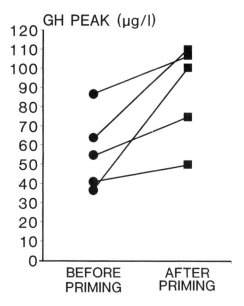

FIGURE 22.2. Peak GH responses to iv bolus injection of hexarelin (2 μg/kg) in five males with constitutional growth delay before and after priming with testosterone (testosterone enanthate, 100 mg im one week before the test).

tal girls = 1,916 ± 326 μg.min.L, $p < .02$ vs. GHRH; pubertal boys = 2,877 ± 415 μg.min.L, $p < .01$ vs. GHRH; pubertal girls = 2,757 ± 509 μg.min.L, $p < .005$ vs. GHRH). The GH responses to GHRH and to Hex were similar between boys and girls, either prepubertal or pubertal, as well as between prepubertal and pubertal subjects (Fig. 22.1). After priming with testosterone the GH response to Hex was increased in all five subjects studied (Fig.

TABLE 22.1. Mean baseline and peak cortisol and prolactin concentrations following intravenous administration of hexarelin (2 μg/kg) in 45 short normal, in 10 obese, and in 6 GHD children.

	Cortisol (nmol/L)		Prolactin (μg/L)	
	Baseline	Peak	Baseline	Peak
Short normal children				
Mean	439	646[a]	8.0	16.9[a]
SEM	30	27	0.9	1.8
Obese children				
Mean	420	617[a]	12.3	24.2[c]
SEM	22	23	3.0	5.7
GHD children				
Mean	274	439[b]	5.8	15.7[b]
SEM	102	95	1.2	2.8

[a] $p < .001$, [b] $p < .02$, [c] $p < .05$ vs. baseline.

22.2). Mean GH peak and AUC responses to Hex before priming were 57.4 ± 8.7 μg/L and 2,448 ± 379 μg.min.L, respectively, and rose to 89.2.1 ± 11.3 μg/L ($p < .05$) and to 4,143 ± 590 μg.min.L ($p < .05$) one week after testosterone administration.

In all children Hex administration caused a significant increase from baseline of both cortisol (from 439 ± 30 to 646 ± 27 nmol/L, $p < .001$) and prolactin (PRL) (from 8.0 ± 0.9 to 16.9 ± 1.8 μg/L, $p < .001$) concentrations, which returned to the baseline values within 2 hours. As no differences were found between boys and girls as well as between prepubertal and pubertal subjects, results of cortisol and PRL measurements after Hex administration have been pooled together and are shown in Table 22.1.

Obese Children

Ten obese children (seven boys and three girls, ages 7.5 to 12.0 years, excess body weight 47% to 86.2% above their ideal body weight derived from Tanner standards, body mass index 23.0 to 30.5, all prepubertal and of normal stature) were studied according to the same protocol described above. In the obese children the GH response to GHRH and to Hex were significantly lower than in the prepubertal children (GHRH: peak = 5.8 ± 0.8 μg/L, $p < .02$, AUC = 402 ± 51 μg.min.L, $p < .01$; Hex: peak = 19.7 ± 4.4 μg/L, $p < .001$, AUC = 1,043 ± 239 μg.min.L, $p < .001$) (Fig. 22.3). The GH response to Hex was significantly higher than that observed after GHRH evaluated both as maximum GH peak ($p < .01$) and as AUC ($p < .02$), and was similar to the response observed in the short normal children after GHRH. Also in the obese children Hex administration caused a significant increase over baseline of both cortisol and PRL concentrations.

Growth Hormone Deficiency

Five subjects (four males and one female, ages 8.4 to 21 years) had organic hypopituitarism as a result of surgical operation for craniopharyngioma. At the time of the study they were on replacement therapy with hydrocortisone, L-thyroxine, and deamino-8-D-arginine vasopressin (DDAVP). Six subjects (four boys and two girls, ages 6.0 to 15.8 years) had isolated GHD, which was idiopathic in three and associated with anatomical abnormalities in the others. In particular, one patient had an arachnoid cyst, one had empty sella syndrome, and one had pituitary stalk interruption syndrome. The latter was diagnosed on magnetic resonance imaging on the basis of lack of a visible pituitary stalk and ectopic posterior pituitary lobe. The diagnosis of GHD was made by the classic criteria, i.e., short stature with subnormal growth rates and delayed bone age, plasma GH levels <10 μg/L after two pharmacologic tests (clonidine, insulin hypoglycemia or L-dopa),

25 352   S. Loche et al.

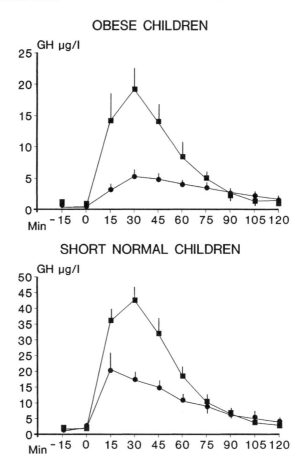

FIGURE 22.3. Mean serum GH responses to iv bolus injection of GHRH (1 μg/kg) (●) and Hexarelin (2 μg/kg) (■) in 10 prepubertal obese children. For comparison, the mean GH responses of 24 short prepubertal boys and girls is shown in the lower panel. Please note the different scale.

and subnormal insulin-like growth factor-I (IGF-I) concentrations. At the time of the study all GHD patients were on replacement therapy with rGH, which was discontinued 2 to 3 weeks before the experiments.

No GH increase after either GHRH or Hex administration was observed in the hypopituitary subjects (data not shown). Individual peak GH responses to Hex and to GHRH in the patients with isolated GHD is shown in Figure 22.4. In two of the three patients with idiopathic GHD (cases 1 and 3 of Fig. 22.4) Hex elicited a sizable GH increase that was of greater magnitude than that observed after GHRH, while the remaining patient had a low GH response (<10 μg/L) to GHRH and a slight response to Hex. In the three patients with GHD associated with anatomical abnormalities (cases 4 to 6 of Fig. 22.4) the GH response to Hex was of lower magnitude than that observed after GHRH.

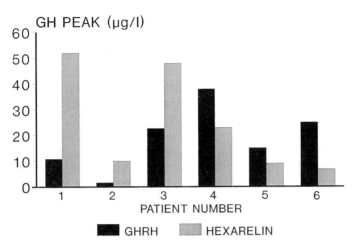

FIGURE 22.4. Individual peak GH responses to iv bolus injection of hexarelin (2 μg/ Kg) and GHRH (1 μg/Kg) in six children with isolated GHD.

As observed in the short normal children and in the obese children, also Hex administration caused a small but significant increase of cortisol and PRL concentrations in the GHD patients (Table 22.1). The increase in cortisol and PRL after Hex administration was absent in the hypopituitary subjects (data not shown).

Discussion

The data presented indicate that Hex is a potent GH-releasing stimulus in short-statured and in obese children, as previously reported in normal adult subjects (4). Hex is also an effective GH secretagogue in patients with idiopathic GHD. As already shown in adults after administration of GHRP-6 (15, 16) and Hex (4), a maximal dose of Hex elicited a greater GH response than a maximal dose of GHRH. The GH response to Hex was similar between males and females as has been previously observed in children after administration of both GHRH (24) and GHRP-6 (18), and in adults after administration of Hex (4). Similarly, we found no difference in the GH response to Hex between prepubertal and early pubertal subjects. However, priming with testosterone in five subjects with constitutional growth delay induced an augmentation of the Hex-induced GH secretion. This discrepancy might be due to the fact that the administration of exogenous testosterone causes a brisk rise of T levels to near adult values, far greater than those observed in the early stages of spontaneous puberty. Alternatively, since the increase in spontaneous GH secretion in boys occurs in late puberty (25), one would not be able to observe an increased GH response to Hex in early pubertal boys.

Hex was not more effective than GHRH in all patients with GHD, as observed in the short normal and obese children. In this regard, the patients with the lower GH responses to Hex than to GHRH had a hypothalamic-pituitary abnormality. It has been recently shown that the effects of Hex on GH synthesis (26) and release (27) are markedly reduced in rats and in sheep after hypothalamic-pituitary disconnection. Furthermore, it is known that the GH-releasing effect of GHRPs is far greater in vivo than in vitro or when the experiments are carried out on hypothalamic-pituitary incubates (11, 12). These findings suggest that the hypothalamus is the principal site of action of GHRPs. It is tempting to speculate that the hypothalamic-pituitary abnormalities present in these GHD patients cause an interruption of the normal connections between hypothalamic influences and the somatotrophs, and thus Hex would not be able to fully express its GH releasing activity. The low GH responses to Hex observed in two patients may reflect direct somatotroph stimulation by the peptide. Whether priming with Hex in these subjects would result in increased GH responses requires further investigation.

In contrast to data in experimental animals (11, 12) but in agreement with previous findings in humans (4, 14, 15, 19), Hex also caused a small but significant increase of both cortisol and PRL concentrations, which returned to the baseline values within 2 hours. As already observed after iv administration of GHRP-1 (19), GHRP-6 (14, 15), and Hex (4), these increases were within the normal limits and were small relative to the GH increases. Interestingly, no effect on PRL or cortisol secretion were observed after oral administration of GHRP-6 and Hex in either normal men or short statured children (3, 4). The absent cortisol response in our patients operated for craniopharyngioma indicates that the cortisol-releasing effect of Hex is exerted at the hypothalamic/pituitary level.

Hex markedly stimulated GH release also in obese children, who reportedly have blunted GH responses to GHRH (28, 29). Also in obese children the GH response to Hex was significantly higher than that after GHRH, although of lower magnitude than that observed in the short normal children. These results compare favorably to those of Cordido et al. (30), who found that in obese adults GHRP-6 administration elicited a marked GH response, which was even increased when GHRP-6 and GHRH were administered simultaneously. However, both the GH response to GHRP-6 and to GHRP-6 plus GHRH, although of considerable magnitude, were lower than those observed in normal-weight subjects (30). It has been previously shown that in obese children the GH response to GHRH can be normalized by pretreatment with drugs capable of inhibiting endogenous somatotropin release inhibiting hormone (SRIH) release such as pyridostigmine (31), galanin (32), or atenolol (33), implying that an increased endogenous SRIH tone may be at least partially responsible for the reduced GH secretion. A chronic somatotroph inhibition by excess SRIH would explain why, although normalized, the GH response to GHRH after pretreatment with the aforementioned drugs still remains lower than that observed in normal-weight children given the same treatment. In agreement with the findings of Cordido et al. (30), these findings reinforce the hypothesis that the reduced GH secretion typical of obesity is a functional, potentially reversible state. Whether Hex administration would eventually restore a normal spontaneous GH secretion in obese children warrants further investigation.

GH hyposecretory states may result from a primitive abnormality of the somatotrophs or, more frequently, from reduced GHRH synthesis and/or release (34). Based on this assumption several trials have been carried out to investigate the therapeutic effectiveness of long-term GHRH administration (see ref. 35 for review). The best results were obtained when the peptide was administered in a pulsatile manner (35), thus demonstrating that normal growth in GHD patients can be obtained by stimulating their endogenous GH secretion. However, long-term pulsatile treatment with GHRH is obviously impractical, and, therefore, the availability of long-acting or orally active GH secretagogues would represent a great advance in the treatment of patients with GHD. Although the long-term effectiveness and safety of such a therapeutic approach with GHRPs need to be established, several data indicate that this could indeed be feasible. First, GHRPs are effective GH secretagogues after oral administration in normal volunteers and short-statured children (2–4), and in most patients with GHD acute intravenous administration of Hex does stimulate GH secretion. Second, both GHRP-6 and Hex have been shown to be potent stimulators of GH gene expression in the GHRH-deprived infant rat (26, 36). Third, prolonged GHRP-6 administration in men results in augmentation of pulsatile GH release as well as increased production of IGF-I (37, 38).

Conclusion

We have shown that Hex stimulates GH secretion in short normal and obese children as well as in children with GHD. The ability of GHRPs to increase GH synthesis and release and to augment spontaneous GH secretion coupled to their effectiveness after oral administration make this family of peptides potentially useful in the diagnosis and treatment of growth disorders.

Acknowledgments. The authors wish to thank Prof. R. Deghenghi for his helpful suggestions, and Mrs. Valentina Bianco and Mrs. Gabriella De Barba for their expert assistance.

References

1. Bowers CY. GH releasing peptides—structure and kinetics. J Pediatr Endocrinol 1993;6:21–31.

2. Hartman ML, Farello G, Pezzoli S, Thorner MO. Oral administration of growth hormone (GH)-releasing peptide stimulates GH secretion in normal men. J Clin Endocrinol Metab 1992;74:1378–84.

3. Bowers CY, Alster DK, Frenz JM. The growth hormone releasing activity of a synthetic hexapeptide in normal men and short-statured children after oral administration. J Clin Endocrinol Metab 1992;74:292–8.

4. Ghigo E, Arvat E, Gianotti L, et al. GH-releasing activity of hexarelin, a new synthetic hexapeptide, after intravenous, subcutaneous, intranasal and oral administration in man. J Clin Endocrinol Metab 1994;78:693–8.

5. Cheng K, Chan WWS, Barreto A, Convey EM, Smith RG. The synergistic effect of His-D-Trp-Ala-Trp-D-Phe-Lys-NH$_2$ on growth hormone (GH)-releasing factor-stimulated GH release and intracellular adenosine 3′,5′-monophosphate accumulation in rat primary pituitary cell culture. Endocrinology 1989;124:2791–8.

6. Goth MI, Lyons CE, Canny BJ, Thorner MO. Pituitary adenylate cyclase activating polypeptide, GH-releasing peptide and GH-releasing hormone stimulate GH release through distinct pituitary receptors. Endocrinology 1992;130:939–44.

7. Codd EE, Shu AYL, Walker RF. Binding of a growth hormone releasing hexapeptide to specific hypothalamic and pituitary binding sites. Neuropharmacology 1989;28:1139–44.

8. Sethumadhavan K, Veeraragavan K, Bowers CY. Demonstration and characterization of the specific binding of growth hormone-releasing peptide to rat anterior pituitary and hypothalamic membranes. Biochem Biophys Res Commun 1991;178:31–7.

9. Blake AD, Smith RG. Desensitization studies using perifused rat pituitary cells show that growth hormone-releasing hormone and His-D-Trp-Ala-Trp-D-Phe-Lys-NH$_2$ stimulate growth hormone release through distinct receptor sites. J Endocrinol 1991;129:11–19.

10. Cheng K, Chan WWS, Butler B, Barreto A, Smith RG. Evidence for a role of protein kinase-C in His-D-Trp-Ala-Trp-D-Phe-Lys-NH$_2$-induced growth hormone release from rat primary pituitary cells. Endocrinology 1991;129:3337–42.

11. Bowers CY, Momany FA, Reynolds GA, Hong A. On the in vitro and in vivo activity of a new synthetic hexapeptide that acts on the pituitary to specifically release growth hormone. Endocrinology 1984;114:1537–45.

12. Bowers CY, Sartor AO, Reynolds GA, Badger TM. On the actions of the growth hormone-releasing hexapeptide, GHRP. Endocrinology 1991;128:2027–35.

13. Malozowski S, Hao EH, Ren SG, et al. Growth hormone (GH) response to the hexapeptide GH-releasing peptide and GH-releasing hormone (GHRH) in the cynomolgus macaque: evidence for non-GHRH-mediated responses. J Clin Endocrinol Metab 1991;73:314–7.

14. Ilson BE, Jorkasky DK, Curnow RT, Stote RM. Effect of a new synthetic hexapeptide to selectively stimulate growth hormone release in healthy human subjects. J Clin Endocrinol Metab 1989;69:212–4.

15. Bowers CY, Reynolds GA, Durham D, Barrera CM, Pezzoli SS, Thorner MO. Growth hormone (GH)-releasing peptide stimulates GH release in normal men and acts synergistically with GH-releasing hormone. J Clin Endocrinol Metab 1990;70:975–82.

16. Robinson BM, DeMott Friberg R, Bowers C, Barkan AL. Acute growth hormone (GH) response to GH-releasing hexapeptide in humans is independent of endogenous GH-releasing hormone. J Clin Endocrinol Metab 1992;75:1121–4.
17. Penalva A, Carballo A, Pombo M, Casanueva F, Dieguez C. Effect of growth hormone (GH)-releasing hormone (GHRH), atropine, pyridostigmine, or hypoglycemia on GHRP-6-induced GH secretion in man. J Clin Endocrinol Metab 1993;76:168–71.
18. Penalva A, Pombo M, Carballo A, Barreiro J, Casanueva FF, Dieguez C. Influence of sex, age and adrenergic pathways on the growth hormone response to GHRP-6. Clin Endocrinol (Oxf) 1993;87–91.
19. Laron Z, Bowers CY, Hirsch D, Almonte AS, Pelz M, Keret R, Gil-Ad I. Growth hormone-releasing activity of growth hormone-releasing peptide-1 (a synthetic heptapeptide) in children and adolescents. Acta Endocrinol (Copenh) 1993;129:424–6.
20. Deghenghi R, Cananzi M, Battisti C, Locatelli V, Müller EE. Hexarelin (EP23905): a superactive growth hormone releasing peptide. J Endocrinol Invest 1992;15(suppl 4):45(abstr).
21. Deghenghi R, Cananzi MM, Torsello A, Battisti C, Müller EE, Locatelli D. GH-releasing activity of hexarelin, a new growth hormone releasing peptide, in infant and adult rats. Life Sci 1994;54:1321–8.
22. Loche S, Cambiaso P, Carta D, et al. The GH-releasing activity of hexarelin, a new synthetic peptide, in short normal and obese children, and in hypopituitary subjects. J Clin Endocrinol Metab 1995;80:674–8.
23. Imbimbo BP, Mant T, Edwards M, et al. Growth hormone releasing activity of hexarelin in humans: a dose response study. Eur J Clin Pharmacol 1994;46:421–5.
24. Gelato MC, Malozowski S, Caruso-Nicoletti M, et al. Growth hormone (GH) responses to GH-releasing hormone during pubertal development in normal boys and girls: comparison to idiopathic short stature and GH deficiency. J Clin Endocrinol Metab 1986;63:174–9.
25. Rose S, Municchi G, Barnes KM, et al. Spontaneous growth hormone secretion increases during puberty in normal girls and boys. J Clin Endocrinol Metab 1991;73:428–35.
26. Torsello A, Locatelli V, Ghigo MC, Grilli R, Luoni M, Deghenghi R, Müller EE. Study on the mechanism of action of hexarelin, a new growth hormone-releasing peptide. Eur J Endocrinol 1994;130(suppl 2):125(abstr).
27. Fletcher TP, Thomas GB, Willoughby JO, Clarke IJ. Constitutive growth hormone secretion in sheep after hypothalamo-pituitary disconnection and the direct in vivo pituitary effect of growth hormone releasing peptide 6. Neuroendocrinology 1994;60:76–86.
28. Pertzelan A, Keret R, Bauman B, et al. Responsiveness of pituitary hGH to GHRH 1-44 in juveniles with obesity. Acta Endocrinol (Copenh) 1986;111:151–3.
29. Loche S, Cappa M, Borrelli P, et al. Reduced growth hormone response to growth hormone-releasing hormone in children with simple obesity: evidence for somatomedin C mediated inhibition. Clin Endocrinol (Oxf) 1987;27:145–53.
30. Cordido F, Penalva A, Dieguez C, Casanueva F. Massive growth hormone (GH) discharge in obese subjects after the combined administration of GH-

releasing hormone and GHRP-6: evidence for a marked somatotroph secretory capability in obesity. J Clin Endocrinol Metab 1993;76:819–23.

31. Loche S, Pintor C, Cappa M, et al. Pyridostigmine counteracts the blunted growth hormone response to growth hormone releasing hormone of obese children. Acta Endocrinol (Copenh) 1989;120:624–8.

32. Loche S, Pintus S, Cella SG, et al. The effect of galanin on baseline and GHRH-induced growth hormone secretion in obese children. Clin Endocrinol (Oxf) 1990;33:187–92.

33. Loche S, Pintus S, Carta D, et al. The effect of atenolol on the growth hormone response to growth hormone-releasing hormone in obese children. Acta Endocrinol (Copenh) 1992;126:124–7.

34. Frohman LA, Jansson O. Growth hormone-releasing hormone. Endocr Rev 1986;7:223–53.

35. Thorner MO, Hartman ML, Gaylinn BD, Aloi JA, Kirk SE, Pezzoli SS, Vance ML. Current status of therapy with growth hormone-releasing neuropeptides. In: Savage MO, Bourguignon JP, Grossman AB, eds. Frontiers in pediatric neuroendocrinology. Cambridge: Blackwell Scientific, 1994:161–7.

36. Locatelli V, Grilli R, Torsello A, Cella SG, Wehrenberg WB, Müller EE. Growth hormone-releasing hexapeptide is a potent stimulator of GH gene expression and release in the GHRH-deprived infant rat. Pediatr Res 1994;36:169–74.

37. Huhn WC, Hartman ML, Pezzoli SS, Thorner MO. Twenty-four-hour growth hormone (GH)-releasing peptide (GHRP) infusion enhances pulsatile GH secretion and specifically attenuates the response to a subsequent GHRP bolus. J Clin Endocrinol Metab 1993;76:1202–8.

38. Jaffe CA, Ho PJ, Demott-Friberg R, Bowers CY, Barkan AL. Effects of prolonged growth hormone (GH)-releasing peptide infusion on pulsatile GH secretion in normal men. J Clin Endocrinol Metab 1993;77:1641–7.

23

Lymphocytes and Hypothalamic Peptides

Douglas A. Weigent and J. Edwin Blalock

It has previously been shown that pituitary hormones and hypothalamic releasing hormones have direct effects upon cells of the immune system (1). Because cells of the immune system also produce pituitary hormones, we and others hypothesized that cells of the immune system might also produce hypothalamic releasing factors. In fact, there are now data to support the extrahypothalamic immune synthesis of corticotropin releasing hormone (CRH), growth hormone releasing hormone (GHRH), luteinizing hormone releasing hormone (LHRH), and somatostatin (SS) (1–3). Further, functional receptors for these same hypothalamic hormones have been identified on cells of the immune system (4). The results discussed in this chapter show that hypothalamic hormones produced in the immune system are structurally similar to, and are produced in some respects similarly to, their hypothalamic counterparts, whereas important differences also exist. The data taken together support the existence of hypothalamic signal molecules and their receptors in the immune system, and suggest they play important roles in lymphocyte function.

Materials and Methods

Materials

GHRH, substance P, CRH, LHRH, and vasoactive intestinal peptide (VIP) were obtained from the Sigma Chemical Co. (St. Louis, MO). [125I]-labeled hGHRH was purchased from Amersham. Bovine serum albumin (BSA) and RPMI 1640 were obtained from Irvine Scientific (Santa Ana, CA). Nutridoma-SP was obtained from Boehringer-Mannheim Biochemicals (Indianapolis, IN). We obtained rat GHRH (1-43) from Peninsula Laboratories (Belmont, CA) and Sigma. [3H]TdR, 32PdCTP, 32P-ATP were purchased from New England Nuclear. First strand cDNA synthesis and PCR supplies were obtained from Amersham and Perkin Elmer, respectively. For antibody and antisense studies, unmodified deoxyribonucleotides were purchased from a core DNA synthesizer facility located at University of Alabama at Birmingham (UAB), and antibodies to GHRH were purchased from Peninsula Labs.

Cell Preparations

Adult Sprague-Dawley male rats (approximately 180 g) were obtained from Harlan Laboratories (Prattville, AL). Animals were sacrificed by cervical dislocation and their spleens and thymuses were placed into sterile plastic dishes containing RPMI 1640 medium and 20 mM Hepes (pH 7.4). Cell suspensions were prepared by teasing the tissues and washing two times in RPMI 1640 followed by centrifugation on a Ficoll-Hypaque density gradient to remove red blood cells. The leukocytes (2×10^6 cells/ml) were suspended in RPMI 1640, supplemented with 5% fetal calf serum or 1% Nutridoma. Unless otherwise indicated, cell cultures were treated with hormones and/or chemicals for binding, and incubated as described in the figure legends.

Proliferation Studies

Cell suspensions (2×10^6/ml) of lymphoid cells were prepared in RPMI 1640 supplemented with 1% Nutridoma-SP containing 100 U of penicillin and streptomycin (GIBCO Laboratories, Grand Island, NY) per milliliter. These suspensions were incubated in the presence of [^{3}H]thymidine (16 hours; 1 µC/ml, New England Nuclear) in plastic tissue culture dishes with and without various concentrations of GHRH as indicated elsewhere. The cultures were harvested using a Whittaker Mini-Mash Harvester and the glass fiber filters counted in a Tracor Analytic Model 6892 liquid scintillation system.

Binding of ^{125}I-labeled GHRH

Binding assays were carried out in the presence of ^{125}I-labeled hGHRH (0.1–20 nM) in a total volume of 0.1 ml. Nonspecific binding was calculated using 10 µM unlabeled hGHRH in the control tube. The incubations were carried out in microfuge tubes treated with Sigmacote in RPMI 1640 containing 25 mM HEPES (pH 7.4) and 0.1% BSA at 4°C for 60 min. At the end of the incubations, the cells were centrifuged for 3 min in a Beckman 12 microfuge; duplicate 100 µl aliquots of supernate were removed for the measurement of free. The pellets were washed three times in ice-cold buffer, and the tip of the tube was cut off and counted in a gamma counter. Specific binding (SB) fraction is defined as total binding (TB) fraction minus nonspecific binding (NSB) fraction. Radioactivity not removed in the presence of an excess of unlabeled hGHRH (1-44 or 1-29) at $10\,\mu M$ was considered nonspecific. Nonspecific binding was 25% to 30% of total cell binding.

Competition Binding and Scatchard Analysis

The inhibition of the specific binding of ^{125}I-labeled hGHRH (0.136 nM approximately 100,000 cpm) was measured as a function of increasing cold hGHRH (1-44) concentration (10^{-11}–10^{-6} M) under the assay condition described above. The polypeptides CRH, LHRH, and GH were used at a concentration of $1\,\mu M$; VIP was used at concentrations of 10^{-10} to 10^{-6} M. The results were plotted as saturation and inhibition curves and subsequently were analyzed by the Scatchard method to determine the binding capacity of the cells.

Radioimmunoassay (RIA)

The RIA for GHRH was done with a sensitive and specific kit purchased from Peninsula Laboratories (cat. no. RIK-8068) according to the manufacturer's protocol. Lower limits of sensitivity for GHRH was 16 pg/tube.

RNA Isolation

Total RNA was prepared by homogenizing leukocytes in 5 M guanidine thiocyanate, 1% sarkosyl, 20 mM ethylenediaminetetraacetic acid (EDTA), 1% 2-mercaptoethanol, and 50 mM Tris-HCl (pH 7.5), with subsequent protease K digestion and extraction with phenol and chloroform. Polyadenylated messenger RNA (mRNA) was isolated on oligo(dT)-cellulose, as described previously (5). Southern gel analysis (5) was performed on 0.8% agarose gels and followed by transfer to nitrocellulose paper. Nitrocellulose paper was hybridized with GH and GHRH probes.

GH and GHRH cDNA and Hybridization

A rat pituitary GH cDNA was kindly provided by Drs. John Baxter and Fran Denoto (Neurochemistry Laboratories, Veterans Administration Medical Center, Sepulveda, CA). An 800-base pair HindIII GH cDNA insert was obtained from this plasmid by standard techniques (5), and it was labeled with [^{32}P]dCTP by nick translation (Bethesda Research Laboratories, Rockville, MD) to a specific activity of 1 to 2×10^8 cpm/μg. A GHRH oligodeoxynucleotide probe corresponding to amino acids 13–27 (48 bases; 5′-CCTGTTCATGATTTCGTGCAGCAGTTTGCGGGCATATA-ATTGGCCCAG-3′) was designed and synthesized in our laboratory. The

GHRH probe was end-labeled with T_4 polynucleotide kinase (Boehringer Mannheim, Indianapolis, IN) using standard procedures (5). Nitrocellulose filters from Southern analysis were prehybridized for 4 hours at 42°C and hybridized for 18 hours at 42°C, according to the procedure of Maniatis et al. (5). After hybridization, the membranes were washed by standard techniques. The nitrocellulose papers were exposed to Kodak AR film at −70°C with Dupont Cronex Lightning Plus intensifying screens for 2 to 3 days.

Reverse Transcription (RT) and Amplification by Polymerase Chain Reaction (PCR)

First-strand cDNA synthesis was performed using a commercially available reagent kit (Amersham) that is based on the procedure of Okayama and Berg (6). Target sequences from this cDNA were amplified in a PCR using the antisense primer (corresponding to amino acids 180–187) (1 μM) and a sense primer (corresponding to amino acids 4–11) (1 μM) by standard procedures. The primers were prepared and purified in our laboratory. PCR reactions were performed in a Perkin Elmer DNA thermal cycler. Generally, reactions were performed in a total volume of 0.1 ml containing 200 μM of each dNTP, 1 μM of each primer, 1 to 10 ng of template DNA, and 2.5 units of Taq polymerase. The final reaction mixture was overlaid with 0.1 ml of mineral oil (Perkin Elmer) to prevent evaporation. A usual cycle consists of 1 min at 94°C (denaturation), 2 min at 52°C (annealing of primer), and 3 min at 72°C (extension). Twenty-five to forty cycles (7 min total/cycle) were usually run over a 3- to 5-hour period. Amplified samples were run on a 2% agarose gel and stained with ethidium bromide to determine efficiency and size.

Results

Structure of Lymphocyte-Derived Hypothalamic Peptides

The production and release of most pituitary hormones are regulated by peptides released from the hypothalamus as a result of central nervous system stimulation. These releasing hormones act on cells of the pituitary gland to release a particular hormone. The first evidence that cells of the immune system could produce a hypothalamic peptide was published for corticotropin releasing factor (CRF) several years ago (7). Since then, four different hypothalamic peptides have been shown to be produced by lymphoid cells (Table 23.1). A summary of the evidence for CRF, GHRH, LHRH, and SS production in the immune system compared to their hypothalamic counterparts is shown in Table 23.1. The data taken together support the idea that the hypothalamic hormones found in leukocytes are

TABLE 23.1. Comparison of the structural analysis of hypothalamic hormones produced in the immune system and in the hypothalamus.

Strategy[a]	CRF	GHRH	LHRH	SS
Immunofluorescence	ND	+	ND	+
Immunoaffinity purification	ND	+	ND	ND
Gel filtration	ND	S	ND	ND
Reverse phase (HPLC)	S/D	S	S	ND
RIA	S/D	S	S	S
Bioactivity	S	S	S	ND
Northern analysis	S/D	S	S	S
Southern analysis	S	S	S	ND
Nucleotide sequence	ND	ND	S	ND
Content (pg/mg)	5–10	50–200	ND	30–80
References	7–9	17, 18	26, 29, 53	25

[a] More detail on the experimental strategy can be obtained in the references as indicated.
+, positive cells identified or de novo synthesized immunoreactive protein isolated; S, similar; D, different; ND, not done.

structurally similar if not identical to hypothalamic hormones originally identified in the hypothalamus.

CRF was first identified in unfractionated human peripheral leukocytes (7) and later confirmed in human T cells (8) and in the rat thymus and spleen (9). The CRF-like mRNA was also detected in neutrophils, a T-cell clone, and an Epstein-Barr virus (EBV)-transformed B cell line. Northern blot analysis using the same CRF probe detected a CRF-like mRNA of around 1.7 kb compared with the 1.5 kb mRNA species associated with the human hypothalamus. The peptide did not appear to be identical with hypothalamic CRF since RIA curves were not parallel and the elution profiles on HPLC were different (7). In human T cells, the CRF-like material appeared identical to hypothalamic CRH and a significant increase in CRF mRNA levels was observed after treatment with phytohemagglutinin and 12-0-tetradecanoylphorbol-13-acetate (8). In the rat, RT-PCR gave a product of the expected size in the spleen and thymus (9). Furthermore, CRF could be secreted from lymphoid cells and the levels increased by a lipoxygenase pathway inhibitor but not interleukin-1 (IL-1). These data suggest that although the structure of lymphocyte CRF may be similar, the regulation of spleen CRF may differ from that in the hypothalamus. This appears to be the case where IL-1 has been shown to be a potent secretagogue for hypothalamic CRF (10).

GH releasing hormone (GHRH) is a 43 amino acid hypothalamic peptide that regulates the synthesis and secretion of GH in the anterior pituitary gland (11). The amino acid sequence of GHRH from several species have been determined and with one exception they exhibit nearly complete homology with hGHRH. Molecular cloning and sequence analysis of GHRH has also been performed. Only a single copy of the gene has been found in each of two tumors studied, suggesting that ectopic GHRH is derived from the same gene as is hypothalamic GHRH (12). The rat GHRH gene is 10 kb in length and contains five exons that encode a 104 amino acid precursor to the 5.2-kd rat GHRH peptide. In the rat, GHRH-like immunoreactivity is localized to cell bodies in the arcuate nucleus and the medial perifornical region of the lateral hypothalamus.

Similar to other neuropeptides, GHRH is also present in extra-hypothalamic tissues including the gastrointestinal tract (13), placenta (14), testis (15), tumors (16), and cells of the immune system (17, 18). The characterization and sequence analysis of the cDNA encoding the precursor peptides have only been reported in the hypothalamus and placenta. The placental and hypothalamic GHRH transcripts are similar in size (~750 nucleotides), as are their peptide products (5.2 kd) (19). Despite these similarities, it was suggested that GHRH gene expression may be differentially regulated in the two tissues, reflecting different sites of transcription initiation (20). A subsequent study has shown that placental and hypothalamic GHRH mRNAs differ in the region corresponding to the first exon as a consequence of the use in placenta of an alternative promoter located 10 kb upstream from the hypothalamic promoter (21). Multiple transcription initiation sites were found in the placental GHRH mRNA that correlated to the lack of a consensus TATA box in the promoter region (21). In addition, a larger mRNA species (1750) that hybridizes to GHRH cDNA has been observed in the testis of postpubertal rats, where immunoreactive GHRH is detectable in Leydig cells. A GHRH-like peptide has also been demonstrated in human ovarian tissue by immunoperoxidase staining and in human follicular fluid (22).

Because we knew that cells of the immune system could produce GH, we investigated whether the GH in these cells might be subject to a similar GHRH regulatory mechanism described in the neuroendocrine system (18). To test the idea, we isolated RNA from lymphoid tissues and probed for the presence of specific GHRH RNA with a ^{32}P-end-labeled oligonucleotide specific for rat GHRH (amino acids 13–27, 42-mer). The data showed that we were able to detect GHRH message in hypothalamus, spleen, thymus, and peritoneal exudate cells. In addition, we could detect GHRH RNA in the transformed mouse macrophage cell line P-388, whereas rat pituitary and muscle cells were negative.

To determine the kinetics of GHRH RNA synthesis, we cultured rat mononuclear spleen and thymus cells for various periods of time after removal from animals and isolated, slotted, and probed for RNA. The data from both spleen and thymus showed that maximal levels were obtained after 8 and 16 hours of in vitro culture for thymus and spleen cells, respectively. The expression of GHRH RNA was also detectable after 128 hours but at reduced levels. We next determined whether mononuclear leukocytes from rat thymus showed an increase in GHRH-specific RNA by Northern blot analysis. The data showed broad RNA staining from the thymus and hypothalamus with a molecular weight in the range of 500 to

900 base pairs. Similar data were observed with RNA obtained from the spleen and analyzed by Northern analysis.

We have also selectively enriched the concentration of GHRH-specific nucleotide sequences by RT-PCR so that they can be observed by conventional gel electrophoresis (23). Total RNA was extracted from the spleen, thymus, and hypothalamus and poly-A containing RNA purified by poly dT-cellulose chromatography. The mRNA was reverse transcribed and then amplified by PCR using two oligonucleotide primers specific for rat hypothalamic GHRH. A sample of the PCR product was size fractionated by electrophoresis in a 1% agarose gel, and was then stained with ethidium bromide and examined under ultraviolet light. The data showed a single major DNA band corresponding in length to the distance between two GHRH-specific primers (260 base pairs) in the hypothalamus and the lymphoid cells. We confirmed the specific amplification of GHRH by performing the Southern blotting analysis and probing with an authentic hypothalamic GHRH oligodeoxynucleotide probe (Fig. 23.1). Thus, after cDNA synthesis only one band was present that was of the expected size that hybridized with the GHRH-specific probe. Taken together, these data confirm and strongly support the identity of the lymphocyte PCR product as the GHRH sequence.

To analyze the GHRH protein made by lymphocytes, we purified the material over sepharose affinity columns coupled with antibodies to rat GHRH. Culture fluids were passed twice over an antibody affinity column

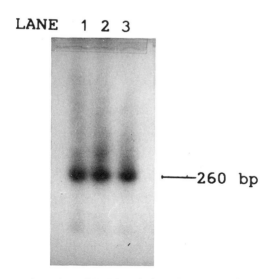

FIGURE 23.1. Detection of rat GH RNA in lymphocytes by Southern gel transfer. The transfer and probing methods are described in the methods. Lane 1, spleen; lane 2, thymus; lane 3, hypothalamus. Reprinted with permission from Weigent et al. (23) by Blackwell Science, Inc., 1991.

(GHRH 1–43 antibody) and the eluted material dialyzed and concentrated. A high-performance liquid chromatography (HPLC) profile of immuno-affinity purified leukocyte-derived fluids showed the presence of a 5,000-dalton peak migrating exactly as bona fide rat GHRH 1–43. In another study, we were able to confirm the de novo synthesis of this material by radiolabeling the protein by culturing leukocytes with ^{3}H-labeled amino acids. In addition, the data in Figure 23.2 obtained with a commercial RIA kit for rat GHRH confirms the immunologic identity of the material isolated from leukocytes. According to Peninsula Laboratories, the antibody to rat GHRH used for this assay was raised against synthetic rat GHRH and shows no cross-reactivity with human GHRH 1–44, hGHRH 1–40, porcine GHRH, PH1, or VIP and has a sensitivity of 15 pg/tube. Leukocytes secrete approximately 76 pg of GHRH per 10^6 leukocytes in a 24-hour culture period. This value is approximately 10-fold lower than that observed for hypothalamic cells (24).

GHRH RNA and protein has also been detected in human lymphocytes where differences with the hypothalamic peptide are more apparent (17). Although binding of lymphocyte-derived irGHRH to a $GHRH_{1-44}$ antibody diluted in parallel with standard $GHRH_{1-44}$, irGHRH eluted earlier on gel

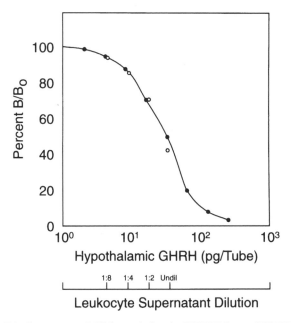

FIGURE 23.2. Displacement of ^{125}I-hypothalamic GHRH from GHRH antibody by synthetic hypothalamic GHRH (●) and irGHRH in lymphocyte culture supernatant fluids (○). A standard RIA procedure was used to measure the displacement of GHRH as described by the suppliers (Peninsula Labs). The immune complexes were pelleted by centrifugation and the precipitated material counted.

chromatography with an apparent molecular weight of approximately 50 kd. On reverse-phase HPLC, irGHRH eluted later than $GHRH_{1-44}$. Northern blot analysis also detected the presence of two GHRH mRNA transcripts in peripheral blood lymphocyte (PBL), one similar in size to the hypothalamic species of 0.75 kb and another larger transcript of approximately 10 kb. Despite these differences in physicochemical behavior, both the lymphocyte-derived irGHRH and $GHRH_{1-44}$ were biologically active as demonstrated by enhancing the transcription of GH mRNA both in dispersed rat pituitary cells and in PBL. It is unclear why differences have been detected between the human and rat lymphocyte GHRH species and what the biologic significance might be.

The release of GH from the hypothalamus is negatively regulated by somatostatin (SS). This hormone not only has potent effects on cells of the immune system but has been localized in lymphoid tissue (25). SS-specific mRNA has been identified in spleen and thymus by an S1 nuclease protection assay similar in size to that seen in the hypothalamus. These studies were corroborated by immunocytochemical studies identifying SS positive cells (25).

The first identification of an immunoactive and bioactive LHRH in purified lymphocytes was made in the spleens of rats (26). Lymphocyte LHRH-like immunoactivity coeluted on reverse phase columns and displaced ^{125}I-LHRH from LHRH antibody in a manner similar to synthetic hypothalamic LHRH. Lymphocyte lysates applied to rat anterior pituitary cells in culture caused secretion of LH in a dose-dependent manner and could be inhibited by an LHRH antagonist (27). The cDNA of rat thymus lymphocyte LHRH has been synthesized and sequenced and is identical to hypothalamic LHRH (28, 29). A recent report in the rat Nb2 T cell line suggests that the lymphocyte LHRH receptor and an alternatively spliced LHRH mRNA are produced at various times during the cell cycle regulated by PRL (30). In the placenta, the extrahypothalamic production of LHRH involves a transcriptional start site upstream from that in the hypothalamus, and mature placental LHRH mRNA retains intron 1 while intron 1 is removed in the hypothalamus (31).

Lymphocyte Receptors for Hypothalamic Peptides

The production of pituitary hormones by cells of the immune system was initially observed in response to mitogens and viruses (1). Since this time, cells of the immune system have also been shown to produce pituitary hormones in response to hypothalamic hormones. Thus far, lymphocytes have been shown to produce ACTH, TSH, LH, and GH in response to CRF, thyrotropin releasing hormone (TRH), LHRH, and GHRH, respectively (1, 3). In addition and along with SS, cells of the immune system have hypothalamic peptide receptors on their membranes and respond to these peptides with an altered immune activity. Some of the findings are summarized in Table 23.2 and suggest that lymphocyte hypothalamic receptors may be similar to their neuroendocrine counterparts.

The CRF receptor mediates a variety of effects on cells in the immune system that take place through specific cell receptors. Radioligand binding studies in membrane homogenates and autoradiographic studies in tissue sections have identified, characterized, and localized CRF receptors in the brain, pituitary, and spleen (32). The relative density of CRF binding was highest in the anterior lobe of the pituitary but detectable binding was evident on spleen macrophages and B cells. A number of similarities have been identified between the pituitary and spleen binding of CRF, including affinity, binding in the presence of cations and guanine nucleotides, stimulation of adenylate cyclase, and apparent subunit molecular weight (32). High-affinity LHRH receptors have also been identified on thymocytes (33). The release of LH stimulated by LHRH apparently is mediated through an LHRH receptor since antagonist to the LHRH receptor blocked the increase (34). The inhibition was dose dependent. Structural studies on this receptor using immunoaffinity chromatography techniques have shown that the binding site is a single polypeptide with a molecular weight of 51 kd, similar to one of the two LHRH binding proteins from pituitary cells (27). Specific receptors for SS and TRH have also been found in lymphocytes (4). Recent work has shown the presence of two receptor types for TRH present on T cells (35). One of these sites satisfies the criteria for a classical TRH receptor and is involved in the release of interferon-γ (IFN-γ) from T cells (36). The existence of distinct subsets of SS receptors have been described on the Jurkat line of human leukemic T cells and U266 immunoglobulin G (IgG)-producing human myeloma cells (37). This study showed that these cells have both high- and low-affinity receptors with K_d values in the picomolar and nanomolar range, respectively. The authors suggest that two subsets of receptors may account for the biphasic concentration-dependent nature of the effects of SS in some systems.

Specific receptors for GHRH have also been identified on lymphocytes (38, 39). Our results show that the binding of ^{125}I-GHRH to spleen and thymic cells was saturable and of a high affinity, approximately 3.5 nM and 2.5 nM for thymus and spleen cells, respectively (Fig. 23.3). The Scatchard

TABLE 23.2. Receptors for hypothalamic hormones on cells of the immune system.

Hormone receptor	Cell type	K_d; B_{max}	References
GHRH	Rat spleen	2.5 nM; 2.2 × 10⁴ sites/cell	38, 39
	Rat thymus	3.5 nM; 3.2 × 10⁴ sites/cell	
CRF	Human PBL	0.2 nM; 8.7 fmol/mg	32
TRH	T cell lines	2.5 nM; 48 fmol/mg	35, 36
LHRH	Rat thymocytes	84 nM; 14 fmol/mg	27, 33
SS	Human PBL	11 nM; 7 × 10⁵ sites/cell	4, 37

Examples of the cell type, K_d and B_{max} of various hypothalamic hormone receptors on cells of the immune system. See text for details.

analysis revealed a binding capacity of approximately 54 fmol and 35 fmol per 10^6 cells on thymus and spleen, respectively. The binding of GHRH was not blocked by 10^{-5} M GH, CRF, vasoactive intestinal peptide (VIP), substance P, or luteinizing hormone releasing hormone. We have observed that after treatment of leukocytes with GHRH (10 nM) a rapid increase in intracellular free calcium concentrations occurs from a basal level of 70 ± 20 nM to a plateau value of 150 ± 20 nM in 6 min after stimulation. We have studied the functional activity of GHRH receptors by measuring lymphocyte proliferative responses and the level of cytoplasmic GH RNA after GHRH treatment (Table 23.3). The presence of GHRH resulted in a twofold increase in thymidine incorporation and a twofold increase in the

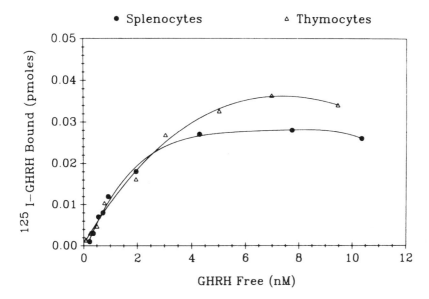

FIGURE 23.3. Thymocytes and splenocytes possess specific, high-affinity sites for GHRH. Cells (1×10^6) (thymocytes [open triangles] or splenocytes [closed circles]) in RPMI 1640 (pH 7.3) with 0.1% bovine serum albumin, 25 mM HEPES (binding buffer) were incubated (60 min at 4°C) in the presence of increasing concentrations of ^{125}I-labeled GHRH (0.1–12 nM) with or without excess (10 μM) unlabeled GHRH (1–29). Incubated, labeled cells were washed with binding buffer (three times) and free and bound ligand were separated by centrifugation. The curves represent the amount ^{125}I-labeled GHRH specifically bound to the cells at increasing doses of labeled GHRH (0.1–12 nM). Cell-associated radioactivity was determined by a Tm Analytical gamma counter. Nonspecific binding (determined by the amount remaining after the addition of excess, unlabeled ligand) was subtracted from total binding to obtain specifically bound ligand; 50% to 70% of the labeled ligand could be blocked with excess, unlabeled GHRH. Each point represents the mean of three determinations. Reprinted with permission from Guarcello et al. (38) by Academic Press, 1991.

TABLE 23.3. Cytoplasmic GH RNA expression and thymidine incorporation in rat thymocytes after GHRH treatment.

Treatment GHRH (M)	^{3}H-thymidine incorporation (cpm)[a]	GH RNA[b] Slot	Relative intensity (z)
None	$4,258 \pm 209$		100 ± 6
10^{-6}	$6,719 \pm 417$		160 ± 7
10^{-7}	$7,551 \pm 496$		184 ± 10
10^{-8}	$10,063 \pm 904$		229 ± 14
10^{-9}	$6,122 \pm 528$		139 ± 18
10^{-10}	$4,781 \pm 372$		110 ± 10

[a] Control and experimental cultures contained 2×10^6 thymus cells maintained in RPMI 1640 plus Nutridoma for 16 hours. After this incubation, cells were pulsed with [3H]thymidine for 14 hours and counted as described. Values represent the mean $\pm$ SEM of six replicate cultures.
[b] Analysis of cytoplasmic GH RNA expression after GHRH treatment. Lymphocytes from thymus (2×10^6/ml) were cultured for 24 hours in the presence of GHRH. Cell cultures were harvested and total cytoplasmic RNA isolated from the different cultures, slotted onto nitrocellulose, and probed as described in the text section Materials and Methods. After a 48-hour exposure, the autoradiograph was densitometrically scanned and graphed as percentage above the nontreated control as shown above.
Reprinted with permission from Guarcello et al. (38) by Academic Press, 1991.

levels of GH RNA in the cytoplasm of treated thymus cells. To determine if leukocyte-derived irGHRH was biologically active, we tested the ability of leukocyte-derived irGHRH to stimulate the expression of GH RNA in primary pituitary cell cultures in vitro. On the average, the amount of GH RNA in the pituitary after treatment increased approximately two- to threefold, supporting the notion that a substance produced by leukocytes can stimulate rat primary pituitary cell cultures to increase transcription of the gene for GH. The data support the production of a bioactive GHRH-like substance from leukocytes that is similar in activity to bona fide hypothalamic-derived GHRH and support the idea that GHRH may be active in a local immune response.

Effects of Hypothalamic Peptides on Immune Function

The concept that hypothalamic hormones influence cells of the immune system is now well established (Table 23.4). The functions of immune-derived hypothalamic peptides compared with exogenous hypothalamic peptides in an immune response are unknown. The accumulating data, however, support the idea that the function of leukocyte-derived hypothalamic peptides in the immune system can be both similar to and different from the role these peptides serve in the hypothalamus (40).

CRH was originally named for its property to stimulate adrenocorticotropic hormone (ACTH) secretion, the systemic hormone that regulates

TABLE 23.4. Modulation of immune responses by hypothalamic peptides.

Hormone	Modulating effect	References
CRF	IL-1 production	43, 46
	Enhanced NK activity	
	Immunosuppression	
TRH	Increased antibody synthesis	52
	Release of IFN-γ	36
GHRH	Stimulates proliferation	38
	Inhibits NK activity	48
	Inhitits chemotactic response	47
SS	Inhibits proliferation	50
	Reduces IFN-γ production	49
LHRH	Increases IL-2 receptor	55
	Stimulates proliferation	56
	Inhibits NK activity	54

Examples of the effect of hypothalamic hormones on activities
of the immune system. See text for details.

production of glucocorticoids by the adrenal cortex. By activating glucocorticoid and catecholamine secretion, central nervous system CRF participates in the suppressive effects of stress on the immune or inflammatory
system. CRF can directly stimulate leukocytes to produce immuno-
regulatory proopiomelanocortin (POMC)-related peptides (β-endorphin,
ACTH, and α-melanocyte-stimulating hormone) (41, 42). CRF stimulates
lymphocyte proliferation, enhances the proliferative response of leukocytes
to lectins, and increases the expression of the IL-2 receptor on T lymphocytes (43). The effect of systemic immunoneutralization of CRF has been
tested in an experimental model of chemically induced aseptic inflammation in rats (44). Intraperitoneal administration of rabbit antiserum to CRF
caused suppression of both inflammatory exudate volume and cell concentration by approximately 50% to 60%. CRF could be detected in the
inflamed area but not in the systemic circulation. Immunoreactive CRF is
therefore produced in peripheral inflammatory sites where, in contrast to its
systemic indirect immunosuppressive effects, it apparently acts as an
autocrine or paracrine inflammatory cytokine (44).

The specific humoral regulators of immune CRF and its secretion are yet
to be completely discerned. However, lymphocyte CRF secretion may be
augmented or suppressed by agents that activate or inhibit inflammation,
respectively. For example, streptococcal cell wall peptidoglycanpoly-
saccharide, which induces arthritis in Lewis rats, also causes the expression
of immune CRF in the arthritic joints (44). In contrast, SS or glucocorticoids
have proportional inhibitory effects on expression of immune CRF and
inflammation (45). IL-1 has been shown to stimulate CRF secretion from
the hypothalamus directly (10). Alternatively, the secretory roles of
CRF and IL-1 may be reversed in immune cells, since CRF has been

shown to stimulate IL-1 production in monocytes (46). Also, increased plasma levels of IL-1 and IL-2 have been reported after intravenous administration of CRF in humans. Another difference between hypothalamic and thymic or splenic CRF secretory responses is the effect of arachidonic acid metabolites on CRF secretion (9). Blocking cyclooxygenase and to a lesser degree the lipoxygenase pathway can result in increased secretion of CRF from a hypothalamic preparation. In this study, the cyclooxygenase blocker indomethacin was without any significant effect, but the lipoxygenase inhibitor nordihydroguaiaretic acid (NDGA) was a very effective stimulator of splenic and thymic CRF secretion. The findings suggest that the direct effects of CRF on immune cells are mediated by thymic and splenic CRF, which may act as an autocrine or paracrine hormone, whereas the more long-term effects of stress are mediated by hypothalamic CRF acting via the neuroendocrine system and glucocorticoids.

The hypothalamic neuropeptide GHRH is the primary stimulus for GH release from somatotrophs of the anterior pituitary. In addition, our previous studies (38) (Table 23.4) show that the addition of GHRH to leukocytes causes a significant increase in thymidine incorporation, GH-specific RNA, and a stimulation of calcium efflux. Several findings by others also show that exogenous GHRH can have a direct influence on immune responses. GHRH has been shown to stimulate proliferation in vitro (38), inhibit natural killer cell activity (47), and inhibit a chemotactic response to zymosan-activated human serum (48). Very little has been done to understand the function of lymphocyte-derived GHRH. In a recent study, we have specifically measured the effect of antibodies to GHRH and antisense oligonucleotides to GHRH on the expression of GH by cells of the immune system (Table 23.5). Basically, spleen and thymus cells were cultured overnight with antibodies and antisense oligos to GHRH. The cells were harvested and spotted onto glass slides to detect changes in the percentage of cells producing the GH protein. The results show that antibodies to GHRH had no effect on GH expression by lymphocytes, whereas GHRH antisense oligodeoxynucleotides blocked the expression of GH by 60%. These data are consistent with an intracellular role for GHRH in the production of lymphocyte-derived GH within the immune system.

There are numerous data to suggest that SS plays a role in regulating the activity of cells of the immune system (40). SS can either inhibit or stimulate lymphocyte proliferation, inhibit the release of colony-stimulating factor from activated lymphocytes, and suppress immunoglobulin synthesis, whereas it enhances leukocytes migration-inhibiting factor formation in activated lymphocytes (49, 50). These actions are probably mediated by a combination of SS with its specific receptors, which have been localized on the surface of lymphocytes and monocytes (37). It is tempting to speculate that GHRH and SS might have antagonistic paracrine actions on immune cells just as they affect the release of GH from the pituitary gland in an opposing manner.

TABLE 23.5. Abundance of GH protein in lymphoid cells from normal rats incubated with antibodies and antisense oligonucleotides to GHRH.

Cell type	Treatment[a]	Percent positive for GH by immunofluorescence
Spleen	None	6.9 ± 0.3
	Antibody GHRH	8.1 ± 1.2
	Antisense GHRH ODN	4.3 ± 0.6*
	Sense GHRH ODN	9.1 ± 0.2
Thymus	None	8.5 ± 1.3
	Antibody to GHRH	8.9 ± 1.3
	Antisense GHRH ODN	5.5 ± 0.6*
	Sense GHRH ODN	9.6 ± 0.3

[a] Rats were sacrificed and the spleen and thymus teased into single cells and cultured for 16 hours with antibody to GHRH (10,000 U/ml) or GHRH antisense oligodeoxynucleotides (ODN) (20 μM). Cells were removed and spotted on glass slides for immunofluorescence as previously described. Each value represents the mean ± SEM of four replicate cultures. * Significant; $p < .05$ (spleen) and $p < 0.1$ (thymus).

There are also some data to support an interaction between TRH and LHRH and the immune system. In the case of TRH, it has been shown to stimulate the release of IFN-γ (36) and enhance antibody synthesis in vitro (51). The data suggest that TRH specifically enhances the in vitro antibody response via production of lymphocyte TSH since antibody to the TSH-β subunit blocked the enhancement (52). It has been shown that rat thymocytes have specific binding sites for LHRH and that LHRH or LHRH analogues can potentiate the concanavalin A–induced stimulation of thymidine incorporation in thymocytes. Thus, some of the effects of LHRH may be directly mediated at the thymocyte level (33). LH is produced by lymphocytes in response to LHRH (53). An LHRH agonist has been shown to increase IL-2 receptor expression, stimulate T-cell proliferation, and diminish natural killer cell activity (27, 54–56).

Summary

Collectively the evidence is mounting that many hypothalamic peptides are found in the immune system. The data gathered to date suggest that the hypothalamic molecules synthesized by cells of the immune system are structurally identical to the gene product of the hypothalamus. The regulation of leukocyte hypothalamic peptide synthesis and secretion is just beginning to be investigated and exhibits both similarities to and differences from hypothalamic mechanisms. The important signal molecules—CRF, GHRH, SS, and LHRH—and their receptors are present in the immune systems. The available data suggest that hypothalamic molecules from lymphocytes are involved in hormone synthesis by lymphocytes, similar to what has been observed for the pituitary. Recent evidence also suggests that hypothalamic peptides modulate cells involved in the immune response. The activity of all major immune cell types, including T cells, B cells, natural killer (NK) cells, and macrophages, can be altered by these hormones. Taken together, these observations have begun to provide a biochemical basis for a mechanism of networking between cells of the immune system via neuroendocrine hormones and peptide neurotransmitters.

Acknowledgments. This work was supported by NIH grant RO1 NS24636. The authors are indebted to Diane Weigent for typing this manuscript and for excellent editorial assistance.

References

1. Blalock JE. Production of peptide hormones and neurotransmitters by the immune system. In: Blalock JE, ed. Neuroimmunoendocrinology, chemical immunology, vol. 52, 2nd ed. Basel: Karger, 1992:1–19.
2. Blalock JE. The syntax of immune neuroendocrine communication. Immunology Today 1994;15:504–11.
3. Weigent DA, Blalock JE. Associations between the neuroendocrine and immune systems. J Leukocyte Biology 1995;58:137–50.
4. Carr DJJ. Neuroendocrine peptide receptors on cells of the immune system. In: Blalock JE, ed. Neuroimmunoendocrinology, chemical immunology, vol. 52, 2nd ed. Basel: Karger, 1992:49–83.
5. Maniatis T, Fritsch EF, Sambrook J. Molecular cloning: a laboratory manual. Cold Spring Harbor, NY: Cold Spring Harbor Laboratory, 1982.
6. Okayama H, Berg P. High efficiency cloning of full-length cDNA. Mol Cell Biol 1982;2:161–70.
7. Stephanou A, Jessop DS, Knight RA, Lightman SL. Corticotropin-releasing factor-like immunoreactivity and mRNA in human leukocytes. Brain Behav Immunol 1990;4:67–73.
8. Ekman R, Servenius B, Castro MG, Lowry PJ, Cederlund AS, Bergman O, Sjogren HO. Biosynthesis of corticotropin-releasing hormone in human T lymphocytes. J Neuroimmunol 1993;44:7–13.
9. Aird F, Clevenger CV, Prystowsky MB, Rebei E. Corticotropin-releasing factor mRNA in rat thymus and spleen. Proc Natl Acad Sci USA 1993;90:7104–8.
10. Sapolsky R, Rivier C, Yamamoto G, Plotsky P, Vale W. Interleukin-1 stimulates the secretion of hypothalamic corticotropin-releasing factor. Science 1987;238:522–4.
11. Frohman LA, Jansson J. Growth hormone-releasing hormone. Endocr Rev 1986;7:223–53.
12. Mayo KE, Cerelli GM, Rosenfeld MG, Evans RM. Characterization of cDNA and genomic clones encoding the precursor to rat hypothalamic growth hormone releasing factor. Nature 1985;314:464–7.

13. Bruhn TO, Mason RT, Vale W. Presence of growth hormone releasing factor-like immunoreactivity in rat duodenum. Endocrinology 1985;117:1710–2.
14. Meigan G, Sasaki A, Yoshinaga K. Immunoreactive growth hormone-releasing hormone in rat placenta. Endocrinology 1988;123:1098–102.
15. Berry SA, Pescovitz OH. Ontogeny and pituitary regulation of testicular growth-hormone releasing hormone-like messenger ribonucleic acid. Endocrinology 1988;127:1404–10.
16. Rivier J, Spiess J, Thorner M, Vale W. Characterization of a growth hormone-releasing factor from a human pancreatic islet tumour. Nature 1982;300: 276–8.
17. Stephanou A, Knight RA, Lightman SL. Production of a growth hormone-releasing hormone-like peptide and its mRNA by human lymphocytes. Neuroendocrinology 1991;53:628–33.
18. Weigent DA, Blalock JE. Immunoreactive growth hormone-releasing hormone in rat leukocytes. J Neuroimmunol 1990;29:1–13.
19. Margioris AN, Brockmann G, Bohler HCL, Grino M, Vamvakopoulos N, Chrousos GP. Expression and localization of growth hormone releasing hormone messenger ribonucleic acid in rat placenta: in vitro secretion and regulation of its peptide product. Endocrinology 1990;126:151–8.
20. Mizobuchi M, Frohman MA, Downs TR, Rohman LA. Tissue specific transcription initiation and effects of growth hormone (GH) deficiency on the regulation of mouse and rat GH-releasing hormone gene in hypothalamus and placenta. Mol Endocrinol 1991;5:476–84.
21. Gonzalez-Crespo S, Boronat A. Expression of the rat growth hormone-releasing hormone gene in placenta is directed by an alternative promoter. Proc Natl Acad Sci USA 1991;88:8749–53.
22. Moretti C, Fabbri A, Gnessi L, Forni L, Fraioli F, Frajese G. Immunohistochemical localization of growth hormone releasing hormone in human gonads. J Endocrinol Invest 1990;13:301–5.
23. Weigent DA, Riley JE, Galin FS, LeBoeuf RD, Blalock JE. Detection of GH and GHRH-related messenger RNA in rat lymphocytes by the polymerase chain reaction. Proc Soc Exp Biol Med 1991;198:643–8.
24. Frohman LA. Regulation of growth hormone-releasing hormone gene expression and biosynthesis. Yale J Biol Med 1989;62:427–33.
25. Aguila MC, Dees WL, Haensly WE, McCann SM. Evidence that somatostatin is localized and synthesized in lymphoid organs. Proc Natl Acad Sci USA 1991;88:11485–9.
26. Emanuele NV, Emanuele MA, Tentler J, Kirsteins L, Azad N, Lawrence AM. Rat spleen lymphocytes contain an immunoreactive and bioactive luteinizing hormone-releasing hormone. Endocrinology 1990;126:2482–6.
27. Costa O, Mulchahey JJ, Blalock JE. Structure and function of luteinizing hormone-releasing hormone (LHRH) receptors on lymphocytes. Prog Neuroendocrinol Immunol 1990;3:55–60.
28. Maier CC, Marchetti B, LeBoeuf RD, Blalock JE. Thymocytes express a mRNA that is identical to hypothalamic luteinizing hormone-releasing hormone mRNA. Cell Mol Neurosci 1992;12:447–54.
29. Azad N, Emanuelle NV, Halloran MM, Tentler J, Kelley MR. Presence of luteinizing hormone-releasing hormone (LHRH mRNA in spleen lymphocytes). Endocrinology 1992;128:1679–81.

30. Wilson TM, Yu-Lee L, Kelley MR. Coordinate gene expression of luteinizing hormone-releasing hormone (LHRH) and the LHRH-receptor following prolactin stimulation in the rat Nb2 T cell line: implications for a role in immunomodulation and cell-cycle gene expression. Mol Endocrinol 1995;9:44–53.
31. Radovick S, Wondisford FE, Nakayama Y, Yamada M, Cutler GBJ, Weintraub BD. Isolation and characterization of the human gonadotropin-releasing hormone gene in the hypothalamus and placenta. Mol Endocrinol 1990;4:476–80.
32. De Souza EB. Corticotropin-releasing factor and interleukin-1 receptors in the brain-endocrine-immune axis. Ann NY Acad Sci 1993;697:9–27.
33. Marchetti B, Guarcello V, Morale MC, Bartoloni G, Raiti F, Palumbo GJ, Farinella Z, Cordaro S, Scapagnini U. Luteinzing hormone-releasing hormone (LHRH) agonist restoration of age-associated decline of thymus weight, thymus LHRH receptors, and thymocyte proliferative capacity. Endocrinology 1989;125:1037–45.
34. Morale MC, Batticane N, Bartoloni G, Guarcello V, Farinella Z, Galasso MG, Marchetti B. Blockade of central and peripheral luteinizing hormone-releasing hormone (LHRH) receptors in neonatal rats with a potent LHRH-antagonist inhibits the morphofunctional development of the thymus and maturation of the cell-mediated and humoral responses. Endocrinology 1990;128:1073–85.
35. Harbour DV, Leon S, Keating C, Hughes TK. Thyrotropin modulates B-cell function through specific bioactive receptors. Prog Neuroendocrinol Immunol 1990;3:266–76.
36. Harbour DV, Hughes TK. Thyrotropin releasing hormone (TRH) induces gamma interferon release. FASEB J 1991;A5884(abstr).
37. Hiruma K, Nakamura KH, Sumida T, Maeda T, Tomioka H, Yoshida S, Fujita T. Somatostatin receptors on human lymphocytes and leukaemia cells. Immunology 1990;71:480–5.
38. Guarcello V, Weigent DA, Blalock JE. Growth hormone releasing factor receptors on lymphocytes. Cell Immunol 1991;136:291–301.
39. Weigent DA, Baxter JB, Guarcello V, Blalock JE. Growth hormone production and growth hormone releasing hormone receptors in the rat immune system. Ann NY Acad Sci 1990;594:432–4.
40. Johnson HM, Downs MO, Pontzer CH. Neuroendocrine peptide hormone regulation of immunity. In: Blalock JE, ed. Neuroimmunoendocrinology, chemical immunology, 2nd ed. Basel: Karger, 1992:49–83.
41. Galin FS, LeBoeuf RD, Blalock JE. A lymphocyte mRNA encodes the adrenocorticotropin/β-lipotropin region of the pro-opiomelanocortin gene. Prog Neuroendocrinol Immunol 1990;3:205–8.
42. Galin FS, LeBoeuf RD, Blalock JE. Corticotropin-releasing factor upregulates expression of two truncated pro-opiomelanocortin transcripts in murine lymphocytes. J Neuroimmunol 1991;31:51–8.
43. Jain R, Zwickler D, Hollander CS, Brand H, Saperstein A, Hutchinson B, Brown C, Audhya T. Corticotropin-releasing factor modulates the immune response to stress in the rat. Endocrinology 1991;128:1329–36.
44. Karalis K, Sano H, Redwine J, Listwak S, Wilder RL, Chrousos GP. Autocrine or paracrine inflammatory actions of corticotropin-releasing hormone in vivo. Science 1991;254:421–3.
45. Kavelaars A, Berkenbosch F, Croiset G, Ballieux RE, Heijnen CJ. Induction of beta-endorphin secretion by lymphocytes after subcutaneous administration of corticotropin-releasing factor. Endocrinology 1990;126:759–64.

46. Kavelaars A, Ballieux RE, Heijnen CJ. The role of IL-1 in the corticotropin-releasing factor and arginine-vasopressin-induced secretion of immunoreactive beta-endorphin by human peripheral blood mononuclear cells. J Immunol 1989;142:2338–42.
47. Pawlikowski M, Zelazowski P, Dohler KD, Stepien H. Effects of two neuropeptides: somatoliberin (GRF) and corticoliberin (CRF) on human lymphocyte natural killer activity. Brain Behav Immunol 1988;2:50–6.
48. Zelazowski P, Dohler KD, Stepien H, Pawlikowski M. Effect of growth hormone releasing hormone on human peripheral blood leukocyte chemotaxis and migration in normal subjects. Neuroendocrinology 1989;50:236–9.
49. Stanisz AM, Befus D, Bienenstock J. Differential effects of vasoactive intestinal peptide, substance P and somatostatin on immunoglobulin synthesis and proliferation by lymphocytes from Peyer's patches, mesenteric lymph nodes and spleen. J Immunol 1986;136:152–6.
50. Muscettola M, Grasso G. Somatostatin and vasoactive intestinal peptide reduce interferon-gamma production by human peripheral blood mononuclear cells. Immunobiology 1990;180:419–30.
51. Blalock JE, Johnson HM, Smith EM, Torres BA. Enhancement of the in vitro antibody response by thyrotropin. Biochem Biophys Res Commun 1985;125:30–4.
52. Kruger TE, Smith LR, Harbour DV, Blalock JE. Thyrotropin: an endogenous regulator of the in vitro immune response. J Immunol 1989;142:744–7.
53. Ebaugh MJ, Smith EM. Human lymphocyte production of immunoreactive luteinizing hormone. FASEB J 1988;2:7811(abstr).
54. Marchetti B, Guarcello V, Morale MC, Bartoloni G, Farinella Z, Cordaro S, Scapagnini U. Luteinizing hormone-releasing hormone binding sites in the rat thymus: characteristics and biological function. Endocrinology 1989;125:1025–36.
55. Batticane N, Morale MC, Gallo F, Farinella Z, Marchetti B. Luteinizing hormone-releasing hormone signalling at the lymphocyte involves stimulation of interleukin-2 receptor expression. Endocrinology 1991;129:277–86.
56. Azad N, Paglia NL, Abel K, Jurgens J, Kirsteins L, Emanuele NV, Kelley MR, Lawrence AM, Mohagheghpour N. Immunoactivation enhances the concentration of luteinizing hormone-releasing hormone peptide and its gene expression in human peripheral T lymphocytes. Endocrinology 1993;133:215–23.

24

Growth Hormone Releasing Peptide—Hexarelin—in Children: Biochemical and Growth Promoting Effects

Zvi Laron, Jenny Frenkel, and Aviva Silbergeld

Hexarelin is a synthetic hexapeptide (His-D-2-methyl-Trp-Ala-Trp-D-Phe-Lys-NH$_2$) that stimulates the release of pituitary growth hormone both in vitro (1) and in vivo (2). It has proved active also in adult men (3) even when administered by the oral or intranasal route (4, 5). We report forthwith our experience with the acute, short, and longer administration of this peptide to children.

Subjects

All the children included in the investigation were short (below the third percentile for age) and had a normal growth hormone (GH) response to oral clonidine (6), and/or insulin hypoglycemia tests (7). The cause of their short stature was either intrauterine growth retardation (IUGR) or familial short stature (FSS), i.e., one or both parents had a height below the third centile.

Three studies were performed:

Study 1: Comparative effect of one intravenous (0.1 U/kg) and intranasal bolus of hexarelin (20 µg/kg). Twelve children, aged $5\frac{1}{2}$ to $15\frac{1}{2}$ years (nine boys, three girls) were investigated. The two tests were performed at an interval of one week between them in a randomized order.

Study 2: Three one-week trials with increasing multiple daily doses of intranasal hexarelin (20 µg/kg b.i.d.; 20–20–40 µg/d; and 40 µg/kg t.i.d.). Between the different doses, there was a washout period. Seven prepubertal children aged $5\frac{1}{2}$ to $11\frac{1}{2}$ years (six boys and one girl) were included in this study.

Study 3: Three months' administration of intranasal hexarelin (60 μg/kg t.i.d.) to five prepubertal children, aged $5\frac{1}{2}$ to $11\frac{1}{2}$ years (four boys, one girl).

The pertinent clinical data of the 12 children enrolled in the investigations is shown in Table 24.1. Five children participated in all three studies, two in only two studies, and five in only the first acute study.

The Drug

The synthetic hexapeptide hexarelin was supplied by Europeptide, Argenteuil, France. Study 1: batches BAD-0001, BIC-9320, 9401. Study 2: batches BIC-9320, 9401. Study 3: batches 9403-6, 9408. The intravenous dose was administered in one bolus. The intranasal spray was administered while the child was in the recumbent position.

Tests

All the tests were performed after an overnight fast. Blood samples were withdrawn via an indwelling catheter in a forearm vein at 0, 15, 30, 60, 90, and 120 minutes.

Blood samples for general chemistry (by Hitachi Autoanalyzer), insulin-like growth factor-I (IGF-I), thyrotropin stimulating hormone (TSH), free thyroxine (T_4), total triiodothyronine (T_3), were drawn in the fasting state on days 0 and 8 in study 2 and in study 3 on days 30, 60 and 90. Serum IGF-I was also sampled on days 15 and 45. A complete blood count was also done periodically.

TABLE 24.1. Pertinent clinical data of 12 children with short stature enrolled in the hexarelin trials.

	Patient		Height		Pubertal	Study		
No.	Age (y:m)	Sex	(cm)	SDS	stage	1	2	3
1	5:9	F	102.0	−2.2	I	+	+	+
2	5:10	M	95.4	−3.7	I	+	+	+
3	5:10	M	104.8	−1.9	I	+	+	(−)
4	7:6	M	113.6	−1.8	I	+	+	+
5	9:1	M	124.0	−1.3	I	+	+	(−)
6	9:11	M	121.3	−2.5	I	+	+	+
7	11:4	M	128.5	−2.3	I	+	+	+
8	12:4	F	135.0	−2.4	IV	+	(−)	(−)
9	13:3	M	136.8	−2.3	II	+	(−)	(−)
10	15:3	M	153.5	−1.9	II	+	(−)	(−)
11	15:5	F	140.7	−3.5	IV	+	(−)	(−)
12	15:6	M	152.3	−2.3	II	+	(−)	(−)

At each visit, the children underwent a complete physical examination, including anthropometric measurements (using a Harpender Stadiometer), and skinfold measurements using a Harpender Caliper. A hand and wrist x-ray was performed at the initiation of the trials to establish the state of skeletal maturation. Plasma hGH was measured by a double-antibody modification of radioimmunoassay (RIA) described by Laron and Mannheimer (8). Serum IGF-I was measured by RIA after acid ethanol extraction followed by cryoprecipitation as previously described (9). The sensitivity of the assay was 25 pmol/L; the within assay coefficient of variation for a concentration of 20 nM/L of IGF-I was 47%. TSH and free T_4 were determimed by routine RIA. All samples were determined in a single assay. Statistical evaluations were performed by Student's paired t-test.

The investigations were approved by the hospital's Ethical Committee and written informed consent was obtained from the parents.

Results

Study 1

Figure 24.1 shows the comparative response of plasma hGH after one iv or intranasal bolus of hexarelin in 12 children with constitutional short stature (CSS).

The mean peak response between the two tests for the whole group was similar for the iv test [79.2 ± 49.3 mU/L (mean ± SD) and for the intranasal

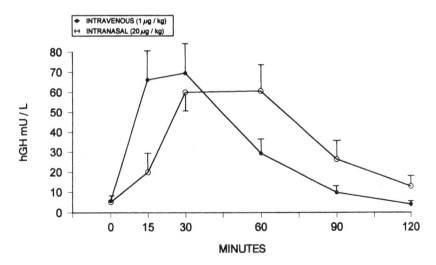

FIGURE 24.1. Plasma hGH response to acute intravenous or intranasal hexarelin administration in 12 children with constitutional short stature (mean ± SEM).

test (69 ± 35.4 mU/L)]. However, the peak response after the iv administration occurred earlier (15–30 min) than that after the intranasal administration (30–60 min).

Figure 24.2 illustrates the individual responses of plasma hGH to the iv or intranasal administration of hexarelin to 11 children with CSS. It is seen that despite having been performed at the same hour after an overnight fast, the responses between the two tests varied in most children. Considering that 20 mU/L is consistent with a normal response of hGH to insulin hypoglycemia or clonidine in healthy subjects, it is evident that both routes of administration of this hexapeptide are potent hGH releasers.

Study 2

The main effects during the 7-day intranasal administration of hexarelin are summarized in Table 24.2. Comparing the basal serum levels of IGF-I, plasma values of alkaline phosphatase, and TSH with the levels on day 8, as well as the day 8 values between the various doses used, it is evident that the serum IGF-I levels increased significantly in four of seven children, this trend being already seen at the lowest dose of 20 μg/kg b.i.d. The plasma alkaline phosphatase rose significantly in four of six children treated by at least two daily doses of hexarelin. Plasma TSH increased significantly in six of seven children within the limits of normal and without any change in plasma T_4 or T_3. No significant changes were observed in the other bio-

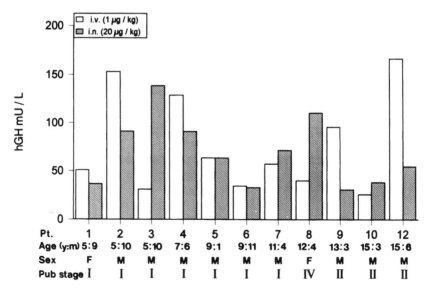

FIGURE 24.2. Individual hGH responses to acute intravenous or intranasal hexarelin administration in 11 children with constitutional short stature.

TABLE 24.2. Effect of intranasal administration of hexarelin in various doses for 7 days to constitutionally short children on IGF-I, plasma alkaline phosphatase, and TSH.

Patient			IGF-I (nM/L)				Alkaline phosphatase (U/L)				TSH (mlU/L)			
			Basal	Day 8			Basal	Day 8			Basal	Day 8		
No.	Sex	CA (y:m)		20 b.i.d.	20–20–40	40 t.i.d.		20 b.i.d.	20–20–40	40 t.i.d.		20 b.i.d.	20–20–40	40 t.i.d.
1	F	5:9	12.7	(−)	13.6	20.7	162	(−)	167	142	1.27	(−)	1.92	1.56
2	M	5:10	7.2	6.3	(−)	8.6	189	190	(−)	208	0.937	1.54	(−)	0.932
3	M	5:10	5.8	8.3	(−)	(−)	135	151	(−)	(−)	1.28	1.66	(−)	(−)
4	M	7:6	14.2	16.9	12.8	16.1	284	259	265	259	1.17	2.17	1.89	1.16
5	M	9:1	8.4	10.1	16.7	18.8	213	210	240	241	0.532	1.61	1.43	1.52
6	M	9:11	7.1	6.8	9.5	14.3	218	217	254	238	0.962	1.53	1.36	1.42
7	M	11:4	15.4	17.4	26.5	22.8	139	143	137	153	0.828	0.659	0.805	0.56

$p < .05$ $p < .05$ $p < .05$ $p < .05$

Doses are in μg/kg.
CA, chronological age; y, years; m, months.

chemical parameters examined such as glucose, Ca, PO₄, electrolytes, and liver enzymes.

Study 3

The principal effects registered during 3 months of intranasal hexarelin (60 μg/kg t.i.d.) administration are summarized in Figure 24.3 and Table 24.3. Table 24.3 shows that the intranasal administration of hexarelin (60 μg/kg t.i.d.) induced a rise in plasma alkaline phosphatase that was significantly higher than the basal levels already after 60 days. A similar significant rise was also registered for plasma inorganic phosphorus but only after 90 days. No significant changes were observed in plasma TSH, free T_4, total T_3, and other biochemical parameters such as glucose, Ca, electrolyte, and liver enzymes, or in the hemogram. The peak morning IGF-I levels registered during 3 months' hexarelin administration to five children with constitutional short stature are illustrated in Figure 24.3. It is evident that the hexarelin treatment raised the serum IGF-I levels in all instances above the basal levels. It is noteworthy that in two children the basal levels were below the normal limits and hexarelin administration normalized the values. A very interesting finding was the significant enhancement of linear growth seen in three out of five children (Fig. 24.4). The body weight increased during these 3 months but no significant change in skinfold thickness in these nonobese children was registered.

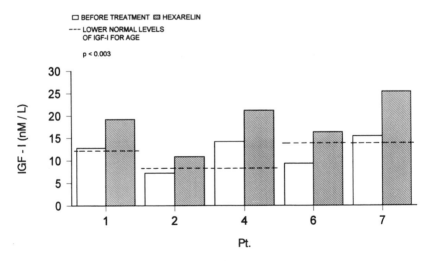

FIGURE 24.3. Serum IGF-I levels before and peak values during 3 months' intranasal hexarelin treatment (60 μg/kg t.i.d.) in five children with constitutional short stature.

TABLE 24.3. Changes in plasma alkaline phosphatase, inorganic phosphorus, and TSH during 90 days of daily intranasal hexarelin administration (60µg/kg t.i.d.) to five children with constitutional short stature.

Patient			Alkaline Phosphatase U/L				Phosphorus (mg/dl)				TSH (mIU/L)			
				Day				Day				Day		
No.	Sex	CA (y:m)	0	30	60	90	0	30	60	90	0	30	60	90
1	F	6:4	156	(−)	206	214	5.2	(−)	6.1	5.7	3.55	1.1	0.839	1.46
2	M	6	195	205	259	221	5.2	5.7	5.6	5.4	0.952	1.29	1.32	1.04
4	M	8	259	317	324	336	4.2	4.8	5.3	5.5	1.85	1.76	1.83	1.73
6	M	10	201	247	258	240	4.8	4.4	4.8	5.3	1.97	1.51	1.48	0.919
7	M	11:8	150	163	170	137	4.3	5.5	4.5	5.3	0.886	0.572	0.742	0.766
				$p < .005$				$p < .05$				N.S.		

CA, chronological age; y, years; m, months.

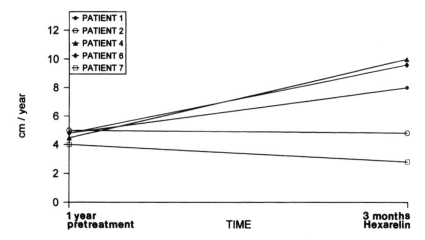

FIGURE 24.4. Growth velocity of five prepubertal children with constitutional short stature before and during 3 months' intranasal hexarelin administration (60 µg/kg t.i.d.).

Drug Tolerance

The drug was well tolerated and no undesirable effects were observed or reported.

Discussion

These investigations are, to the best of our knowledge, the first long-term trial using one of the new synthesized GH-releasing peptides. Although the period of treatment is not very long, it is evident that hexarelin is a very potent GH secretagogue. Its persistent effect is expressed by the evidence of activating the IGF-I release and stimulating the biologic effects of IGF-I, such as the increase in plasma alkaline phosphatase and inorganic phosphorus (9, 10). Of interest was that in three out of five children a definite acceleration of the growth velocity was observed in the first months of treatment. It remains to be seen whether the metabolic and growth promoting effects of hexarelin will be maintained during long-term administration and can compare with those of hGH and IGF-I.

An interesting finding was the rise of plasma TSH observed even after just one week of hexarelin administration. This finding differs from our observations with another synthetic GH secretagogue, the heptapeptide GRP-1, the acute intranasal administration of which induced a fall in plasma TSH (11). The changes in both directions were within the normal limit, and further studies on the possible effect of the GH secretagogues on the thyroid function are indicated.

Acknowledgments. The authors are indebted to Sara Anin, R.N., M.A., for her assistance in the trials and to Drs. R. Deghenghi and V. Lenaerts (Europeptide, Argenteuil, France) for their valuable advice. This investigation was supported by a grant from Europeptide, Argenteuil, France. Z.L. holds the Irene and Nicholas Marsh Chair of Endocrinology and Juvenile Diabetes, Tel Aviv University.

References

1. Deghenghi R, Cananzi M, Battisti C, Locatelli V, Muller E. Hexarelin (EP-23905)—a superactive growth hormone releasing peptide. Abstract #36. J Endocrinol Invest 1992;15(suppl 4):45.
2. Deghenghi R, Cananzi MM, Torsello A, Battisti C, Muller EE, Locatelli V. GH-releasing activity of hexarelin. A new growth hormone releasing peptide, in infant and adult rats. Life Sci 1994;54:1321–8.
3. Imbimbo BP, Mant T, Edwards M, Amin D, Dalton N, Boutignon F, et al. Growth hormone-releasing activity of hexarelin in humans. Eur J Clin Pharmacol 1994;46:421–5.
4. Ghigo E, Arvat E, Gianotti L, Imbimbo BP, Lenaerts V, Deghenghi R, et al. Growth hormone-releasing activity of hexarelin, a new synthetic hexapeptide, after intravenous, subcutaneous, intranasal, and oral administration in man. J Clin Endocrinol Metab 1994;78:693–8.
5. Laron Z, Frenkel J, Gil-Ad I, Klinger B, Lubin E, Wuthrich P, et al. Growth hormone releasing activity by intranasal administration of a synthetic hexapeptide (hexarelin). Clin Endocrinol 1994;41:539–41.
6. Laron Z, Topper E, Gil-Ad I. Oral clonidine—a simple, safe and effective test for growth hormone secretion. In: Laron Z, Butenandt O, eds. Evaluation of growth hormone secretion. Pediatric adolescent endocrinology, vol. 12. Basel: Karger, 1983:103–21.
7. Josefsberg Z, Kauli R, Keret R, Brown M, Bialik O, Greenberg D, et al. Comparison of response of growth hormone to insulin hypoglycemia and to arginine in children with constitutional short stature in different pubertal stages. In: Laron Z, Butenandt O, eds. Evaluation of growth hormone secretion. Pediatric adolescent endocrinology, vol. 12. Basel: Karger, 1983:66–74.
8. Laron Z, Mannheimer S. Measurement of human growth hormone. Description of the method and its clinical applications. Isr J Med Sci 1968;2:115–9.
9. Laron Z, Klinger B. IGF-I treatment of adult patients with Laron syndrome. Clin Endocrinol 1994;41:631–8.
10. Klinger B, Laron Z. Renal function in Laron syndrome patients treated by IGF-I. Pediatr Nephrol 1994;8:684–8.
11. Laron Z, Bowers CY, Hirsch D, Selman-Almonte A, Pelz M, Keret R, et al. Growth hormone-releasing activity of growth hormone-releasing peptide-1 (a synthetic heptapeptide) in children and adolescents. Acta Endocrinol 1993;129:424–6.

25

Growth Hormone Secretagogues in Disease States Associated with Altered Growth Hormone Secretion

Felipe F. Casanueva, Vera Popovic, Alfonso Leal-Cerro,
José L. Zugaza, Manuel Pombo, and Carlos Dieguez

Growth hormone (GH) is necessary for normal linear growth, a fact that conditioned its name and the main focus of research in the last decades (1). Recently, it has become evident that GH exerts other relevant actions over general metabolism, being anabolic, lipolytic, and diabetogenic (2). Furthermore, as a relative GH deficiency is associated with advanced age, this fact has been considered responsible for the deleterious effects of aging on body function and body composition. In this context, understanding the complex mechanisms of GH regulation exerted either by central or peripheral signals is a compelling task.

There is no doubt that in all species, including man, GH is secreted by spontaneous bursts of secretory episodes throughout the 24-hour day, and that growth hormone releasing hormone (GHRH) and somatostatin exert stimulatory and inhibitory actions, respectively, on GH release. Beside these facts is the well-accepted dogma of several components—that such an episodic pattern of release is mandatory to allow GH to exert its biologic actions, and that both spontaneous GH pulses and induced GH discharges are explained by the bihormonal control of the somatotroph with asynchronous periodic release of GHRH and somatostatin. In this scheme a GH surge is the final result of a reduction in the hypothalamic release of somatostatin plus a rise in GHRH secretion (3). This dual hypothalamic regulation has been the Procrustean bed in which we have put our experimental data regarding GH regulation, even if the model may be an oversimplification.

The introduction into the field of a new and artificial class of GH secretagogues developed by Bowers and coworkers may change the accepted paradigm of regulation of somatotroph function in the future. In fact, His-D-Trp-Ala-Trp-D-Phe-Lys-NH$_2$ (so-called GHRP-6), and its cognate peptides suggest the existence of another system that, beside GHRH and somatostatin, regulates somatotroph function (4). The hexapeptide GHRP-6 was originally developed based on its capability to release GH both in vivo and in vitro, and it is now supposed that it might well be the artificial activator of a pituitary receptor naturally operated by an unknown hypothalamic factor. GHRP-6 releases GH in a specific and dose-related manner in all species examined so far, through mechanisms and point of action mostly unknown but different from those of GHRH or somatostatin. Furthermore, the hexapeptide possesses peculiar characteristics; namely, it activates pituitary as well as hypothalamic receptors, it does not act through GHRH receptors and, though debatable, it does not seem to operate through the modulation or action of somatostatin (5, 6). It has also been reported that the combined administration of GHRH plus GHRP-6 elicits a powerful GH discharge (7). This GH release is not additive but synergistic, i.e., the GH secretion after GHRH plus GHRP-6 is significantly higher than the arithmetical addition (sum) of GHRH and GHRP-6 tested separately.

To understand the fine-tuning regulation of GH secretion, it may be useful to study not only physiologic systems but also some natural models of deranged GH secretion, namely Cushing's syndrome, obesity, or aging. In the present work we took advantage of two properties of GHRP-6—its peculiar mechanism of action and its synergistic activity when administered in combination with GHRH—in the exploration of disease states associated with blocked or impaired GH secretion. The two aims of this work were to gain further insight into the disrupted secretory mechanisms of such pathologic states, and to understand GHRP-6's mechanism of action.

Cushing's Syndrome

Cushing's syndrome is a paradigmatic example of a disease state associated with impaired secretion of growth hormone. Glucocorticoids exert a complex action on growth hormone secretion in man. When administered acutely to normal subjects, corticoids possess a consistent and reproducible GH releasing capability 2 or 3 hours after its administration (8). On the other hand, when corticoids are administered for prolonged periods they inhibit somatic growth as well as GH secretion (9, 10). Chronic hypercortisolism, as in Cushing's syndrome, is associated with a blunted GH secretion with respect to all stimuli tested to date. However, the mechanism of this alteration on somatotroph function is unknown (2, 11). Furthermore, the GH hyposecretion of Cushing's syndrome continues for at least a year during convalescence, taking a long time to reach normal levels after surgical cure (12). As the GH deficit may play a contributory role in the catabolic action of cortisol excess on body tissues, it is most relevant to fully understand any altered mechanism of that situation. Although excessively schematic, the glucocorticoid-induced GH blockade may be explained in terms of three mechanisms, either operating alone or in combination: (a) a deficit in the endogenous release of GHRH, (b) a tonic hypersecretion of somatostatin, and (c) a direct inhibitory action upon the somatotroph cell.

The first possibility seems unlikely, because administration of high doses of exogenous GHRH does not release GH in Cushing's syndrome patients compared with control subjects (Fig. 25.1) (13, 14). Even repetitive administration of GHRH (priming) scarcely increases the GHRH-induced GH release (13). Although this is a more indirect pharmacologic approach, there is no support for the second alternative. In fact, administration of pyridostigmine, a drug that presumptively acts by inhibiting endogenous somatostatin (15), does not even minimally modify the blocked GH secretion in Cushing's syndrome patients (16). These facts leave the direct alteration of the somatotroph cell (17) as an open possibility.

As Figure 25.1 shows, administration of GHRP-6 did not induce a positive GH response in Cushing's syndrome patients, and it is highly potent in normal control subjects (18). GHRP-6 can thus be added to the long list of GH stimuli devoid of action in Cushing's syndrome. Although impaired in these patients, GHRP-6 was more effective in releasing GH than GHRH, a finding also observed in control subjects.

The synergistic action of GHRH plus GHRP-6 administered together, probably the most potent stimulus of GH secretion available, was also unable to elicit a significant release of GH in Cushing's syndrome patients

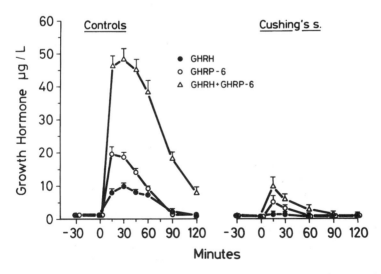

FIGURE 25.1. Mean ± SEM of GH secretion in 10 patients with Cushing's syndrome and 5 normal volunteers. Each subject was tested on three separate occasions with either GHRH 100 μg iv, GHRP-6 100 μg iv, or GHRH plus GHRP-6 at the same doses. Reproduced from Leal et al. (18) with permission.

(Fig. 25.1). Despite being more potent than either GHRH or GHRP-6 administered alone, the combined stimulus induced in only one out of ten patients a GH discharge over 10 μg/L, this being the conventional limit for considering a positive GH response. On the other hand, all the control subjects exhibited a GH peak over 42 μg/L (18). These results indicate that the GH impairment of chronic hypercortisolism is not overridden by this potent combined stimulus, inferentially indicating that glucocorticoids exert permanent inhibition upon the pituitary cells. This lack of response after any GH stimulus is somewhat surprising considering the normal level of insulin-like growth factor-I (IGF-I) observed in Cushing's syndrome patients.

Obesity

An abnormal secretion of GH in obesity and in starvation in humans points to the close interrelationship between nutritional status and GH secretion (19). As with chronic hypercortisolism, obese subjects characteristically present impaired GH secretion when subjected to all stimuli tested so far, namely hypoglycemia, L-dopa, arginine, glucagon, exercise, opioid peptides, clonidine, and the deep sleep nocturnal phase (1, 20–23). As GH is a hormone endowed with lipolytic properties that is also capable of conserving lean mass when individuals are receiving hypocaloric treatments, it has been postulated that the low GH secretion in obesity may contribute to perpetuating the morbid status. Although a partial alteration of GH clearance cannot be ruled out (19), it is undisputed that the main mechanism altered in obesity is a reduced somatotroph responsiveness to stimuli and even to spontaneous secretion (24). Although the blunted GH response can be reversed after weight reduction (25), the first demonstration of a partial and acute reversibility came after short-term caloric restriction (26). However, the mechanism of action of this food restriction is not known, considering that the obese subjects tested became partially responsive for GH secretion before changes in body weight occurred (26).

Following the scheme of reasoning mentioned above for Cushing's syndrome, the fact that even GHRH-stimulated GH secretion is blocked (26) rules out that an alteration of the release of endogenous GHRH could be the explanation for the alterations observed in obesity. The alternatives remaining were an increased somatostatinergic tone, a permanent alteration in the pituitary somatotrophs, or a combination of both mechanisms. As pyridostigmine, a drug that seems to act by inhibiting the hypothalamic release of somatostatin, notably potentiates GHRH stimulation in obese subjects (27), it was suggested that an enhanced somatostatinergic tone was at the root of the altered somatotroph function in this disease state. However, even if the combination of pyridostigmine plus GHRH was able to induce a GH secretion over 20 μg/L in obese individuals, such a response was still considerably lower than that in nonobese counterparts after similar stimuli. Therefore, we hypothesized that obesity was followed by a chronic increase in somatostatin release by the hypothalamus that, in the long run, led to a reduction in the capability of GH synthesis and release by the somatotroph cell (28). This working hypothesis, in part supported by experimental data, may explain why eliminating the presumed high somatostatin tone with pyridostigmine yields a lower response in obese subjects than in normal ones.

GHRP-6 either alone or in combination with GHRH was employed to further understand the altered secretory mechanisms of GH in obesity. As Figure 25.2 demonstrates, obese subjects showed the expected low secretion of GH after GHRH administration. Conversely, GHRP-6 at saturating dose elicited a GH discharge that was lower than that observed in control subjects but nevertheless over the artificial limit of 10 µg/kg (29). GHRP-6 appears to be the most potent stimulus for GH in obesity with a potency higher than hypoglycemia, a stimulus that acts at least partially by reduction of somatostatin release. However, the most striking finding came when obese subjects were tested with GHRH plus GHRP-6 (Figs. 25.2 and 25.3). Under this combined stimulus massive GH secretion was evident (mean peak 40 µg/L), which did not differ greatly from that observed in normal

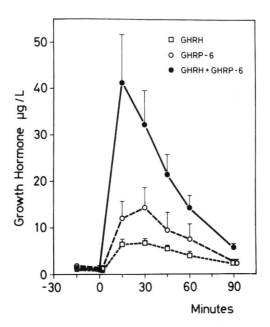

FIGURE 25.2. Mean ± SEM plasma GH levels in a group of obese patients after the administration on three separate days of either GHRH (100 µg iv), GHRP-6 (100 µg iv), or GHRH plus GHRP-6 at the same doses. Reproduced from Cordido et al. (29) with permission.

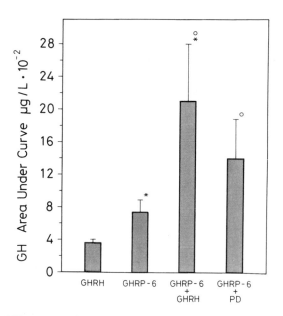

FIGURE 25.3. GH areas under the curve in a group of obese patients after the administration on separate days of either GHRH (100 μg iv), GHRP-6 (100 μg iv), or GHRH plus GHRP-6 at the same doses, and finally pyridostigmine 120 mg po 60 min before GHRP-6. Reproduced from Cordido et al. (29) with permission.

subjects after the same combined stimulus (see Fig. 25.1). This action of the combined stimulus is not attributable to reduction of endogenous somatostatin. In fact, pretreatment with pyridostigmine was able to increase the action of GHRP-6 in a way similar to the increase reported for GHRH (15), but the GH secretion with GHRP-6 and pyridostigmine was lower than GHRP-6 plus GHRH (Fig. 25.3) (29). These results have interesting implications both in the clinical setting and for our understanding of the altered mechanisms of obesity (30). First, this opens a way to pharmacologically induce GH discharge that could help to maintain lean mass in obese subjects under hypocaloric treatments. Second, this is the first demonstration that the somatotroph inhibition of obese subjects is a functional and potentially reversible state. As a side conclusion, it is evident that the altered GH secretion reported in Cushing's syndrome is not attributable to the increased body mass index (BMI) of those patients. It is very interesting in obesity that a somatotroph cell that has been under low GH secretion for years is able to discharge a near normal amount of GH when challenged with the appropriate stimulus.

Aging

Both spontaneous and stimulated GH secretion in middle and late adulthood decline with age, and the diminished GH secretion plus IGF-I reduction may contribute to the change in body mass composition that occurs with aging. In fact, the expansion of adipose tissue mass and the thinning of skin that occur in old age can be reversed to some extent by exogenous GH administration (31–33). Although the GH secretion decline becomes more evident after the decade of 30 to 40 years in humans, that decrease is in fact a continuum, being evident at any age interval provided that a large number of subjects are studied (34). It has been calculated that each decade of increasing age attenuates the GH production by about 14% (35).

Despite being physiologically striking and clinically relevant, the mechanisms responsible for this age-related impairment are not yet fully understood. Impairment in GH synthesis or a decrease in the GH releasable pool has been blamed for the decrease in GH secretion (36). However, the finding that arginine infusion, either alone or in combination with GHRH, restored it to near normality powerfully suggested a chronic increase in somatostatinergic tone as the basis of the altered GH secretion in aged subjects (37, 38).

In the present work, nine healthy young subjects (ages 22 $\pm$ 1.1 years) and nine healthy subjects in their late adulthood (59.5 $\pm$ 1.7 years) were challenged on three separate occasions with either GHRH (100 μg iv), GHRP-6 (90 μg iv), or GHRH plus GHRP-6 at the above doses. As Figure 25.4 shows, administration of GHRH at saturating doses induced a GH release in the late adulthood individuals that was severely reduced as compared with young subjects. Interestingly enough, no differences with regard to age were observed in GHRP-6–stimulated GH secretion in both groups (Fig. 25.4) (39). The administration of GHRH and GHRP-6 in combination elicited in the control young population the expected large GH discharge of a synergistic nature, and a similar high response was observed in the late adulthood group (39). Although the subjects were under 65 years of age, the official limit for being considered old, there was an age gap of 30 years between the groups, making the results relevant. There is no possibility of knowing at present if a more dramatic reduction in GH secretion could take place in older groups. In any case, this result—a normal GH secretion after GHRP-6 either alone or in combined stimulus in aged subjects—suggests that the GH derangement of aging is a reversible state. This finding opens new avenues of therapy for aging, because GHRP-6 is very active by the oral route.

Discussion

Developed originally on theoretical grounds, it is now supposed that GHRP-6 might well be the artificial activator of pituitary and hypothalamic receptors naturally activated by an as yet unknown factor. To explain the mechanism of action of GHRP-6, it has been postulated that its main effect is exerted at the hypothalamic level, through inducing the release of an unknown factor with GH releasing capabilities (3, 6). In brief, the experi-

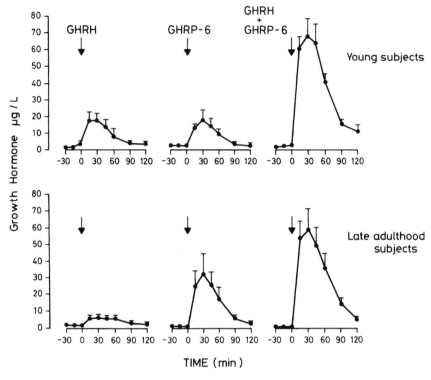

FIGURE 25.4. Mean ± SEM GH secretion in nine young subjects and nine subjects in late adulthood (30-year age gap between the groups) after the administration on three separate days of either GHRH (100 µg iv), GHRP-6 (90 µg iv), or GHRH plus GHRP-6 at the same doses. GH secretion was statistically different between groups only after GHRH stimulation. Reproduced from Micic et al. (39) with permission.

mental data leading to this working hypothesis were (1) GHRP-6 is more potent in vivo than in vitro; (2) in experimental animals with hypothalamic ablation, GHRP-6 is less potent than GHRH; (3) GHRP-6 plus GHRH, both at maximal doses, are synergic in vivo but merely additive in vitro; and (4) the synergistic action of GHRH plus GHRP-6 is absent when the somatotrophs are under an impaired hypothalamic control, as in acromegaly (40, 41). Although this set of findings suggests a hypothalamic action for GHRP-6, direct evidence is lacking.

To gain insight into the mechanism and point of action of GHRP-6, the GH releasing activity of the hexapeptide was assessed in patients with hypothalamopituitary disconnection. Although there were different pathologies affecting the hypothalamopituitary area, the common finding in this group of patients was the presence of a secondary hypopituitarism requiring variable degrees of end-organ hormone replacement therapy

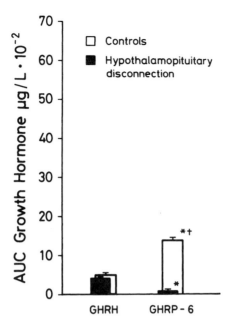

FIGURE 25.5. Mean ± SEM area under the secretory curve (AUC) of GH in normal volunteers (white bars) and patients with hypothalamopituitary disconnection (black bars) after the administration on separate occasions of GHRH (100 µg iv) or GHRP-6 (90 µg iv). *$p < .05$ vs. GHRH; †$p < .01$ vs. the hypothalamopituitary disconnection patients group. Reproduced from Popovic et al. (42) with permission.

prior to testing. In these patients, any component of a GH releasing factor operating through hypothalamic structures should be absent. As Figure 25.5 shows, although slightly reduced as compared with control subjects, patients with hypothalamopituitary disconnection showed a normal GHRH-induced GH secretion, evidencing a well-preserved somatotroph pool (42). As expected, in the control group the GHRP-6–stimulated GH release was three times greater than after GHRH. Conversely, and as a new finding, in patients with hypothalamopituitary disconnection, GHRP-6–stimulated GH secretion was severely reduced, either when compared with control subjects or internally compared with GHRH in the same group of patients (42). These data are the first direct evidence that, in man, the main action of GHRP-6 is exerted at the hypothalamus. This finding is quite surprising when dealing with a peptide whose development was guided by its in vitro GH releasing ability.

    Despite the fact that considerable uncertainties remain in the GHRP-6 mechanism and point of action, the above-described results underscore the fact that the hexapeptide is a valuable tool in neuroendocrinology. The different GH secretory pattern after its administration in disease states as different as obesity, Cushing's syndrome, aging, and pathologies leading to hypothalamopituitary disconnection would provide information about both the normal and the altered regulation of growth hormone secretion in man.

Acknowledgments. This work is supported by a research grant from the Fundacion Ramon Areces, Spain.

References

1. Dieguez C, Page MD, Scanlon MF. Growth hormone neuroregulation and its alterations in disease states. Clin Endocrinol 1988;28:109–43.
2. Casanueva FF. Physiology of growth hormone secretion and action. Endocrinol Metab Clin North Am 1992;21:483–517.
3. Casaneuva FF, Popovic V, Leal-Cerro A, Alvarez CV, Zagaza JL, Dieg EZ. The physiology of growth hormone secretion. In: Melmed S, ed. Molecular and clinical advances in pituitary disorders—1993. Proceedings of the Third International Pituitary Congress. Los Angeles: Endocrine Research and Education, 1993:145–51.
4. Bowers CY, Sartor AO, Reynolds GA, Badger TM. On the actions of the growth hormone-releasing hexapeptide, GHRP-6. Endocrinology 1991;128:2027–35.
5. Bowers CY, Momany FA, Chang D, Hong A, Chang K. Structure-activity relationships of a synthetic pentapeptide that specifically releases GH in vitro. Endocrinology 1980;106:663–7.
6. McCormick GF, Millard WJ, Badger TN, Bowers CY, Martin JB. Dose-response characteristics of various peptides with growth hormone-releasing activity in the unanesthetized male rat. Endocrinology 1985;117:97–105.
7. Bowers CY, Reynolds GA, Durham D, Barrera CM, Pezzoli SS, Thorner MO. Growth hormone (GH)-releasing peptide stimulates GH release in normal men and acts synergistically with GH-releasing hormone. J Clin Endocrinol Metab 1990;70:975–82.
8. Casanueva FF, Burguera B, Muruais C, Dieguez C. Acute administration of corticoids: a new and peculiar stimulus of growth hormone secretion in man. J Clin Endocrinol Metab 1990;70:234–7.
9. Boldgett FM, Burgin L, Iezzoni D, Gribetz D, Talbot ND. Effects of prolonged cortisone therapy on the statural growth, skeletal maturation and metabolic status of children. N Engl J Med 1956;254:636–40.
10. Hartog M, Gaafar MA, Fraser R. Effects of corticosteroids on serum growth hormone. Lancet 1964;1:376–8.
11. Demura R, Demura H, Nunokawa T, Baba H, Miura K. Responses of plasma ACTH, GH, LH and 11-hydroxycorticosteroids to various stimuli in patients with Cushing's syndrome. J Clin Endocrinol Metab 1972;34:852–9.
12. Magiakou MA, Mastorakos G, Gomez MT, Rose SR, Chrousos GP. Suppressed spontaneous and stimulated growth hormone secretion in patients with Cushing's disease before and after surgical cure. J Clin Endocrinol Metab 1994;78:131–7.

13. Leal A, Pumar A, Villamil F, Astorga R, Dieguez C, Casanueva FF. Growth hormone releasing hormone priming increases growth hormone secretion in patients with Cushing's syndrome. Clin Endocrinol 1993;38:399–403.
14. Smals AM, Pieters GFM, Smals AG, Benraad TJ, Kloppenborg PW. Human pancreatic growth hormone releasing hormone fails to stimulate human growth hormone both in Cushing's disease and in Cushing's syndrome due to adreno-cortical adenoma. Clin Endocrinol 1986;24:401–7.
15. Ghigo E, Mazza E, Imperiale E, Molinatti P, Bertagna A, Camanni F, Massara F. Growth hormone responses to pyridostigmine in normal adults and in normal and short children. Clin Endocrinol 1987;27:669–73.
16. Leal A, Pereira JL, Garcia-Luna PP, Astorga R, Cordido F, Casanueva FF, Dieguez C. Effect of the enhancement of endogenous cholinergic tone with pyridostigmine on growth hormone (GH) responses to GH-releasing hormone in patients with Cushing's syndrome. Clin Endocrinol 1990;33:291–5.
17. Chrousos GP, Gold PW. The concept of stress and stress system disorders: overview of physical and behavioural homeostasis. JAMA 1992;267:1244–52.
18. Leal A, Pumar A, Garcia E, Dieguez C, Casanueva FF. Inhibition of growth hormone release after the combined administration of GHRH and GHRP-6 in patients with Cushing's syndrome. Clin Endocrinol 1994;41:649–54.
19. Veldhuis JD, Iranmanesh A, Ho KK, Waters MJ, Johnson ML, Lizarralde G. Dual defects in pulsatile GH secretion and clearance subserve the hyposomatotropism of obesity in man. J Clin Endocrinol Metab 1991;72:51–9.
20. Sims EAH, Danforth E, Horton E, Bray GA, Glennon JA, Salans LB. Endocrine and metabolic effects of experimental obesity in man. Recent Prog Horm Res 1973;29:457–63.
21. Copinschi G, Wegienka LC, Hane S, Forsham PH. Effect of arginine on serum levels of insulin and growth hormone in obese subjects. Metabolism 1967;16:485–91.
22. Glass AR, Burman KD, Dahms WT, Bohem TM. Endocrine function in human obesity. Metabolism 1981;30:89–104.
23. Finer H, Price P, Grossman A, Besser GM. The effects of enkephalin analogue on pituitary hormone release in obesity. Horm Metab Res 1987;19:68–70.
24. Tannenbaum GS, Lapointe M, Gurd W, Finkelstein JA. Mechanisms of impaired GH secretion in genetically obese Zucker rat; roles of GH-releasing factor and somatostatin. Endocrinology 1990;127:3087–95.
25. Williams T, Berelowitz M, Joffe SN, Thorner MO, Rivier J, Vale W, Frohman L. Impaired growth hormone responses to GH-releasing factor in obesity. N Engl J Med 1984;311:1403–7.
26. Kelijman M, Frohman LA. Enhanced growth hormone responsiveness to GH-releasing hormone after dietary manipulation in obese and non obese subjects. J Clin Endocrinol Metab 1988;66:489–94.
27. Cordido F, Casanueva FF, Dieguez C. Cholinergic receptor activation by pyridostigmine restores growth hormone (GH) responsiveness to GH-releasing hormone administration in obese subjects. J Clin Endocrinol Metab 1989;68:290–3.
28. Cordido F, Dieguez C, Casanueva FF. Effect of central cholinergic neurotransmission enhancement by pyridostigmine on the growth hormone secretion elic-

ited by clonidine, arginine or hypoglycemia in normal and obese subjects. J Clin Endocrinol Metab 1990;70:1361–70.

29. Cordido F, Peñalva A, Dieguez C, Casanueva FF. Massive growth hormone (GH) discharge in obese subjects after the combined administration of GH-releasing hormone and GHRP-6: evidence for a marked somatotroph secretory capability in obesity. J Clin Endocrinol Metab 1993;76:819–23.

30. Bowers C. Editorial: a new dimension on the induced release of growth hormone in obese subjects. J Clin Endocrinol Metab 1993;76:817–8.

31. Rudman D, Feller AG, Nagraj HS, Gergans GA, Lalitha PY, Goldberg AF, Schlenker RA, Cohn L, Rudman IW, Mattson DE. Effects of human growth hormone in men over 60 years. N Engl J Med 1990;323:1–6.

32. Corpas E, Harman SM, Blackman S. Human growth hormone and human aging. Endocr Rev 1993;14:20–39.

33. Vance ML. Metabolic status and growth hormone secretion in man. J Paediatric Endocrinol 1993;6:267–72.

34. Ghigo E, Goffi S, Arvat E, Nicolosi M, Procopio M, Bellone J, Imperiale E, Mazza E, Baracchi G, Camanni F. Pyridostigmine partially restores the GH responsiveness to GHRH in normal aging. Acta Endocrinol 1990;123:169–71.

35. Iranmanesh A, Lizarralde G, Veldhuis JD. Age and relative adiposity are specific negative determinants of the frequency and amplitude of growth hormone (GH) secretory bursts and the half life of endogenous GH in healthy men. J Clin Endocrinol Metab 1991;73:1081–8.

36. Iovino M, Monteleone P, Steardo L. Repetitive growth hormone-releasing hormone administration restores the attenuated growth hormone (GH) response to GH-releasing hormone testing in normal aging. J Clin Endocrinol Metab 1989;69:910–3.

37. Ghigo E, Goffi S, Nicolosi M, Arvat E, Valente F, Mazza E, Ghigo MC, Camanni F. Growth hormone (GH) responsiveness to combined administration of arginine and GH-releasing hormone does not vary with age in man. J Clin Endocrinol Metab 1991;71:1481–5.

38. Ghigo E, Arvat E, Goffi S, Bellone J, Valente F, Procopio M, Ghigo MC, Camanni F. Repetitive GHRH and arginine administration to explore the maximal secretory capacity of somatotrope cells during lifespan. Exp Clin Endocrinol 1992;10:191–8.

39. Micic D, Popovic V, Kendereski A, Macut D, Casanueva FF, Dieguez C. Growth hormone (GH) secretion after the administration of GHRP-6 or GHRH plus GHRP-6 does not decline in late adulthood. Clin Endocrinol 1995;42:191–4.

40. Bowers C, Veeraragavan K, Sethumadhavan K. Atypical growth hormone releasing peptides. In: Bercu BB, Walker RF, eds. Growth hormone: basic and clinical aspects. Serono Symposia. New York: Springer-Verlag, 1994:203–22.

41. Popovic V, Damjanovic S, Micic D, Petakov M, Dieguez C, Casanueva FF. Growth hormone (GH) secretion in active acromegaly after the combined administration of GH-releasing hormone and GH-releasing peptide-6. J Clin Endocrinol Metab 1994;79:456–60.

42. Popovic V, Damjanovic S, Micic D, Djurovic M, Dieguez C, Casanueva FF. Blocked growth hormone-releasing peptide (GHRP-6)-induced GH secretion

and absence of the synergic action of GHRP-6 plus GH-releasing hormone in patients with hypothalamopituitary disconnection: evidence that GHRP-6 main action is exerted at the hypothalamic level. J Clin Endocrinol Metab 1995;80:942–7.

26

Growth Hormone Releasing Hormone (GHRH) Effects in Healthy Aging Men

M.R. Blackman, J. Vittone, E. Corpas, J. Busby-Whitehead,
T. Stevens, K. Stewart, J. Tobin, M. Rogers, M.F. Bellantoni,
J. Roth, A. Schwartz, P.L. Smith, R.G.S. Spencer,
and S.M. Harman

Studies of diurnal growth hormone (GH) release have shown 15% to 70% reductions in most GH secretory parameters in middle-aged (40–65 years) and in elderly (>60 years) men and women (1–5). (Age terminology varies in the literature.) In one investigation performed in healthy, nonobese men, aged 21 to 71, there was a 14% reduction in the GH production rate with each advancing decade (6). Aging is also associated with decreases in baseline plasma insulin-like growth factor-I (IGF-I) levels (1, 4, 7–9), but with unaltered responsivity of serum IGF-I levels to GH administered exogenously (10), or to GH increased endogenously in response to GHRH (11). These data suggest that age-related decreases in IGF-I reflect decreased GH secretion, rather than tissue resistance to the effect of circulating GH.

Aging in humans is also associated with reduced muscle and bone mass and increased body fat (12), and it has been suggested that age-related decrements in GH axis activity may be partly responsible for these changes in body composition. GH release is decreased with obesity (13, 14), particularly with intraabdominal adiposity (6). There is a reduction in the frequency of GH secretory bursts and a significant shortening of the circulatory half-life of GH, both of which reduce integrated daily plasma GH concentrations (6, 15). By comparison with normal weight controls, obese men and women also have diminished GHRH-stimulated GH secretion (9, 14, 16, 17), which is partly reversible by exogenous administration of GHRH, arginine, and pyridostigmine (18–20). These observations suggest that the derangement of GH secretion in obesity may be partly a consequence of increased somatostatinergic tone. Combined administration of GHRH and the GH secretagogue GHRP-6 in obese subjects elicits a synergistic increase in GH secretion (20), suggesting that obesity-associated hyposecretion of GH is functional and potentially reversible.

Both healthy (21, 22) and unhealthy (23) older adults treated with GH for short periods (up to 2 weeks) exhibited dose-dependent increases in nitrogen, phosphate and sodium balance, IGF-1, mid-arm circumference, and/or body weight. By enhancing the metabolic efficiency of parenteral nutrition, GH administration has produced similar beneficial effects in severely malnourished older patients with chronic obstructive pulmonary disease (24), and in middle-aged and elderly acutely ill and perioperative patients (25–27). GH treatment of elderly postsurgical patients or middle-aged overweight men with GH also increased fat oxidation (28, 29), and decreased total and low-density lipoprotein (LDL) cholesterol, with no change in high-density lipoproteins (HDL) (30). There were early increases in plasma insulin and glucose levels, and subsequent persistent hyperinsulinemia with a return of glucose to normal. A hyperinsulinemic euglycemic clamp study indicated no change in glucose disposal.

Administration of GH for periods up to 6 months leads to substantial changes in body composition. In one study, when GH was administered subcutaneously three times per week for 6 months to 12 healthy older men with reduced plasma IGF-1 levels, an 8.8% increase in lean body mass, a 14.4% decrease in adipose tissue mass, and a significant increase in skin thickness were observed (31). A follow-up report on these men revealed that the changes in body composition were greater in those men whose intratreatment levels of IGF-1 were at the lower end of the normal range (32). In another 6-month GH intervention study conducted in elderly women, there was a 9% decrease in body fat, as assessed by skinfold thickness, and a decrease in LDL cholesterol only in women cotreated with estrogen (33). No significant change in nitrogen balance occurred, although creatinine clearance increased by 9.2%. The physiologic importance of GH-mediated changes in body composition is as yet unclear, as emphasized by a recent report in which older men underwent resistance training for 14 weeks, followed by 10 weeks of GH versus placebo treatment. As compared with the placebo-treated group,the GH-treated men exhibited increased lean body mass and decreased fat mass, but no greater gains in muscle strength than in the control men (34).

Concurrent increases of plasma osteocalcin, a marker of bone formation, and urinary markers of bone resorption such as calcium, hydroxyproline, pyridinoline, and deoxypyridinoline result from short-term treatment of healthy older people with GH (21) or GHRH (35). In a 12-week study in which the effects of GH and GHRH were compared in postmenopausal osteoporotic women, elevated levels of markers of bone formation and resorption were detected only in the GH treated group, with no changes in bone mineral density in either treatment group (36). Administration of GH to healthy older men and women (21) and osteoporotic patients (37) increases plasma levels of parathyroid hormone and calcitriol, suggesting that

GH augments bone turnover in both osteoporotic and nonosteoporotic older people.

Administration of GH to adults can cause supraphysiologic increases in IGF-I levels, and may be accompanied by impaired glucose tolerance, fluid retention, hypertension, and nerve entrapment syndromes (38–40). The extent to which these adverse effects result from a nonphysiologic plasma GH profile versus excess GH administration is currently unknown.

Long-term (18 months) GHRH administration accelerates growth velocity in GH-deficient children (41–44), with no subsequent desensitization of acute GHRH-stimulated GH responses. Short-term (12–24 hour) continuous intravenous infusions of GHRH to normal adult men augment nocturnal GH release in proportion to the GHRH dose (45), with preservation of the GH response to bolus intravenous GHRH testing (46). Continuous intravenous infusions of GHRH to young adult men for 14 days led to significant increases in IGF-I levels, augmentation of pulsatile GH release, and preservation of the response to supramaximal bolus doses of GHRH (47). These findings suggest that administration of GHRH to children and nonelderly adults preserves the normal, pulsatile pattern of endogenous GH release, without desensitizing the somatotroph to subsequent stimulation by GHRH.

The acute secretory response of GH to direct pituitary stimulation with GHRH is reduced in healthy old women (48, 49), and in some (50–52), but not other (5, 9) studies of old men. This reduction may be due to a progressive decrease in the secretion or action of GHRH, and/or to a rise in somatostatinergic tone with age, as the releasable pool of pituitary GH remains constant (53). Evidence favoring a decrease in GHRH activity derives from studies examining the effects on GH secretion of "priming" doses of GHRH. Alternate day intravenous injections of GHRH for 12 days restore the acute GH responses to GHRH of older men to levels comparable with those of young men (51). Similar acute GH responses, accompanied by significant increase in IGF-I levels, also occur after administration of intravenous GHRH for 8 days to postmenopausal women (35).

We assessed 24-hour pulsatile GH secretion, acute GH responses to intravenous bolus administration of GHRH, and plasma IGF-I measurements in healthy older men, before and after 2 weeks of either twice daily subcutaneous GHRH injections (5) or constant subcutaneous GHRH infusions by portable pump (54), each repeated at cumulative doses of 1 or 2 mg daily. Spontaneous pulsatile GH secretion was lower in old versus young men, and increased into the range seen in untreated normal young men after daily administration of 2 mg of GHRH by injection and either 1 or 2 mg GHRH by constant infusion. GH responses to evening GHRH injections were significantly greater than to those in the morning (5). Diurnal pulsatile secretory variation appeared more physiologic with GHRH injections than during infusions, in that infusion tended to raise the interpeak

GH levels and to diminish amplitude of nocturnal GH secretory peaks. Before GHRH injections, the mean IGF-I level was lower in old men than in young men (5). Treatment evoked a dose-responsive increase in IGF-I to levels not significantly different from the mean for young men at the high dose. At both GHRH doses the mean IGF-I level was significantly greater than the pretreatment level in old men. Two weeks after GHRH treatment was discontinued IGF-I levels had fallen but remained significantly elevated compared with basal conditions. Constant subcutaneous infusions of GHRH gave similar responses of IGF-I, with significant increases at both doses (54). Thus, the IGF-I responses appeared to be independent of the method of GHRH administration. There were no adverse effects. These data revealed that twice daily subcutaneous administration of GHRH to healthy old men for 2 weeks can reverse age-related changes in GH secretory dynamics and IGF-I levels and suggest that more prolonged treatment may beneficially affect alterations in body composition in older people.

Taken together, the above observations are consistent with the hypothesis of diminished secretion or action of GHRH with age.

The GH secretagogue arginine, administered by intravenous infusion, elicits a similar GH response in young and old men (55–57), but enhances the response to GHRH only in old men (57). Given that arginine is a known inhibitor of somatostatin, the latter results suggest that augmented somatostatinergic tone contributes to age-related diminution in GH release. In one study, a single oral dose of 8 g of arginine also potentiated GHRH-stimulated GH release in elderly men (30), whereas in another study, administration of an oral arginine/lysine preparation (6 g of each amino acid daily) for 2 weeks did not increase spontaneous or GHRH-stimulated GH secretion (58). These investigations emphasize that age differences in regulation of GH release may vary with the experimental paradigm, and have implications for the effects of nutritional interventions on GH release in the frail elderly.

GH releasing peptide-6 (GHRP-6), one member of a family of GH secretagogues, is a synthetic hexapeptide acting through a non-GHRH, nonsomatostatin receptor to enhance GH production. It is effective in humans whether administered orally, intranasally, subcutaneously, or intravenously (59–62), and it elicits a large GH response in obese subjects (63). When administered orally, GHRP-6 elicits a similar GH secretory response in young and elderly subjects (64). However, when combined with oral arginine (8 g) administration, the GH response in old men exceeds that of young men (64). The GH secretory response to intravenous hexarelin, a methylated analogue of GHRP-6, is reduced in older men, but is greater than the maximal response to GHRH, and is augmented by coadministration with arginine only in elderly men, suggesting a mechanism of action independent of somatostatin (65). Similarly, the GH responses to acute intravenous infusions of the nonpeptide GHRP analogue L-692,429 are reduced in elderly men and women, but are greater than the maximal responses to GHRH (66, 67). Whether the effects of these GH secretagogues can be sustained chronically, and their effects on outcome measures of physiologic relevance, remain to be determined.

The bulk of GH secretion occurs during delta slow-wave (stages 3 and 4) sleep (68). Elderly people and GH-deficient younger adults have decreased delta sleep (68–70). Although earlier studies suggested that there was no significant correlation between age-related decreases in GH levels and these sleep changes (68), more recent work demonstrates concomitant reductions with age in GH and slow-wave sleep (71). Taken together with a recent report that GHRH administration in young men increased slow-wave sleep and decreased nocturnal wakefulness (72), these data suggest that the sleep pattern changes in old men might be related to their decreased GH (or GHRH) and that GHRH could potentially improve their sleep profile.

We extended our previous studies by assessing, in 11 healthy, ambulatory, nonobese, elderly men, the effects of single nightly subcutaneous injections of GHRH (2 mg) for 6 weeks on the GH–IGF-I axis; certain metabolic indices of fat and bone metabolism; body composition; muscle histology, strength, and bioenergetics; and sleep patterns. After 6 weeks, mean nocturnal GH release, area under the GH peaks (AUPGH), and GH peak amplitude all increased significantly, with no change in GH pulse frequency. There were no significant changes in levels of IGF-I, IGFBP-3, or GH-binding protein (GH-BP), or in weight, body mass index (BMI), waist/hip ratio, DEXA measures of muscle and fat, or muscle histology. However, two of six measures of muscle strength and a test of muscle endurance all improved. There were no alterations in glucose, insulin, and GH responses to oral glucose tolerance test (OGTT) challenge, in circulating lipids, or in urinary creatinine (UCr) excretion. Urinary calcium (UCa) increased, whereas the UCa/UCr ratio, urine deoxypyridoline cross-links, serum osteocalcin, and procollagen peptide remained unchanged. GHRH antibodies were detected in three subjects, and did not affect GH responsivity to GHRH. There were no significant adverse effects.

The latter observations suggest that age-related reductions in GH and IGF-I may contribute to decreased muscle mass and strength in older persons. To determine whether ^{31}P nuclear magnetic resonance spectroscopy (NMRS) can detect effects of treatment with GHRH on metabolism of high-energy phosphate compounds and intracellular pH in forearm muscle during isometric exercise, we assessed the normalized phosphocreatine abundance, PCr/[PCr + Pi], and intracellular pH in forearm muscle by NMRS during both sustained and ramped intermittent exercise protocols, before and at the end of 6 weeks of GHRH treatment. Nuclear magnetic resonance spectra at 32.5 MHz were obtained with a 1.9-T, 31-cm Bruker Biospec. Metabolites were quantified by integration of spectral resonances and corrected for saturation factors. The normalized phosphocreatine abundance, PCr/[PCr+Pi], and intracellular pH declined appropriately during both sustained and ramped intermittent exercise protocols. There were no significant effects of GHRH treatment on descriptive variables derived from the time course of changes in high-energy phosphate and pH. Before GHRH treatment there were significant inverse correlations between reductions in PCr/[PCr + Pi] induced by the sustained exercise protocol and three of six muscle strength measurements, and between reductions in pH and five of six strength measurements. After GHRH treatment, there were no significant correlations between measures of PCr/[PCr+Pi] and muscle strength, while change in pH was related to only one of six strength measures. No significant correlations of strength with PCr/[PCr+Pi] or pH were seen during ramped exercise before or after GHRH administration. These data suggest that in healthy elderly men muscle strength and muscle bioenergetics are related. The changes we observed in these relationships are consistent with the hypothesis that nightly low-dose GHRH treatment reduced the need for anaerobic metabolism during stressful exercise.

We also measured various sleep parameters in nine of the men. At baseline, each subject underwent acclimatization to the sleep laboratory on night 1, followed on night 2 by serial blood sampling for GH and concurrent polysomnographic assessments of EEG activity, REM sleep, and wakefulness. Sleep and GH studies were repeated at 6 weeks, without an acclimatization night. Slow-wave sleep was decreased (compared with that in healthy young men studied in the same laboratory) at baseline, and did not change significantly after GHRH. Similarly, there were no significant changes in total sleep time, stages 1, 2, or REM sleep, time of sleep onset, or movement arousals. Integrated nocturnal GH secretion was positively related to total sleep time before, but not after, GHRH administration. There was a significant positive relationship between AUPGH and slow-wave sleep at baseline, but not after GHRH treatment. In contrast, there were no significant relationships between AUPGH and REM sleep or total sleep before or after GHRH. It remains to be determined whether other paradigms of nocturnal GHRH administration will significantly alter sleep architecture in healthy elderly men or women.

Taken together, the data suggest that optimized methods of administration of GHRH or other GH secretagogues may provide an alternative, and perhaps physiologic, means of attenuating the effects of reduced GH and IGF-I in some older persons.

References

1. Corpas E, Harman SM, Blackman MR. Human growth hormone and human aging. Endocr Rev 1993;14:20–39.

2. Finkelstein J, Roffwarg H, Boyar P, Kream J, Hellman L. Age-related changes in the twenty-four hour spontaneous secretion of growth hormone in normal individuals. J Clin Endocrinol Metab 1972;35:665–70.
3. Zadik Z, Chalew SA, McCarter RJ, Meistas M, Kowarski AA. The influence of age on the 24-hour integrated concentration of growth hormone in normal individuals. J Clin Endocrinol Metab 1985;60:513–6.
4. Ho KY, Evans WS, Blizzard RM, et al. Effects of sex and age on 24-hour profile of growth hormone secretion in men: importance of endogenous estradiol concentrations. J Clin Endocrinol Metab 1987;64:51–8.
5. Corpas E, Harman SM, Piñeyro MA, Roberson R, Blackman MR. Growth hormone (GH)-releasing hormone-(1-29) twice daily reverses the decreased GH and insulin-like growth factor-1 levels in old men. J Clin Endocrinol Metab 1992;75:530–5.
6. Iranmanesh A, Lizarralde G, Veldhuis JD. Age and relative adiposity are specific negative determinants of the frequency and amplitude growth hormone (GH) secretory bursts and the half-life of endogenous GH in healthy men. J Clin Endocrinol Metab 1991;73:1081–8.
7. Florini JR, Prinz PN, Vitrello MV, Hintz RL. Somatomedin-C levels in healthy young and old men. Relationships to peak and 24-hour integrated levels of growth hormone. J Clin Endocrinol Metab 1985;40:2–7.
8. Vermeulen A. Nyctohemoral growth hormone profiles in young and aged men: correlations with somatomedin-C levels. J Clin Endocrinol Metab 1987;64: 884–8.
9. Pavlov EP, Harman SM, Merriam GR, Gelato MC, Blackman MR. Responses of growth hormone and somatomedin-C to GH-releasing hormone in healthy aging men. J Clin Endocrinol Metab 1986;62:595–600.
10. Johanson AJ, Blizzard RM. Low somatomedin-C levels in older men rise in response to growth hormone administration. Johns Hopkins Med J 1981;149:115–7.
11. Pavlov EP, Harman SM, Merriam GR, Gelato MC, Blackman MR. Responses of growth hormone and somatomedin-C to growth hormone releasing hormone in healthy aging men. J Clin Endocrinol Metab 1986;62:595–600.
12. Forbes G. The adult decline in lean body mass. Hum Biol 1976;48:161–73.
13. Rudman D, Vintner MH, Rogers CM, Lubin MF, Fleming GH, Raymond PB. Impaired growth hormone secretion in the adult population. Relation to age and adiposity. J Clin Invest 1981;67:1361–9.
14. Williams T, Berelowitz M, Jaffe SN, et al. Impaired growth hormone (GH) responses to GH-releasing factor (GRF) in obesity: a pituitary defect reversed with weight reduction. N Engl J Med 1984;311:1403–7.
15. Veldhuis JD, Iranmanesh A, Ho KY, Waters MJ, Johnson ML, Lizzaralde G. Dual defects in pulsatile growth hormone secretion and clearance subserve the hyposomatotropism of obesity in man. J Clin Endocrinol Metab 1991;72:51–9.
16. Kelijman M, Frohman LA. Enhanced growth hormone (GH) responsiveness to GH releasing hormone after dietary manipulation in obese and nonobese subjects. J Clin Endocrinol Metab 1988;66:489–94.
17. DeMarinis L, Folli G, D'Amico C, et al. Differential effects of feeding on the ultradian variation of the growth hormone (GH) response to GH releasing hormone in normal subjects and patients with obesity and anorexia nervosa. J Clin Endocrinol Metab 1988;66:598–604.

18. Ghigo E, Procopio M, Boffano GM, et al. Arginine potentiates but does not restore the blunted growth hormone response to growth hormone-releasing hormone in obesity. Metabolism 1992;41(5):560–3.
19. Tanaka K, Inoue S, Shiraki J, et al. Age-related decrease in plasma growth hormone: response to growth hormone releasing hormone, arginine and L-dopa in obesity. Metabolism 1991;40(12):1257–62.
20. Cordido F, Penalva A, Dieguez C, Casanueva FF. Massive growth hormone (GH) discharge in obese subjects after the combined administration of GH releasing hormone and GHRP-6: evidence for a marked somatotroph secretory capability in obesity. J Clin Endocrinol Metab 1993;76(4):819–23.
21. Marcus R, Butterfield G, Holloway L, et al. Effects of short term administration of recombinant human growth hormone to elderly people. J Clin Endocrinol Metab 1990;70:519–27.
22. Kaiser FE, Silver AJ, Morley JE. The effect of recombinant human growth hormone on malnourished older individuals. J Am Geriatr Soc 1991;39: 235–40.
23. Binnerts A, Wilson JHP, Lamberts SWJ. The effects of human growth hormone administration in elderly adults with recent weight loss. J Clin Endocrinol Metab 1988;67:1312–6.
24. Suchner U, Rothkopf MM, Stanislaus G, Elwyn DH, Kvetan V, Askanazi J. Growth hormone and pulmonary disease. Metabolic effects in patients receiving parenteral nutrition. Arch Intern Med 1990;150:1225–30.
25. Ziegler TR, Young LS, Ferrari-Baliviera E, Demling RH, Wilmore DW. Use of human growth hormone combined with nutritional support in a critical care unit. JPEN 1990;14:574–81.
26. Mjaaland M, Unneberg K, Hotvedt R, Revhaug A. Nitrogen retention caused by growth hormone in patients undergoing gastrointestinal surgery with epidural analgesia and parenteral nutrition. Eur J Surg 1991;157:21–7.
27. Ziegler TR, Rombeau JL, Young LS, et al. Recombinant human growth hormone enhances the metabolic efficacy of parenteral nutrition: a double-blind, randomized, controlled study. J Clin Endocrinol Metab 1992;74:865–73.
28. Ponting GA, Halliday D, Teale JD, Sim AJW. Postoperative positive nitrogen balance with intravenous hyponutrition on growth hormone. Lancet 1988;2:438–9.
29. Ward HC, Halliday D, Sim AJ. Protein and energy metabolism with biosynthetic human growth hormone after gastrointestinal surgery. Ann Surg 1987;206:56–61.
30. Oscarsson J, Ottosson M, Wiklund O, et al. Low dose continuously infused growth hormone results in increased lipoprotein A and decreased low density lipoprotein cholesterol concentrations in middle aged men. Clin Endocrinol (Oxf) 1994;41(1):109–16.
31. Rudman D, Feller AG, Nagraj HS, et al. Effect of human growth hormone in men over 60 years old. N Engl J Med 1990;323:1–6.
32. Cohn L, Feller AG, Draper MW, Rudman IW, Rudman D. Carpal tunnel syndrome and gynecomastia during growth hormone treatment of elderly men with low circulating IGF-1 concentrations. Clin Endocrinol (Oxf) 1993;39(4):417–25.
33. Holloway L, Butterfield G, Hintz R, Geusundheit N, Marcus R. Effects of recombinant human growth hormone on metabolic indices, body composition,

and bone turnover in healthy elderly women. J Clin Endocrinol Metab 1994;79:470–9.

34. Taafe DR, Pruitt L, Reim J, et al. Effect of recombinant human growth hormone on the muscle strength response to resistance exercise in elderly men. J Clin Endocrinol Metab 1994;79:1361–6.

35. Franchimont P, Urbain-Choffray D, Lambelin P, Fontaine MA, Frangin G, Reginster JY. Effects of repetitive administration of growth hormone-releasing hormone on growth hormone secretion, insulin-like growth factor I, and bone metabolism in postmenopausal women. Acta Endocrinol (Copenh) 1989;120:121–8.

36. Clemmesen B, Overgaard K, Riis B, Christiansen C. Human growth hormone and growth hormone releasing hormone: a double-masked, placebo-controlled study of their effects on bone metabolism in elderly women. Osteoporosis Int 1993;3:330–6.

37. Aloia JF, Vaswani A, Kappor A, Yeh JK, Cohn SH. Treatment of osteoporosis with calcitonin, with and without growth hormone. Metabolism 1985;34:124–9.

38. Jorgensen JOL, Thuesen L, Ingemann-Hansen T, et al. Beneficial effects of growth hormone treatment in GH-deficient adults. Lancet 1989;1:1221–5.

39. Salomon F, Cuneo RC, Hesp R, Sonksen PH. The effects of treatment with recombinant human growth hormone on body composition and metabolism in adults with growth hormone deficiency. N Engl J Med 1989;321:1797–803.

40. Rudman D, Feller AG, Cohn L, Shetty KR, Rudman IW, Draper MW. Effects of human growth hormone on body composition in elderly men. Horm Res 1991;36 Suppl 1:73–81.

41. Gelato MC, Ross JL, Malozowski S, et al. Effects of pulsatile administration of growth hormone (GH)-releasing hormone on short term linear growth in children with GH deficiency. J Clin Endocrinol Metab 1985;61:444–50.

42. Rochiccioli PE, Tauter MT, Conde FS, et al. Results of 1 year growth hormone (GH)-releasing hormone (1-44) treatment on growth, somatomedin-C, and 24-hour GH secretion in six children with partial GH deficiency. J Clin Endocrinol Metab 1987;65:268–74.

43. Low LCK, Wang PT, Cheung PT, et al. Long term pulsatile growth hormone (GH)-releasing hormone therapy in children with GH deficiency. J Clin Endocrinol Metab 1988;66:611–7.

44. Martha PM, Blizzard RM, McDonald JA, Thorner MO, Rogol AD. A persistent pattern of varying pituitary responsivity to exogenous growth hormone (GH) releasing hormone in GH deficient children: evidence supporting periodic somatostatin secretion. J Clin Endocrinol Metab 1988;67:449–54.

45. Sassolas G, Garry J, Cohen R, et al. Nocturnal continuous infusion of growth hormone (GH) releasing hormone results in a dose-dependent accentuation of episodic GH secretion in normal men. J Clin Endocrinol Metab 1986;63:1016–22.

46. Hulse JA, Rosenthal SN, Cuttler L, Kaplan SL, Grumbach NM. The effect of pulsatile administration, continuous infusion, and diurnal variation on the growth hormone (GH) response to GH releasing hormone in normal men. J Clin Endocrinol Metab 1986;63:872–8.

47. Vance ML, Kaiser DL, Martha PM, et al. Lack of in vivo somatotroph desensitization or depletion after 14 days of continuous growth hormone (GH) releas-

ing hormone administration in normal men and a GH-deficient boy. J Clin Endocrinol Metab 1989;68:22–8.

48. Lang I, Schernthaner G, Pietschmann P, Kurz R, Stephenson JM, Templ M. Effects of sex and age on growth hormone response to growth hormone releasing hormone in healthy individuals. J Clin Endocrinol Metab 1987;65:535–40.

49. Bellantoni MF, Harman SM, Cho D, Blackman MR. Effects of progestin-opposed transdermal estrogen administration on GH and IGF-I in postmenopausal women of different ages. J Clin Endocrinol Metab 1991;72:172–8.

50. Shibasaki T, Shizume K, Nakahara M, et al. Age-related changes in plasma growth hormone response to growth hormone-releasing factor in man. J Clin Endocrinol Metab 1984;58:212–4.

51. Iovino M, Monteleone P, Steardo L. Repetitive growth hormone-releasing hormone administration restores the attenuated growth hormone (GH) response to GH releasing hormone testing in normal aging. J Clin Endocrinol Metab 1989;69:910–13.

52. Coiro V, Volpi R, Cavazzini U, et al. Restoration of normal growth hormone responsiveness to GHRH in normal aged men by infusion of low amounts of theophylline. J Gerontol 1991;46:M155–8.

53. Muller EE, Cocchi D, Ghigo E, Arvat E, Locatelli VFC. Growth hormone response to GHRH during lifespan. J Pediatr Endocrinol 1993;6(1):5–13.

54. Corpas E, Harman SM, Piñeyro MA, Roberson R, Blackman MR. Continuous subcutaneous infusions of GHRH 1-44 for 14 days increase GH and IGF-I levels in old men. J Clin Endocrinol Metab 1993;76:134–8.

55. Dudl J, Ensinck J, Palmer E, Williams R. Effect of age on growth hormone secretion in man. J Clin Endocrinol Metab 1973;37:11–16.

56. Blichert-Toft M. Stimulation of the release of corticotrophin and somatotrophin by metyrapone and arginine. Acta Endocrinol (Copenh) 1975;195:65–85.

57. Ghigo E, Goffi E, Nicolosi M, et al. Growth hormone (GH) responsiveness to combined administration of arginine and GH-releasing hormone does not vary with age in man. J Clin Endocrinol Metab 1990;71:1481–5.

58. Corpas E, Blackman MR, Roberson R, Scholfield D, Harman SM. Oral arginine/lysine does not increase growth hormone and insulin-like growth factor-I secretion in old men. J Gerontol 1993;48:M128–33.

59. Bowers CY, Reynolds GA, Barrera CM . A second generation of GH releasing heptapeptide. 73rd Annual Meeting of American Endocrine Society, Washington, DC, 1991:422.

60. Bowers CY, Alster DK, Frentz JM. The growth-releasing activity of a synthetic hexapeptide in normal men and in short statured children after oral administration. J Clin Endocrinol Metab 1992;74:292–8.

61. Bowers CY, Reynolds GA, Durham D, Barrera CM, Pezzoli SS, Thorner MO. Growth hormone (GH)-releasing peptide stimulates GH release in normal men and acts synergistically with GH-releasing hormone. J Clin Endocrinol Metab 1990;70:975–82.

62. Hartman ML, Farello G, Pezzoli SS, Thorner MO. Oral administration of growth hormone (GH) releasing peptide (GHRP) stimulates GH secretion in normal men. J Clin Endocrinol Metab 1992;74:1378–84.

63. DaCosta AP, Lenham JE, Ingram RL, Sonntag WE. Moderate caloric restriction increases type-1 IGF receptors and protein synthesis in aging rats. Mech Ageing Dev 1993;71:59–71.

64. Lieberman SA, Mitchell AM, Marcus R, Hintz RL, Hoffman AR. The insulin-like growth factor I generation test—resistance to growth hormone with aging and estrogen replacement therapy. Horm Metab Res 1994;26:229–33.
65. Arvat E, Gianotti L, Grottoli S, et al. Arginine and growth hormone-releasing hormone restore the blunted growth hormone releasing activity of hexarelin in elderly subjects. J Clin Endocrinol Metab 1994;79(5):1440–3.
66. Aloi JA, Gertz BJ, Hartman ML, et al. Neuroendocrine responses to a novel growth hormone secretagogue, L-692,429, in healthy older subjects. J Clin Endocrinol Metab 1994;79(4):943–9.
67. Gertz BJ, Barrett JS, Eisenhandler R, et al. Growth hormone response in man to L-692,429, a novel non-peptide mimic of growth hormone releasing peptide (GHRP-6). J Clin Endocrinol Metab 1993;77:1393–7.
68. Prinz PN, Weitzman ED, Cunningham GR, Karacan I. Plasma growth hormone during sleep in young and aged men. J Gerontol 1983;38(5):519–24.
69. Astrom C, Lindholm J. Growth hormone deficient young adults have decreased deep sleep. Neuroendocrinology 1990;51:82–4.
70. Astrom C, Jochumsen PL. Decrease in delta sleep in growth hormone deficiency assessed by a new power spectrum analysis. Sleep 1989;12(6):508–15.
71. van Coevorden A, Mockel J, Laurent E, et al. Neuroendocrine rhythms and sleep in aging men. Am J Physiol 1991;260(4 Pt 1):E651–61.
72. Kerkhofs M, Van Cauter E, Van Onderbergen A, Caufriez A, Thorner MO, Copinschi G. Sleep-promoting effects of growth hormone-releasing hormone in normal men. Am J Physiol 1993;264:E594–8.

27

Aging and Growth Hormone Releasing Peptides

Ezio Ghigo, Emanuela Arvat, Laura Gianotti, Silvia Grottoli,
Guido Rizzi, Giampaolo Ceda, Romano Deghenghi,
and Franco Camanni

It is widely accepted that growth hormone (GH) secretion undergoes an age-related decrease in all species (1, 2). There is agreement that in man 24-hour integrated GH concentration is low in elderly subjects (3, 4) and similar to that usually observed in patients with GH deficiency (4). On the other hand, while some authors found that both day- and nighttime mean GH pulse amplitude and duration but not pulse frequency are reduced (3, 5), others found significant reduction in GH secretory burst frequency and half-life but not in other parameters (6). In agreement with the existence of a GH hyposecretory state, insulin-like growth factor-I (IGF-I) levels have been found reduced in aging (1, 2). Decreased IGF-I levels in the elderly could be due to changes in nutrition and/or lifestyle, particularly in physical exercise (1, 2). However, IGF-I response to exogenous GH administration has been reported preserved in aging (7), pointing to hypothalamopituitary impairment as cause of the reduced activity of the GH–IGF-I axis.

In addition to the reduction of spontaneous GH secretion, in aging there is a decrease in the somatotroph responsiveness to several stimuli acting via the central nervous system such as hypoglycemia, clonidine, physostigmine, pyridostigmine, levodopa, met-enkephalin, and galanin (1, 2, 8, 9), a notable exception being arginine (1, 2, 10, 11). Moreover, even the GH response to growth hormone releasing hormone (GHRH), the specific hypophysiotropic neurohormone, has been found reduced in elderly subjects by the majority of authors (1, 2, 8–11). This evidence could suggest a pituitary pathogenesis of GH-insufficiency in aging, thus pointing to exogenous GH as the unique treatment able to restore IGF-I levels. However, much data in animal and in man strongly suggest that an hypothalamic pathogenesis underlies GH insufficiency in aging (8–19).

In man, there is evidence that treatment with exogenous GHRH restores spontaneous GH secretion and IGF-I to young levels (12, 13); these data agree with the hypothesis that the somatotroph hyposecretion in aging depends on the hypoactivity of GHRH secreting neurons or receptors (14–16). Above all, it has also been demonstrated in elderly subjects that the somatotroph hyporesponsiveness to GHRH is totally restored to young levels by arginine (10, 11), which likely acts via inhibition of hypothalamic somatostatin (20). These data show that the pituitary GH releasable pool is preserved in aging. In this condition GH insufficiency is due, at least partly, to somatostatinergic hyperactivity, in agreement with many studies in animals (17–19).

The evidence that in aging the GH insufficiency is due to hypothalamic pathogenesis and that a marked GH releasable pool is available to be secreted indicates that GH releasing substances may be able to restore the activity of the GH–IGF-I axis. Among GH releasing substances known so far, GH releasing peptides (GHRPs) (21–23) represented the beginning of a new era that now could continue with some nonpeptidyl GH releasing compounds, which have been discovered based on the knowledge provided by GHRPs (24). All these substances are potent GH secretagogues that display their activity even after oral administration (21, 23, 25, 26).

This chapter focuses on GHRPs in aging, presenting our results that contribute to better understanding of the effects of GHRPs in elderly subjects, and verifies the possible clinical utility of GHRPs for diagnostic and therapeutic purposes.

There is evidence suggesting that the age-related body changes may, at least partly, depend on GH insufficiency as indicated by the demonstration that treatment with exogenous GH counteracts them (7). However, it is still unclear whether in aging it is useful to restore the activity of the GH–IGF-I axis to young levels. Were this the case, based on the availability of the large GH releasable pool in the aged pituitary, treatment with GH releasing substances would be more appropriate than that with exogenous GH.

Fundamentals of GHRPs

GHRPs were invented rather than isolated (27, 28), and until now they have had to be considered synthetic, nonnatural compounds that, however, have specific receptors both at the pituitary and the hypothalamic level (29, 30). These receptors are non-GHRH, nonsomatostatin, and nonopioid receptors (29, 31–35), although GHRPs were derived from met-enkephalins (27, 28). The specificity of GHRP receptors led to the hypothesis of the existence of a natural GHRP-like ligand that, however, has not been found until now.

GHRP-6 is the first hexapeptide that has been extensively studied in humans (36–38). More recently, a heptapeptide, GHRP-1(21), and two other hexapeptides, GHRP-2 (21) and hexarelin (23, 39, 40) have been synthesized and are now available for human studies.

In humans, more than in animals, all these peptides have been shown to possess high GH releasing activity that is dose related (21, 23, 36, 37). The

GH releasing effect of 1 µg/kg iv of these peptides is generally higher than that elicited by 1 µg/kg iv GHRH, which has been shown to be the maximal effective dose (21, 37). Higher doses of iv GHRPs have rarely been studied, but, more recently, 2 µg/kg iv hexarelin has been shown to elicit a further GH rise (23); thus, this should be considered its maximal effective dose.

It is noteworthy that GHRPs are strongly active by subcutaneous, intranasal, and even oral route of administration (21, 23, 25, 26, 41). The mechanisms underlying the GH releasing effect of GHRPs are still unclear. A large amount of animal and human data favor the hypothesis that these peptides act concomitantly at the pituitary and the hypothalamic level (21, 22) where they have their own receptors (29, 30).

At the pituitary level GHRPs stimulate GH release (29, 31–35) and, possibly, its synthesis (42). There are data indicating that, at the pituitary level, GHRPs act to antagonize the inhibitory activity of somatostatin on GH release (29). This hypothesis could explain why the GH releasing effect of GHRPs seems partly resistant to inhibition by exogenous somatostatin as well as by substances that increase hypothalamic somatostatin release such as pirenzepine, glucose, and free fatty acids (43, 44). These latter substances could also act directly at the pituitary level (44). The GH releasing effect of GHRPs on somatotroph cells in vitro is, on the other hand, clearly smaller than that in vivo (34). Also in vivo their positive interaction with GHRH is strongly potentiated (34, 37, 43). These data point to a hypothalamic action. At this level, GHRPs do not seem to negatively influence somatostatin release (22, 34, 38, 43). On the other hand, there are data suggesting that the action of GHRPs is either dependent on (34, 45, 46) or independent of (31–33, 35) hypothalamic GHRH; these possibilities are not mutually exclusive. It has been also hypothesized that GHRPs could act via a hypothalamic unknown U-factor (22, 34). Although much data indicate that the GHRPs known so far act via a common mechanism (29, 35, 47, 48), it has recently been reported that GHRP-2 could have an unusual action (49).

GH Releasing Effect of GHRPs in Aging

The GH releasing effect of GHRPs undergoes a reduction in aging in both animals and humans (22, 50–54). In humans we studied the age-related effect of the maximal effective dose of hexarelin (2 µg/kg iv). In 16 normal elderly subjects (eight men and eight women, ages 61–84 years) hexarelin induced a GH response (AUC ± SEM: 890.3 ± 187.6 µg/L/hour) lower ($p < .001$) than that recorded in 20 normal young adults (11 men and 8 women, ages 20–35 years) (2069.5 ± 188.8 µg/L/hour) (Fig. 27.1). As in young adults (23), in the elderly the GH response to hexarelin was not dependent on sex. In aging, the somatotroph responsiveness to hexarelin was not inversely related to age.

A GH response to hexarelin similar to that recorded in elderly subjects was also shown in 11 postmenopausal women (ages 51–65 years) (552.1 ±

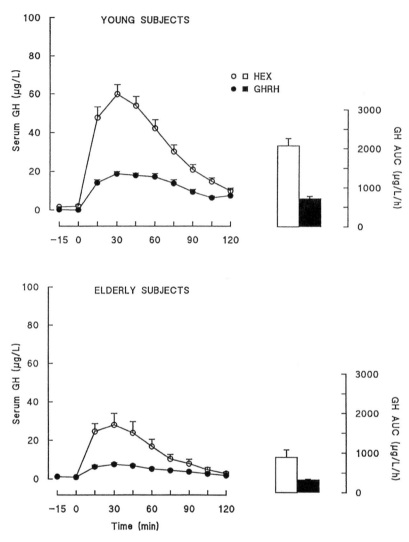

FIGURE 27.1. Serum GH levels (mean ± SEM) after hexarelin (2μg/kg iv) or GHRH (2μg/kg iv) in 20 young and 16 elderly subjects.

62.0μg/L/hour). Thus, these data indicate that the GH releasing effect of GHRPs is reduced by the sixth decade of life, although it remains still marked. Also the somatotroph responsiveness to GHRH is reduced in aging (1, 2, 8, 10, 11). However, it has to be emphasized that, as in young adults (23), even in the elderly hexarelin releases more GH than GHRH (890.3 ± 187.6 vs. 331.1 ± 34.0μg/L/hour, $p < .02$) (Fig. 27.1). Thus, the somatotroph responsiveness of either GHRH or GHRPs decreases from youth to aging and one could argue that this agrees with the hypothesis that

GHRPs act via a GHRH-mediated mechanism (22, 34, 45, 46). However, our preliminary data indicate that at birth, when the GHRH-induced GH rise is maximal (55), the GH response to hexarelin is clearly lower than that in childhood when, in turn, the GH releasing effect of GHRH decreases (personal unpublished results). Similar findings have been observed in the rat (56). These data indicate that the age-related pattern of the somatotroph responsiveness to hexarelin differs from that to GHRH, and that the activity of GHRPs is, at least partly, independent of the activity of GHRH-secreting neurons (31–33, 35).

As in young adults (21, 23, 25, 26, 41), in elderly subjects GHRPs are active also by the subcutaneous, intranasal, and oral routes of administration. The efficacy of subcutaneous administration of GHRP-6 and GHRP-1 in aging has been demonstrated by Bowers (21), who showed that a 3 µg/kg sc dose elicits a clear-cut GH release, which, however, is lower than that in young adults.

We studied the GH releasing effect of the intranasal administration of 1.25 mg (about 18 µg/kg) hexarelin in six elderly subjects (four men and two women, ages 60–78 years). By this route, hexarelin elicited a GH rise that was higher, although not significantly, than that observed after intravenous administration of 1 µg/kg GHRH (219.4 ± 40.8 vs. 136.8 ± 30.0 µg/L/hour) (Fig. 27.2).

In another nine elderly subjects (one man and eight women, ages 64–93 years) we studied the effects of 20 and 40 mg oral hexarelin, comparing

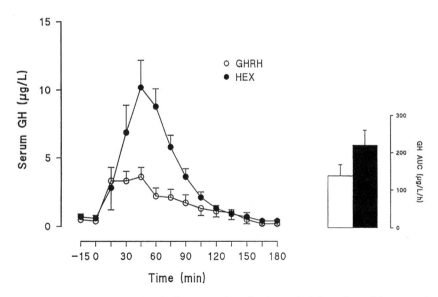

FIGURE 27.2. Serum GH levels (mean ± SEM) after administration of intranasal hexarelin (1.25 mg = about 18 µg/kg) or intravenous GHRH (1 µg/kg) in six elderly subjects.

them to those observed in 12 young adults (six men and six women, ages 20–35 years). Elderly and young subjects had similar mean body weight (58.8 ± 1.5 vs. 63.0 ± 9.0 kg) and, in both groups, 20 and 40 mg represented a dose of about 300 and 600 μg/kg, respectively. In the elderly, 20 mg oral hexarelin induced a significant GH rise which was lower than that in young subjects (358.1 ± 74.0 vs. 762.0 ± 148.7 μg/L/hour, $p < .05$) but higher ($p < .02$) than that observed in the same elderly subjects after 1 μg/kg iv GHRH (166.8 ± 44.5 μg/L/hour). The GH response to 40 mg oral hexarelin was reduced in aging (526.0 ± 157.4 vs. 1360.0 ± 171.4 μg/L/hour, $p < .001$) and it was higher than that after 20 mg, although the difference attained statistical significance ($p < .01$) only in young adults (Fig. 27.3).

Other studies were performed to clarify the mechanisms underlying the GH releasing effect of GHRPs and, particularly, their reduced activity in aging. It has been demonstrated that the reduced GHRH-induced GH response in postmenopausal women is restored by treatment with estrogens (57). Based on this evidence, we were interested in clarifying whether the reduced activity of GHRPs in aging was due to a reduction in gonadal steroids. Thus, in seven postmenopausal women (ages 51–58 years) we studied the GH response to intravenous administration of 2 μg/kg hexarelin before and after 3 months of treatment with transdermal estradiol (50 μg/day). Our preliminary results demonstrate that the GH response to hexarelin in postmenopausal women is not significantly modified by treatment with estradiol and persists lower than that in young women (personal unpublished results). Based on these findings, one could conclude that,

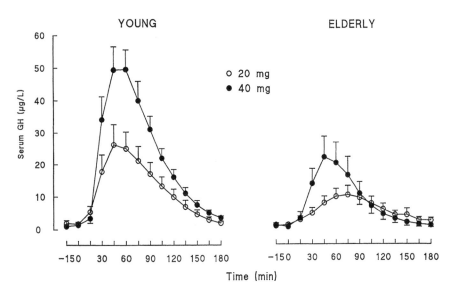

FIGURE 27.3. Serum GH levels (mean ± SEM) after oral hexarelin (20 and 40 mg = about 300 and 600 μg/kg, respectively) in 12 young and nine elderly subjects.

differently from GHRH, the GH releasing effect of hexarelin is independent of the influence of estrogens. This assumption is supported by the evidence that hexarelin has no sex-related effect in young adults (23). However, our recent studies indicated that the effect of hexarelin increases at puberty, when it is more marked in girls than in boys, and is positively related to estradiol levels (58). Thus, it appears that, in agreement with data in animals (50, 59, 60), estrogens influence the activity of GHRPs during puberty while they are probably not involved in the reduction of the GH-releasing activity of these peptides in aging.

To further clarify the mechanisms underlying the age-related reduction in the GH releasing effect of GHRPs, we studied the interactions of hexarelin, GHRH, and arginine, which likely acts via inhibition of hypothalamic somatostatin (20). In the first study (54), eight elderly men (ages 65–78 years) and seven young men (ages 24–29 years) were administered hexarelin ($2\,\mu g/kg$ iv) and GHRH ($2\,\mu g/kg$ iv) alone or in combination. In the elderly, the GH response to hexarelin was higher than that to GHRH (1056.0 ± 141.0 vs. $281.3\,\mu g/L/hour, p < .02$) and both were lower ($p < .002$) than those in young subjects (2424.3 ± 300.0 vs. $727.5 \pm 51.0\,\mu g/L/hour, p < .02$). The combined administration of hexarelin and GHRH enhanced the somatotroph responsiveness in both groups but in elderly subjects ($1947.7 \pm 306.0\,\mu g/L/hour, p < .05$ vs. hexarelin alone) it remained lower ($p < .001$) than in young adults ($3862.3 \pm 251.0\,\mu g/L/hour, p < .03$ vs. hexarelin alone). These results agree with others obtained in animals and humans, where the interaction of GHRP-6 or GHRP-1 with GHRH was studied (21, 22, 37, 51). On the one hand, these data strengthen the view that GHRPs and GHRH act, at least partly, by different mechanisms (31–35, 43). On the other hand, considering that the combined administration of GHRP and GHRH is probably the most powerful stimulus of GH secretion known so far, these data also indicate that the GH-releasable pool in aging, although still marked, is reduced. As there is evidence that normal GHRH activity is needed to allow the GH releasing effect of GHRPs (22, 34, 45), their reduced stimulatory effect in elderly could be due to an impairment of GHRH neurons or receptors, in agreement with other data in animals and humans (12–16).

In a second study (54), another eight elderly men (ages 69–84 years) and six young men (ages 26–30 years) were administered hexarelin ($2\,\mu g/kg$ iv) alone or combined with arginine ($0.5\,g/kg$ iv infused over 30 min). Although in young adults arginine failed to enhance the GH response to hexarelin (2028.2 ± 218.9 vs. $2409.6 \pm 334.0\,\mu g/L/hour$), in elderly subjects the amino acid clearly increased the reduced somatotroph response to the hexapeptide (2069.5 ± 528.7 vs. $824.6 \pm 229.7\,\mu g/L/hour, p < .005$), restoring it to young levels. Considering that the GH-releasing effect of arginine is likely mediated by hypothalamic somatostatin release (20), these results in aging indicate that hexarelin acts by a mechanism independent of the modulation of hypothalamic somatostatin; moreover, the reduction in the

GH releasing effect of GHRPs in the elderly could be explained by the existence of a somatostatinergic hyperactivity, in agreement with other data in animals and humans (8–11, 17–19).

In six elderly women (ages 65–80 years) we found that coadministration of hexarelin, GHRH, and arginine induced a GH rise higher ($p < .05$) than those observed after hexarelin plus arginine or hexarelin plus GHRH. These latter responses were, in turn, higher ($p < .05$) than that to hexarelin alone (personal unpublished results) (Fig. 27.4).

Taken together, these results suggest that hexarelin, GHRH, and arginine act, at least partly, via different mechanisms and that the reduced somatotroph responsiveness to GHRPs in aging is likely due to a concomitant reduction of GHRH activity and an increase of somatostatinergic tone.

We were also interested in studying whether chronic treatment with GHRPs induces desensitization in aging or increases IGF-I levels. We performed studies administering different treatments with GHRP-6 or hexarelin.

First, in seven elderly women (ages 65–82 years) we studied the effect of 4-day oral treatment with 300 µg/kg GHRP-6 twice daily (53). Before treatment the acute oral administration of GHRP-6 induced a clear GH rise (353.1 ± 90.6 µg/L/hour) which was higher ($p < .01$) than that observed after 1 µg/kg iv GHRH (106.5 ± 43.9 µg/L/hour). After the 4-day treatment with the hexapeptide, the somatotroph response to acute oral GHRP-6 was not significantly modified with a trend toward increase (499.8 ± 107.2 µg/L/hour). IGF-I levels were not modified by this GHRP-6 treatment (77.1 ± 8.4 vs. 84.1 ± 12.2 µg/L) (Fig. 27.5).

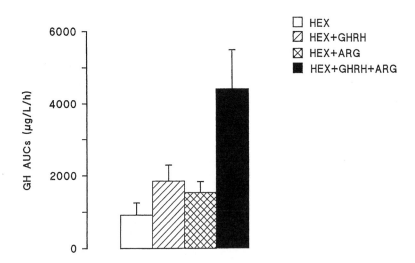

FIGURE 27.4. Serum GH levels (mean ± SEM) after hexarelin (2 µg/kg iv) alone and combined with GHRH (2 µg/kg iv), arginine (0.5 g/kg iv) or both in six normal elderly subjects.

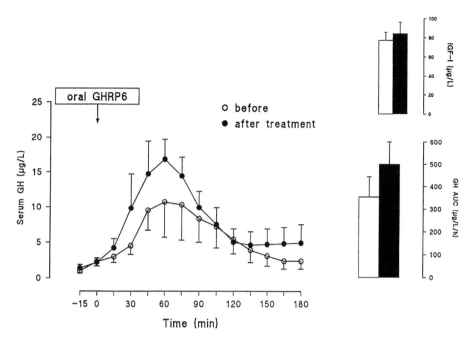

FIGURE 27.5. Effects of 4-day treatment with oral GHRP-6 (300 μg/kg twice daily) on both the acute GH response to the hexapeptide and IGF-I levels in seven normal elderly subjects.

Second, in a group of eight elderly men (ages 66–85 years), we demonstrated (52) that the GH response to an acute oral dose of GHRP-6 (300 μg/kg) is increased even by low-dose oral arginine aspartate (8 g) (649.3 ± 74.3 vs. 378.6 ± 101.7 μg/L/hour, $p < .01$). Based on this evidence, in another eight elderly men (ages 66–85 years), we studied the effects of 4-day treatment with 300 μg/kg oral GHRP-6 combined with 8 g oral arginine aspartate, twice daily. The acute GH response to oral GHRP-6 and arginine did not differ after 4-day treatment (428.5 ± 86.4 vs. 352.8 ± 60.7 μg/L/hour), with a trend toward decrease. Again, IGF-I levels were not modified by this treatment (53.0 ± 5.9 vs. 55.6 ± 8.4 μg/L) (Fig. 27.6).

The results of these two studies indicate that after short-term treatment with oral GHRP-6, the effect of the hexapeptide does not undergo desensitization. This evidence agrees with data showing that intermittent GHRP-6 treatment does not decrease the acute GH response to the hexapeptide in rat (50), while other data in rat (31, 45) and in humans (61–63) indicate that continuous GHRP-6 infusion reduces the somatotroph response to the acute administration of the hexapeptide.

Another interesting finding in our studies is that the effect of oral GHRP-6 in the elderly is acutely enhanced by low-dose oral arginine; however, this effect seems to vanish after a short time. Moreover, IGF-I levels in aging

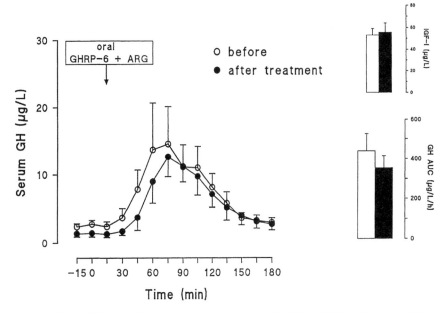

FIGURE 27.6. Effects of 4-day treatment with oral GHRP-6 (300 µg/kg twice daily) and oral arginine aspartate (8 g twice daily) on both the acute GH responses to the hexapeptide and IGF-I levels in eight elderly subjects.

were not increased by these modalities of treatment with GHRP-6 in spite of the persistent somatotroph response to the drug. We think that GH secretion over the day has not been increased sufficiently to enhance somatomedin generation. On the other hand, a peripheral resistance to the GH-induced increase of somatomedin generation in aging could also be hypothesized.

To clarify these points, we studied, in elderly subjects, the effect of more prolonged treatments with more frequent administrations of hexarelin. In one study, the effects of 8-day treatment with intranasal administration of 1.25 mg (about 18 µg/kg) hexarelin thrice daily on the acute GH response to the hexapeptide as well as on IGF-I and IGF binding protein (BP)3 levels were studied in six elderly subjects (four men and two women, ages 60–78 years). In a second study, in six elderly women (ages 65–80 years), the same hormonal parameters were studied, before and after 15-day treatment with oral administration of 20 mg (about 300 µg/kg) hexarelin, thrice daily. Our preliminary data indicate that the GH releasing effect of hexarelin is preserved after treatment with the hexapeptide up to 15 days. Moreover, an increase of IGF-I and IGF-BP3 levels was present in both studies after treatment with hexarelin, indicating that these doses and frequency of administration are needed to obtain an increase of GH secretion sufficient to enhance somatomedin generation.

Taken together, these data demonstrate that the treatment with GHRPs is able to restore the reduced function of the GH–IGF-I axis in aging and suggest that chronic treatment with these compounds could counteract the age-related body changes. According to this hypothesis, there are already data in rats showing that prolonged treatment with GHRPs has somatogenic effects. Particularly in old rats, chronic administration of GHRP-6, although combined with GHRH, has been found to have beneficial effects on body structure and function (51).

Side Effects of GHRP Administration

Acute administration of both GHRP-6 and hexarelin may induce transient mild facial flushing and or sleepiness in a small percentage of subjects (about 10%). Interestingly, the occurrence of facial flushing is more frequent after administration of GHRH-29.

Prolonged GHRP-6 or hexarelin administration did not induce particular side effects; facial flushing as well as sleepiness vanished after 1 to 2 days of treatment. Generally, there was good compliance to intranasal administration of hexarelin and some subjects referred to slight burning of nasal mucosa.

Conclusion

It is well known that the function of the GH–IGF-I axis is reduced in aging; this fact could contribute to the age-related body changes as indicated by evidence that treatment with exogenous GH counteracts them. Moreover, there is evidence that the GH insufficiency in aging is due to hypothalamic alterations and that the GH releasable pool in elderly subjects is similar to that in young adults. This evidence suggests the utility of GH releasing substances to restore the function of the GH–IGF-I axis in aging.

At present there are two major classes of compounds that could serve as GH secretagogues with therapeutic implications. The first is GHRH and its analogues, and the second is GH releasing peptide-6 (GHRP-6) and its analogues (GHRP-1, GHRP-2, hexarelin), including recent development of nonpeptidyl GHRP mimetics.

The primary problem of GHRH is that it needs to be administered parenterally, but the low bioavailability of subcutaneous administration has not been overcome. Synthetic GHRPs have a strong GH releasing effect and are active after subcutaneous, intranasal, and even oral routes of administration in animals and humans. This evidence is prompting studies about effects and possible clinical usefulness of these substances in the states of GH insufficiency such as in aging.

Data available so far indicate that the GH releasing effect of GHRPs is reduced in the elderly of both sexes, although it is still marked and higher than that of GHRH. Also in aging, GHRPs maintain their activity after subcutaneous, intranasal, and oral administration.

Interestingly, the age-dependent decrease of the GHRP effect becomes apparent in the sixth decade of life. However, it does not seem dependent on variations of gonadal steroids; in fact, in contrast to GHRH, the GH response to hexarelin is not restored by the pretreatment with transdermal estradiol in postmenopausal women. The reduced effect of GHRPs in aging is more likely dependent on a concomitant decrease of GHRH activity and increase of somatostatinergic tone. According to this hypothesis, the GH releasing activity of hexarelin is enhanced by GHRH to a lesser extent in elderly than in young subjects, while it is increased by arginine, which likely inhibits hypothalamic somatostatin in aging only. These data agree with the hypothesis that these peptides act, at least partly, independently of modulation of GHRH and/or somatostatinergic neurons. They could antagonize somatostatin activity directly on somatotrophs and concomitantly act within the hypothalamus, directly or indirectly via an unknown factor. These possibilities could apply also for an endogenous GHRP-like ligand, whenever it is discovered.

There are data indicating that the activity of GHRPs may undergo partial, homologous desensitization, but this seems to be true for continuous infusion only. In fact, the intermittent oral treatment with GHRPs up to 15 days does not reduce the acute GH response to these hexapeptides. On the other hand, the treatment with GHRPs seems able to induce an increase of IGF-I levels in elderly subjects. Thus, these data point to the ability of GHRPs to restore the reduced function of the GH–IGF-I axis in aging and suggest that chronic treatment could counteract age-related body changes.

Although it is still unclear whether it is really useful to restore to young levels the activity of the GH–IGF-I axis in aging, treatment with GH secretagogues is likely more appropriate and physiologic than that with biosynthetic GH. Among GH secretagogues, GHRPs appear promising, based on their marked GH releasing effect and oral activity. Many questions and problems remain to be solved but the present state of the art opens interesting perspectives.

Acknowledgments. These studies were supported by CNR (Progetto Invecchiamento contratto n.93.00356. PF40) and Mediolanum/Europeptides. The authors wish to thank G.M. Boffano, M.F. Boghen, L. DiVito, S. Goffi, B.P. Imbimbo, B. Maccagno, J. Ramunni, and M. Talliano for their participation in these studies.

References

1. Ghigo E, Arvat E, Goffi S, Bellone J, Nicolosi M, Procopio M, Maccario M, Camanni. Neural control of growth hormone secretion in aged humans. In:

Muller EE, Cocchi D, Locatelli V, eds. Growth hormone and somatomedins during lifespan. Berlin: Springer-Verlag, 1993:275–87.

2. Corpas E, Harman SM, Blackman S. Human growth hormone and human aging. Endocr Rev 1993;14:20–39.

3. Vermeulen A. Nyctohemeral growth hormone profiles in young and aged men: correlation with somatomedin-C levels. J Clin Endocrinol Metab 1987;64:884–8.

4. Zadik Z, Chalew SA, McCarter RJ, Meistas M, Kowarski AA. The influence of age on the 24-hour integrated concentration of growth hormone in normal individuals. J Clin Endocrinol Metab 1985;60:513–6.

5. Van Coevorden A, Mockel J, Laurent E. Neuroendocrine rhythms and sleep in aging men. Am J Physiol 1991;260:251–61.

6. Iranmanesh A, Lizarralde G, Veldhuis JD. Age and relative adiposity are specific negative determinants of the frequency and amplitude of GH secretory bursts and the half-life of endogenous GH in healthy men. J Clin Endocrinol Metab 1991;73:1081–8.

7. Rudman D, Feller AG, Nagraj HS. Effects of human growth hormone in men over 60 years old. N Engl J Med 1990;323:1–6.

8. Ghigo E, Goffi S, Arvat E, Nicolosi M, Procopio M, Bellone J, et al. Pyridostigmine partially restores the GH responsiveness to GHRH in normal aging. Acta Endocrinol (Copenh) 1990;123:169–74.

9. Giustina A, Bussi AR, Conti C, Doga M, Legati F, Macca C, et al. Comparative effect of galanin and pyridostigmine on the growth hormone response to growth hormone-releasing hormone in normal aged subjects. Horm Res 1992;37:165–70.

10. Ghigo E, Goffi S, Nicolosi M, Arvat E, Valente F, Mazza E, et al. Growth hormone (GH) responsiveness to combined administration of arginine and GH-releasing hormone does not vary with age in man. J Clin Endocrinol Metab 1990;71:1481–5.

11. Ghigo E, Goffi S, Arvat E, Imperiale E, Boffano GM, Valetto MR, et al. A neuroendocrinological approach to evidence of impairment of central cholinergic function in aging. J Endocrinol Invest 1992;15:665–70.

12. Corpas E, Harman M, Pineyro MA, Roberson R, Blakman R. Growth hormone (GH)-releasing hormone-(1-29) twice daily reverses the decreased GH and insulin-like growth factor-I levels in old men. J Clin Endocrinol Metab 1992;75:530–5.

13. Corpas E, Harman MS, Pineyro MA, Roberson R, Blackman MR. Continuous subcutaneous infusions of GHRH 1–44 for 14 days increase GH and IGF-I levels in old men. J Clin Endocrinol Metab 1993;76:134–8.

14. DeGennaro V, Zoli M, Cocchi D. Reduced growth hormone releasing factor (GHRH)-like immunoreactivity and GHRH gene expression in hypothalamus of aged rats. Peptides 1989;10:705–8.

15. Morimoto N, Kawakami F, Makino S, Chihara K, Hasegawa M, Ibata Y. Age-related changes in growth hormone releasing factor and somatostatin in the rat hypothalamus. Neuroendocrinology 1988;47:459–64.

16. Ceda GP, Valenti G, Butturini U, Hoffman AR. Diminished pituitary responsiveness to growth hormone-releasing factor in aging male rats. Endocrinology 1986;118:2109–14.

17. Sonntag WE, Forman LJ, Miki N. Effects of CNS active drugs and somatostatin antiserum on growth hormone release in young and old male rats. Neuroendocrinology 1981;33:73–8.

428   E. Ghigo et al.

18. Ge F, Tsagarakis S, Rees LH, Besser GM, Grossman A. Relationship between growth hormone-releasing hormone and somatostatin in the rat: effects of age and sex on content and in-vitro release from hypothalamic explants. J Endocrinol 1989;123:53–8.
19. Locatelli V, Cella SG, Cermenati P, Panzeri G, Sellan R, Muller EE. Defective growth hormone (GH) secretion in aging mammals: contribution of central and peripheral inhibitory influences and of GH-releasing hormone. In: Valenti G, ed. Psychoneuroendocrinology of aging: basic and clinical aspects. Fidia Research Series. Padova, Italy: Liviana Press, 1989:61–7.
20. Ghigo E. Neurotransmitter control of growth hormone secretion: In: de la Cruz LF, ed. Growth hormone and somatic growth. Amsterdam: Excerpta Medica, 1992:103–36.
21. Bowers CY. GH releasing peptides. Structure and kinetics. J Pediatr Endocrinol 1993;6:21–31.
22. Bowers CY, Veeraragavan K, Sethumadhavan K. Atypical growth hormone releasing peptides. In: Bercu BB, Walker RF, eds. Growth hormone II: basic and clinical aspects. New York: Springer-Verlag, 1993:203–22.
23. Ghigo E, Arvat E, Gianotti L, Imbimbo BP, Lenaerts V, Deghenghi R, et al. Growth hormone-releasing activity of hexarelin, a new synthetic hexapeptide, after intravenous, subcutaneous, intranasal and oral administration in man. J Clin Endocrinol Metab 1994;78:693–8.
24. Smith RG, Cheng K, Schoen WR, Pong SS, Hickey G, Jacks T, et al. A nonpeptidyl growth hormone secretagogue. Science 1993;260:1640–3.
25. Bowers CY, Alster DK, Frentz JM. The growth hormone-releasing activity of a synthetic hexapeptide in normal men and short statured children after oral administration. J Clin Endocrinol Metab 1992;74:292–8.
26. Hartman ML, Farello G, Pezzoli SS, Thorner MO. Oral administration of growth hormone (GH)-releasing peptide stimulates GH secretion in normal men. J Clin Endocrinol Metab 1992;74:1378–84.
27. Bowers CY, Momany FA, Reynolds GA, Hong A. On the in vitro and in vivo activity of a new synthetic hexapeptide that acts on the pituitary to specifically release growth hormone. Endocrinology 1984;114:1537–45.
28. Momany FA, Bowers CY, Reynolds GA, Hong A, Newlander K. Conformational energy studies and in vitro and in vivo activity data on active GH releasing peptides. Endocrinology 1984;114:1531–6.
29. Goth MI, Lyons CE, Canny BY, Thorner MO. Pituitary adenylate cyclase activating polypeptide, growth hormone (GH)-releasing peptide and GH releasing hormone stimulate GH release through distinct pituitary receptors. Endocrinology 1992;130:939–44.
30. Codd EE, Shu AYL, Walker RF. Binding of a growth hormone releasing hexapeptide to specific hypothalamic and pituitary binding sites. Neuropharmacology 1989;28:1139–44.
31. Badger TM, Millard WY, McCormick GF, Bowers CY, Martin JB. The effects of growth hormone (GH)-releasing peptides on GH secretion in perifused pituitary cells of adult male rats. Endocrinology 1984;115:1432–8.
32. Sartor O, Bowers CY, Chang D. Parallel studies of His-D-Trp-Ala-Trp-D-Phe-Lys-NH$_2$ in rat primary pituitary cell monolayer culture. Endocrinology 1985;116:952–7.
33. Blake AD, Smith RG. Desensitization studies using perifused rat pituitary cell show that growth hormone-releasing hormone and His-D-Trp-D-Phe-Lys-NH$_2$

stimulate GH release through different receptor sites. J Endocrinol 1991;129:11–19.
34. Bowers CY, Sartor AO, Reynolds GA, Badger TM. On the action of the growth hormone-releasing hexapeptide, GHRP. Endocrinology 1991;128: 2027–35.
35. Cheng K, Chan WWS, Barreto A, Convey EM, Smith RG. The synergistic effects of His-D-Trp-Ala-Trp-D-Phe-Lys-NH2 on growth hormone (GH)-releasing factor-stimulated GH release and intracellular adenosine 3′,5′-monophosphate accumulation in rat primary cell culture. Endocrinology 1989;124:2791–8.
36. Ilson BE, Jorkasky DK, Curnow RT, Stote RM. Effect of a new synthetic hexapeptide to selectively stimulate growth hormone release in healthy human subjects. J Clin Endocrinol Metab 1989;69:212–4.
37. Bowers CY, Reynolds GA, Durham D, Barrera CM, Pezzoli SS, Thorner MO. Growth hormone (GH)-releasing peptide stimulates GH release in normal men and acts synergistically with GH-releasing hormone. J Clin Endocrinol Metab 1990;70:975–82.
38. Penalva A, Carballo A, Pombo M, Casanueva FF, Dieguez C. Effect of growth hormone (GH)-releasing hormone (GHRH), atropine, pyridostigmine, or hypoglycemia on GHRP-6-induced GH secretion in man. J Clin Endocrinol Metab 1993;76:168–71.
39. Deghenghi R, Boutignon F, Wuthrich P, Lenaerts V, Imbimbo B, Lucchelli PE, et al. GH releasing properties of hexarelin (EP 23905). In: Proceedings of the International Symposium on Growth Hormone II: Basic and Clinical Aspects; 1992, December 3–6; Tarpon Springs (Florida). 1992:31 (abstr).
40. Imbimbo BP, Mant T, Edward M, Amin D, Froud A, Lenaerts U, et al. Growth hormone releasing activity of hexarelin in humans: a dose-response study. Eur J Clin Pharmacol 1994;46:421–5.
41. Hayashi S, Okimura Y, Yagi H, Uchiyama T, Takeshima S, Shakatsui S. Intranasal administration of His-D-Trp-Ala-Trp-D-Phe-Lys-NH$_2$ (growth hormone releasing peptide) increased plasma growth hormone and insulin-like growth factor-I levels in normal men. Endocrinol Jpn 1991;38:15–21.
42. Locatelli V, Torsello A, Grilli R, Ghigo MC, Cella SG, Deghenghi R, et al. GHRP-6 stimulates GH secretion and synthesis independently from endogenous GHRH and SRIF in the infant rat. J Endocrinol Invest 1993;16(suppl 1):106.
43. Arvat E, Gianotti L, Di Vito L, Imbimbo BP, Lenaerts V, Deghenghi R, et al. Modulation of growth hormone-releasing activity of hexarelin in man. Neuroendocrinology 1995;61:51–6.
44. Maccario M, Arvat E, Procopio M, Gianotti L, Grottoli S, Imbimbo BP, et al. Metabolic modulation of the growth-hormone-releasing activity of hexarelin in man. Metabolism Clin Exp 1995;44:134–8.
45. Clark RG, Carisson MS, Trojnar J, Robinson ICAF. The effects of a growth hormone-releasing peptide and growth hormone-releasing factor in conscious and anaesthetized rats. J Neuroendocrinol 1989;1:252–5.
46. Dickson SL, Leng G, Robinson ICAF. Systemic administration of growth hormone-releasing peptide activates hypothalamic arcuate neurons. Neurosci Lett 1993;53:303–7.
47. Akman MS, Girard M, O'Brien L, Ho AK, Chik CL. Mechanisms of action of a second generation growth hormone-releasing peptide (Ala-His-D-βNal-Ala-

Trp-D-Phe-Lys-NH2) in rat anterior pituitary cells. Endocrinology 1993;132:1286–91.

48. Wu D, Chen C, Zhang J, Katoh K, Clark I. Effects in vitro of new growth hormone releasing peptide (GHRP-1) on growth hormone secretion from ovine pituitary cells in primary culture. J Neuroendocrinol 1994;6:185–90.

49. Wu D, Chen C, Katoh K, Zhang J, Clark I. The effect of GH-releasing peptide-2 (GHRP-2 or KP102) on GH secretion from primary cultured ovine pituitary cells can be abolished by a specific GH-releasing factor (GRF) receptor antagonist. J Endocrinol 1994;140:3–9.

50. Sartor O, Bowers CY, Reynolds GA, Momany FA. Variables determining the growth hormone response of His-D-Trp-Ala-Trp-D-Phe-Lys-NH2 in the rat. Endocrinology 1985;117:1441–7.

51. Walker RF, Yang AW, Masuda R, Hu CS, Bercu BB. Effects of growth hormone releasing peptides on stimulated growth hormone secretion in old rats. In: Bercu BB, Walker RF, eds. Growth hormone II, basic and clinical aspects. New York: Springer-Verlag, 1993:167–92.

52. Ghigo E, Arvat E, Rizzi G, Bellone J, Nicolosi M, Boffano GM, et al. Arginine enhances the growth hormone (GH)-releasing activity of a synthetic hexapeptide (GHRP-6) in elderly but not in young subjects after oral administration. J Endocrinol Invest 1994;17:157–62.

53. Ghigo E, Arvat E, Rizzi G, Goffi S, Mucci M, Boghen MF, Camanni F. Growth hormone-releasing activity of growth hormone-releasing peptide is maintained after short-term oral pretreatment with the hexapeptide in normal aging. Eur J Endocrinol 1994;131:1–4.

54. Arvat E, Gianotti L, Grottoli S, Imbimbo BP, Lenaerts V, Deghenghi R, et al. Arginine and growth hormone-releasing hormone restore the blunted growth hormone-releasing activity of hexarelin in elderly subjects. J Clin Endocrinol Metab 1994;79:1440–3.

55. Deiber M, Chatelain P, Naville D, Putet G, Salle B. Functional hypersomatotropism in small for gestational age (SGA) newborn infants. J Clin Endocrinol Metab 1989;68:232–5.

56. Deghenghi R, Cananzi M, Torsello, Battisti C, Muller EE, Locatelli V. GH-releasing activity of hexarelin, a new growth hormone releasing peptide, in infant and adults rats. Life Sci 1994;54:1321–8.

57. Hughes-Dawson B, Stern D, Goldman J, Reichlin S. Regulation of growth hormone and somatomedin-C secretion in postmenopausal women: effect of physiological estrogen replacement. J Clin Endocrinol Metab 1986;63:424–8.

58. Bellone J, Aimaretti G, Bartolotta E, Benso L, Imbimbo BP, Lenaerts V, et al. Growth hormone-releasing activity of hexarelin, a new synthetic hexapeptide, before and during puberty. J Clin Endocrinol Metab 1995;80:1090–4.

59. Bercu BB, Weideman CA, Walker RF. Sex differences in growth hormone (GH) secretion by rats administered GH-releasing hexapeptide. Endocrinology 1991;129:2592–8.

60. Mallo F, Alvarez CV, Benitez L, Burguera B, Coya R, Casanueva FF, et al. Regulation of His-dTrp-Ala-Trp-dPhe-Lys-NH2 (GHRP6)-induced GH secretion in the rat. Neuroendocrinology 1992;57:247–51.

61. Huhn WC, Hartman ML, Pezzoli SS, Thorner MO. Twenty-four-hour growth hormone (GH)-releasing peptide (GHRP) infusion enhances pulsatile GH se-

cretion and specifically attenuates the response to a subsequent GHRP bolus. J Clin Endocrinol Metab 1993;76:1202–8.

62. DeBell WK, Pezzoli, SS, Thorner MO. Growth hormone (GH) secretion during continuous infusion of GH-releasing peptide: partial response attenuation. J Clin Endocrinol Metab 1991;72:1312–6.

63. Jaffe CA, Ho PJ, Demott-Friberg R, Bowers CY, Barkan AL. Effects of a prolonged growth hormone (GH)-releasing peptide infusion on pulsatile GH secretion in normal men. J Clin Endocrinol Metab 1993;77:1641–7.

28

Growth Hormone Relationships to Immune Function in Humans

MARIE C. GELATO AND MARY T. SHEEHAN

The role of growth hormone (GH) in immune function was first established from animal data in the 1960s by the now classic work of Pierpaoli and Sorkin (1) demonstrating that an antiserum to GH induced thymic atrophy in mice and that this effect was reversed by GH. In humans, data have been gathered mostly from patients with GH deficiency (GHD). This chapter briefly reviews some of the data in GHD patients before and after therapy and relates these data to the elderly (adults over the age of 60 years), who also have a high incidence of GHD and immune function abnormalities similar to those seen in young adults and children with GHD.

While patients with GHD do not present with an overt immunodeficient state, specific alterations in immune function have been noted. These include thymic hypoplasia (2), reduced plasma levels of thymulin (3), decreased numbers (4) and activity of natural killer (NK) cells (5), decreased numbers of CD8$^+$ cells (suppressor/cytotoxic T cells) (6), inability to synthesize antibodies in response to clinically important antigens such as tetanus toxoid and pneumococcal polysaccharide, defective antibody and cell-mediated immunity as assessed in vivo (7), and decreased production of important cytokines—the interleukins (IL)-1α and IL-2—that affect T-cell function (8). These data are controversial because there are also reports showing no change in immune function in this same population (9–12).

Similarly, reviewing the literature on the effects of GH therapy in GHD patients yields conflicting results. These studies have shown that GH treatment decreased the percentage of circulating B cells (10). A transient decrease in the CD4/CD8 ratio (T helper/T-suppressor ratio) was observed (13). Short-term GH therapy did not restore NK cell activity (14) but long-term GH administration (1 year) did normalize NK cell function (5). Both an increase (15) and a decrease (9) in phytohemagglutinin (PHA)-stimulated lymphoproliferative responses have been observed with GH therapy. In addition, in vitro production of immunoglobulin M (IgM), which was reported as low in GHD children, was restored to normal after 1 year of therapy with GH (16). While, as pointed out by Wit et al. (17) in a recent review of the immunologic findings in GH-treated patients, there is no clinical evidence to suggest that GHD patients have an increased susceptibility to infection nor other evidence of overt immune dysfunction, there are enough data to suggest that subtle changes are present that by themselves may not have a clinical impact because of the maturational level of the immune system. That is to say, in young adults there is not the dysregulation of the immune system that appears to be present as a part of aging, so that these subtle changes may have more of an impact in the elderly.

In the last several years, the effects of GH administration have been studied in the elderly. This population of patients has received much attention because approximately 30% to 40% of individuals over the age of 60 have biochemical parameters indicative of GHD. Early work by Rudman et al. (18) demonstrated that by age 70 to 80, 53% of subjects have no significant serum immunoreactive GH detectable during a 24-hour period and have serum insulin-like growth factor-I (IGF-I) levels less than 0.03 U/ml, which would classify them as growth hormone deficient. These studies were later expanded by Ho et al. (19), who measured GH secretion over a 24-hour period in young adults (ages 18–33) and older adults (ages 55–76) and found that the number of GH pulses was the same in both groups, but the pulse amplitude, duration, and the fraction of GH secreted in pulses during the 24-hour period were significantly decreased in older adults, both men and women. Thus they substantiated the earlier work by Rudman et al. (18). In addition, the elderly have diminished GH responses to indirect pituitary stimulation. For instance, insulin-induced hypoglycemia, which causes a brisk GH response in young adults (>10 ng/ml), is only partially effective as a GH secretagogue in older individuals (20). However, elderly men and women retain their GH responsiveness to GH releasing hormone (GHRH) (21). In addition, the metabolic clearance of GH does not appear to be altered with age (22). These findings imply that the age-related decrease in 24-hour serum GH concentration is most likely a result of decreased pituitary release, possibly secondary to a decrease in GHRH.

Serum concentrations of IGF-I and its binding protein, IGFBP-3, also decrease with advancing age (23, 24). This is most likely a reflection of diminished GH secretion, since it is well known that GH stimulates the production of both peptides (25). While the levels of these proteins are diminished, the ability of exogenous GH (26) and GHRH (27) to stimulate IGF-I production is normal in the elderly. Thus, GH action is not impaired.

There are both quantitative and qualitative changes that occur in the immune system in the elderly. These changes are similar to some of the defects noted in patients with GH deficiency. Quantitatively, the number of circulating lymphocytes decreases by approximately 15%, primarily due to decreased numbers of T cells (28). Both helper (CD4+) and suppressor (CD8+) cells are decreased to approximately the same extent with age. Some studies have reported decreases in the number of B lymphocytes

(CD19$^+$) and functional abnormalities in NK cell activity (cytotoxic cells) (29). In addition, the thymus decreases in size with age, and there is decreased thymic hormone secretion (30).

Qualitatively, changes can be seen in the surface membrane, cytoplasm, and nucleus of T cells, B cells, and monocytes or macrophages (28). At the surface-membrane level, there are fewer interleukin-2 (IL-2) receptors on T cells from elderly subjects (31). This may account in part for the decline in T-cell proliferation, since IL-2 synthesis diminishes with age as well (32). In addition, there is a loss of B-cell responsiveness to both T-cell–independent and –dependent antigens with age (33). This may be related to the loss of B-cell Fc receptor function (34). At the cytoplasmic level, metabolic changes have been described in aged T cells, such as decreased levels of nicotinamide adenine dinucleotide (NAD), which would imply an impaired ability to generate energy and repair DNA, and alterations in calcium uptake, which could impair second-messenger activities (28, 35). Other second-messenger systems within the T cell may be abnormal as well (36), including defects in G regulatory proteins and defects in the adenylate cyclase catalytic subunit (37).

At the nuclear level, several functional changes have been detected in T cells, such as decreased ability to repair damaged DNA (38), increased breaks in DNA strands (39), and decreased transcription of the IL-2 gene (40). These various changes may be in part responsible for the loss of cytotoxic activity of T cells (28) as well as their decreased responsiveness to mitogens (41). Thus, most of the changes observed in the immune system of the elderly relate to either T-cell dependent processes or alterations in the regulatory mechanisms of the immune system (33). For instance, studies on antibody responses to foreign antigenic stimulation show that primary, but not necessarily secondary, antibody response decreases with age, and the decrease is most pronounced in T-cell dependent processes (42). Additionally, antibody responses do not appear to be sustained in elderly individuals (43–45), and serum concentration of isoagglutinins diminishes with age (46, 47). The proliferative capacity of B cells does not appear to be affected, thus suggesting a change primarily in regulatory T cells. That regulatory mechanisms of the immune system may be altered is also evidenced by an age-related increase in the incidence of monoclonal gammopathies in the absence of B-cell malignancy (33) as well as the increased prevalence of certain autoantibodies (ANA, rheumatoid factor) (48, 49). Thus, the hallmark of aging in the immune system is a reduction in T-cell function and a decrease in the primary immune response of B cells, especially in those responses requiring T-cell interaction. These changes may be relevant to the fact that the elderly have four to five times the case rate of cancer, tuberculosis, and herpes zoster (50) and six to seven times the fatality rate from pneumonia, compared with young adults (51).

Is there a cause-and-effect relationship between the correlated decline in the immune system and the GH/IGF-I axis in the elderly? Clearly, GH and

IGF-I are not the only factors involved in the process of immunosenescence, which is a complex set of events. However, it is possible that the decline in GH/IGF-I may contribute to the process. It has been our working hypothesis that GH and IGF-I can alter immune function, which is impaired in the elderly. To that end, we have proposed a series of investigations to attempt to establish a link between these two systems in the elderly.

First, we sought to establish the incidence of GHD in healthy, elderly adults (24 men and 76 women) ranging in age from 60 to 88 years, who were living at home in the community and were fully ambulatory. The only medications that some of these subjects were taking were for hypertension; none of the medications were those known to alter GH secretion. IGF-I was measured by a specific radioimmunoassay (RIA), and IGF-II levels were assessed by a radioreceptor assay (52). The mean IGF-I level for the group was 130 ± 10 (SE) ng/ml, which was significantly lower ($p < .01$) than the mean of 244 ± 20 ng/ml for younger adults aged 20 to 44 years. Interestingly, women had a significantly ($p < .05$) lower level of IGF-I, 116 ± 10 ng/ml, than men (167 ± 16 ng/ml). There was no correlation of the IGF-I levels with either age or body mass index (BMI). Approximately 37% of these individuals had IGF-I levels less than 100 ng/ml, which would class them as GH deficient based on data published by Rudman et al. (53). The IGF-II levels in the elderly, 338 ± 40 ng/ml, were slightly lower than the levels in younger adults, 450 ± 50 ng/ml.

To further assess the GH status of these individuals, we also measured levels of two of the serum IGFBPs, BP-3 and BP-1. The levels of BP-3 were measured by an immunoradiometric assay (IRMA) (kindly provided by DSL, Webster, Texas) and an enzyme-linked immunosorbent assay (ELISA) for BP-1 measurements (54). The levels of BP-3 were 3.5 ± 0.5 mg/L in the elderly, which were not significantly different from the levels in the younger adults, 4.2 ± 0.4 mg/L. However, IGFBP-1 levels were significantly elevated in the elderly, 70 ± 15 ng/ml ($p < .05$), when compared with values in younger adults, 2.6 ± 0.2 ng/ml. There was a tendency for the IGF-I levels to be negatively correlated with the BP-1 levels, that is, the lower the IGF-I level, the higher the BP-1 level. In addition, using methods previously described (55), we assessed the posttranslational modification of IGFBP-1. In 26% of the elderly individuals tested, IGFBP-1 was present in a highly phosphorylated form (pBP-1) in comparison to no detectable pBP-1 in sera of younger healthy adults. IGFBP-1 is thought to be an inhibitor of IGF-I action, and when BP-1 is phosphorylated it has a sevenfold increase in affinity for IGF-I (56), which may make IGF-I inaccessible to the cell surface receptor. These data suggest that in addition to low IGF-I levels in the elderly, there are changes in the serum BPs for the IGFs, i.e., increased total levels of BP-1 and more circulating pBP-1, which may impair the action of available IGFs at the cellular level. The lower levels of IGF-I in the elderly women may reflect a lack of estrogen since none of these women were on estrogen replacement therapy (19).

In the next phase of the study, we hope to assess the immune status of a subset of these individuals and test the responsiveness of these parameters to short-term therapy with GHRH (kindly supplied by Serono Laboratories Inc., Norwell, MA) 1 mg twice daily, which has been shown in previous work to restore IGF-I levels to normal in older adults (27). Preliminary results in adults over 60 years of age show that there is a decrease in the CD8+ cells (T suppressor/cytotoxic cells) in comparison with young adults. This is similar to what has been previously reported in the elderly (28).

To this end, in elderly men treated with GH for 6 months by Petersen et al. (57), no changes were seen in lymphocyte subset populations nor in their response to mitogenic or antigenic stimulation in response to GH. There was, however, a decrease in NK cell functional activity in these subjects, which returned to normal with GH therapy. It is our belief that what needs to be assessed is more than just quantitative changes or in vitro responses, since these are sometimes poor indicators of the functional capacity of the system. It would be more fruitful to challenge the system and observe whether immune responsiveness changes in vivo. As we learn more about the aging process, there is great hope that the future will offer therapeutic interventions that can decrease human frailty and allow individuals a better quality of life as they age. The ability to delay immunosenescence would be a major step in reducing one of the parameters of aging that accounts for major morbidity in the elderly.

Acknowledgments. The authors wish to thank Mr. Gopal Savjani at DSL for supplying the kits to measure IGFBP-3, UBI for the generous gift of IGF-I and -II, Dr. Robert Frost for measuring phosphorylated BP-1, Kenneth Marsh for his expert technical assistance, and Mrs. Linda Martorana for typing the manuscript. This work was supported in part through grants from Serono Laboratories, Inc. and grant number 2S07RR0573619 from the National Center for Research Resources of the Public Health Service to MCG. Its contents are solely the responsibility of the authors and do not necessarily represent the official views of the NCRR.

References

1. Pierpaoli W, Sorkin E. Hormones and immunologic capacity. I. Effect of heterologous anti-growth hormone (ASTH) antiserum on thymus and peripheral lymphatic tissue in mice. J Immunol 1968;101:1036–43.
2. Sipponen P, Simila S, Callan Y, Autere T, Herva R. Familial syndrome with panhypopituitarism, hypoplasia of the hypophysis and poorly developed sella turcica. Arch Dis Child 1978;53:664–7.

3. Mocchegiani E, Fabris N. Growth hormone influence on thymic endocrine activity in humans. Int J Neurosci 1990;51:253–4.
4. Kiess W, Doers H, Cutenandt O, Celohradsky BH. Lymphocyte subsets and natural killer activity in growth hormone deficiency. N Engl J Med 1986;314:321–4.
5. Bozzola M, Valtorta A, Moretta A, Cisternino M, Biscaldi I, Schimpff RM. In vitro and in vivo effect of growth hormone on cytotoxic activity. J Pediatr 1990;117:596–9.
6. Gupta D, Fikrig SM, Noval MS. Immunological studies in patients with isolated growth hormone deficiency. Clin Exp Immunol 1983;54:87–90.
7. Fleischer TA, White RA, Broder S, Nissley PS, Blaese RM, Mulvihill JJ, et al. X-Linked hypogammaglobulinemia and isolated growth hormone deficiency. N Engl J Med 1980;302:1429–34.
8. Casanova S, Repelin AM, Schimpff RM. Production of interleukin-1α and interleukin-2 by mononuclear cells from children with growth delay in relation to the degree of growth hormone deficiency: effects of substitutive treatment. Horm Res 1990;34:209–14.
9. Church JA, Castin G, Brooks J. Immune functions in children treated with biosynthetic growth hormone. J Pediatr 1989;115:420–4.
10. Petersen BH, Rapaport R, Henry DP, Huseman C, Moore WV. Effect of treatment with biosynthetic human growth hormone (GH) on peripheral blood lymphocyte populations and function in growth hormone-deficient children. J Clin Endocrinol Metab 1990;70:1756–60.
11. Spadoni GL, Rossi P, Ragno W, Galli E, Cianfarani S, Galasso C, Boscherini B. Immune function in growth hormone-deficient children treated with biosynthetic growth hormone. Acta Pediatr Scand 1991;80:75–9.
12. Rapaport R, Petersen B, Skuza KA, Heim M, Goldstein S. Immune functions during treatment of growth hormone-deficient children with biosynthetic human growth hormone. Clin Pediatr 1991;30:22–7.
13. Bozzola MM, Cisternino M, Valtorta A, Moretta A, Biscaldi I, Maghnie M, et al. Effect of biosynthetic methionyl growth hormone therapy on the immune function in GH deficient children. Horm Res 1989;31:153–6.
14. Kiess W, Malozowski S, Gelato M, Butenand O, Doerr H, Crisp B, et al. Lymphocyte subset distribution and natural killer activity in growth hormone deficiency before and during short-term treatment with growth hormone-releasing hormone. Clin Immunol Immunopathol 1988;48:85–94.
15. Abassi V, Bellanti J. Humoral and cell-mediated immunity in growth hormone-deficient children: effect of therapy with human growth hormone. Pediatr Res 1985;19:299–301.
16. Bozzola M, Cisternino M, Valtorta A, Moretta A, Biscaldi I, Maghnie M, et al. Effect of biosynthetic methionyl growth hormone (GH) therapy on the immune function in GH-deficient children. Horm Res 1989;31:153–6.
17. Wit JM, Kooijman R, Rijkers GT, Zegers BJM. Immunological findings in growth hormone-treated patients. Horm Res 1993;39:107–10.
18. Rudman D, Kutner MH, Rogers CM. Impaired growth hormone secretion in the adult population: relation to age and adiposity. J Clin Invest 1981;67:1361–6.
19. Ho KY, Evans WS, Blizzard RM, Veldhuis J, Merriam GR, Samojlik L, et al. Effects of sex and age on 24-hour profile of growth hormone secretion in man:

importance of endogenous estradiol concentrations. J Clin Endocrinol Metab 1987;64:51–8.

20. Muggeo M, Fedele D, Tiengo A. Human growth hormone and cortisol responses to insulin stimulation in aging. J Gerontol 1975;30:546–52.

21. Pavlov EP, Harman SM, Merriam GR, Gelato MC, Blackman MR. Responses of growth hormone (GH) and somatomedin-C to GH-releasing hormone in healthy aging men. J Clin Endocrinol Metab 1986;62:590–600.

22. Blackman MR. Pituitary hormones and aging. Endocrinol Metab Clin 1990;71:575–9.

23. Hammerman MR. Insulin-like growth factors and aging. Endocrinol Metab Clin 1987;16:995–1011.

24. Donahue LR, Hunter SJ, Sherblom AP, Rosen C. Age-related changes in serum insulin-like growth factor-binding proteins in women. J Clin Endocrinol Metab 1990;71:575–9.

25. Rechler MM, Nissley SP. Peptide growth factors and their receptors. In: Sporn MB, Roberts AB, eds. Insulin-like growth factors, vol. 95. Heidelberg: Springer-Verlag, 1990:263–367.

26. Johanson AJ, Blizzard RM. Low somatomedin-C levels in older men rise in response to growth hormone administration. Johns Hopkins Med J 1981;149:115–20.

27. Corpas E, Harman SM, Pineyro MA, Roberson R, Blackman MR. Growth hormone (GH)-releasing hormone-(1-29) twice daily reverses the decreased GH and insulin-like growth factor-I levels in old men. J Clin Endocrinol Metab 1992;75:530–5.

28. Geokas MC, Lakatta EG, Makinodan T, Timiras SP. The aging process. Ann Intern Med 1990;113:456–66.

29. Lehtonen L, Eskola J, Vainio O, Lektonen A. Changes in lymphocyte subsets and immune competence in very advanced age. J Gerontol 1990;45: M108–12.

30. Currie MS. Immunosenescence. Compr Ther 1992;18:26–34.

31. Gillis S, Kozak R, Durante M, Weksler ME. Immunological studies of aging. Decreased production of and response to T cell growth factor by lymphocytes from aged humans. J Clin Invest 1981;67:937–42.

32. Ceuppens JL, Goodwin JS. Regulation of immunoglobulin production in pokeweed mitogen-stimulated cultures of lymphocytes from young and old adults. J Immunol 1982;128:2429–34.

33. Doggett DL, Chang MP, Makinodah T, Strehler BL. Cellular and molecular aspects of immune system aging. Mol Cell Biochem 1981;37:137–56.

34. Scribner DJ, Weiner HL, Moorhead JW. Anti-immunoglobulin stimulation of murine lymphocytes. V. Age-related decline in Fc receptor-mediated immunoregulation. J Immunol 1978;121:377–84.

35. Chapman ML, Zaun MR, Gracy RW. Changes in NAD levels in human lymphocytes and fibroblasts during aging and in premature aging syndromes. Mech Ageing Div 1983;21:157–67.

36. Tam CF, Walford RL. Cyclic nucleotide levels in resting and mitogen-stimulated spleen cells suspensions from young and old mice. Mech Ageing Div 1978;7:309–20.

37. Dax EM. Age-related changes in membrane receptor interactions. Endocrinol Metab Clin 1987;16:947–64.

440   M.C. Gelato and M.T. Sheehan

38. Lambert B, Ringborg U, Skoog L. Age-related decrease of ultra violet light-induced DNA repair synthesis in human peripheral leukocytes. Cancer Res 1979;39:2792–5.
39. Price GB, Modak SP, Makinodan T. Age-associated changes in the DNA of mouse tissue. Science 1971;171:917–20.
40. Wu W, Pahlavani M, Cheung HT, Richardson A. The effect of aging on the expression of interleukin-2 messenger ribonucleic acid. Cell Immunol 1986;100:224–33.
41. Murasko DM, Weiner P, Kaye D. Association of lack of mitogen-induced lympocyte proliferation with increased mortality in the elderly. Aging Immunol Infect Dis 1988;1:1–6.
42. Burns EA, Goodwin JS. Immunology and infectious disease. In: Cassel CK, Riesenberg DE, Sorensen LB, Walsh JR, eds. Geriatric medicine. New York: Springer-Verlag, 1990:312–29.
43. Forrester HL, Jahnigen DW, LaForce FM. Inefficacy of pneumococcal vaccine in a high-risk population. Am J Med 1987;83:425–30.
44. Ammann AJ, Schiffman G, Austrian R. The antibody response to pneumococcal capsular polysaccharides in aged individuals. Proc Soc Exp Biol Med 1980;164:312–6.
45. Kishimoto S, Tomino S, Mitsuya H, Fujiwara H., Tsuda H. Age-related decline in the in vitro and in vivo synthesis of anti-tetanus toxoid antibody in humans. J Immunology 1980;125:2347–52.
46. Czlonkowska A, Korlah J. The immune response during aging. J Gerontol 1979;34:9–14.
47. Gardner ID. The effect of aging on susceptibility to infection. Rev Infect Dis 1980;2:801–10.
48. Goodwin JS, Searles RP, Tung KSK. Immunological responses of a healthy elderly population. Clin Exp Immunol 1982;48:403–10.
49. Delespesse G, Gausset RH, Sarfati M, Dubi-Rucquoy M, Dekisschop MJ, VanHaelst L. Circulating immune complexes in old people and in diabetics: correlation with autoantibodies. Clin Exp Immunol 1980;40:96–102.
50. Saltzman RL, Peterson PK. Immunodeficiency of the elderly. Rev Infect Dis 1982;9:1127–39.
51. Hazzard WR. The biology of aging. In: Braunwald E, ed. Harrison's principles of internal medicine. New York: McGraw-Hill, 1987:447–54.
52. Gelato MC, Rutherford C, San-Roman G, Shmoys S, Monheit A. The serum insulin-like growth factor-II/mannase-6-phosphate receptor in normal and diabetic pregnancy. Metabolism 1993;42:1031–8.
53. Rudman D, Feller AG, Nagraj HS, Gergans GA, Lolitha PY, Goldberg AF, et al. Effects of human growth hormone in men over 60 years old. N Engl J Med 1990;323:1–6.
54. Frost RA, Fuhrer J, Mariuz P, Lang CH, Gelato MC. Wasting in the acquired immune deficiency syndrome is associated with multiple defects in the serum insulin-like growth factor system. Clin Endocrinol 1996 (in press).
55. Frost R, Lang CH, Bereket A, Wilson T, Gelato MC. Insulin-like growth factor (IGF)-binding protein (BP)-1 is phosphorylated in the serum of catabolic patients. J Clin Endocrinol Metab 1994;78:1533–5.
56. Clemmons DR, Camacho-Hubner C, Jones JI, McCusker RH, Busby WH. Insulin-like growth factor binding proteins: mechanisms of action at the cellular

level. In: Spencer EM, ed. Modern concepts of insulin-like growth factors. New York: Elsevier, 1991:475–86.
57. Petersen BH, Steimel LS, Baum LL, Rudman D. Measurement of immune parameters following long-term in vivo treatment with recombinant human growth hormone. 75th Annual Meeting of the Endocrine Society 1993; Abstract #707:227.

Author Index

Subject Index

PROCEEDINGS IN THE SERONO SYMPOSIA USA SERIES

Continued from page ii

FOLLICLE STIMULATING HORMONE: Regulation of Secretion and Molecular Mechanisms of Action
Edited by Mary Hunzicker-Dunn and Neena B. Schwartz

SIGNALING MECHANISMS AND GENE EXPRESSION IN THE OVARY
Edited by Geula Gibori

GROWTH FACTORS IN REPRODUCTION
Edited by David W. Schomberg

UTERINE CONTRACTILITY: Mechanisms of Control
Edited by Robert E. Garfield

NEUROENDOCRINE REGULATION OF REPRODUCTION
Edited by Samuel S.C. Yen and Wylie W. Vale

FERTILIZATION IN MAMMALS
Edited by Barry D. Bavister, Jim Cummins, and Eduardo R.S. Roldan

GAMETE PHYSIOLOGY
Edited by Ricardo H. Asch, Jose P. Balmaceda, and Ian Johnston

GLYCOPROTEIN HORMONES: Structure, Synthesis, and Biologic Function
Edited by William W. Chin and Irving Boime

THE MENOPAUSE: Biological and Clinical Consequences of Ovarian Failure: Evaluation and Management
Edited by Stanley G. Korenman

ISBN 0-387-94707-8

EAN

9 780387 947075 >